The Nationalization of Scientific Knowledge in the
Habsburg Empire, 1848–1918

This page intentionally left blank

The Nationalization of Scientific Knowledge in the Habsburg Empire, 1848–1918

Edited by

Mitchell G. Ash
Professor of Modern History, University of Vienna

and

Jan Surman
Department of History, University of Vienna

palgrave
macmillan

First published 2012 by
PALGRAVE MACMILLAN

Palgrave Macmillan in the UK is an imprint of Macmillan Publishers Limited,
registered in England, company number 785998, of Houndmills, Basingstoke,
Hampshire RG21 6XS.

Palgrave Macmillan in the US is a division of St Martin's Press LLC,
175 Fifth Avenue, New York, NY 10010.

Palgrave Macmillan is the global academic imprint of the above companies
and has companies and representatives throughout the world.

Palgrave® and Macmillan® are registered trademarks in the United States,
the United Kingdom, Europe and other countries

ISBN 978-1-349-33112-3   ISBN 978-1-137-26497-8 (eBook)
DOI 10.1057/9781137264978

This book is printed on paper suitable for recycling and made from fully
managed and sustained forest sources. Logging, pulping and manufacturing
processes are expected to conform to the environmental regulations of the
country of origin.

A catalogue record for this book is available from the British Library.

A catalog record for this book is available from the Library of Congress.

10 9 8 7 6 5 4 3 2 1
21 20 19 18 17 16 15 14 13 12

Contents

Illustrations

Contributors

Mitchell G. Ash is Professor of Modern History, Head of the Working Group in History of Science and coordinator of the PhD programme 'The Sciences in Historical, Philosophical and Cultural Contexts' at the University of Vienna, Austria. He is Full Member of the Berlin-Brandenburg Academy of Sciences and Humanities and was president of the Gesellschaft für Wissenschaftsgeschichte/Society for History of Sciences and Humanities from 2002 to 2008. He is the author or editor of numerous books, articles, and book chapters focusing *inter alia* upon the relations of science, politics and society in the nineteenth and twentieth centuries. His publications include *Gestalt Psychology in German Culture 1867–1967: Holism and the Quest for Objectivity* (1995); (ed.) *German Universities Past and Future: Crisis or Renewal?* (1997), German translation: *Mythos Humboldt: Vergangenheit und Zukunft der deutschen Universitäten* (1999); (ed. with Christian Stifter) *Wissenschaft, Politik und Öffentlichkeit: Von der Wiener Moderne bis zur Gegenwart* (2002); (ed.) *Mensch, Tier und Zoo. Der Tiergarten Schönbrunn im internationalen Vergleich vom 18. Jahrhundert bis heute* (2008); (ed. with Wolfram Nieß and Ramon Pils) *Geisteswissenschaften im Nationalsozialismus: Das Beispiel der Universität Wien* (2010).

Tatjana Buklijas is Research Fellow at the Liggins Institute, University of Auckland, New Zealand. Her research interests include the history of medicine and biology, history of anatomy and embryology, medicine and evolution, and science and medicine in Central Europe. Publications most relevant for this volume include 'Public anatomies in fin-de-siècle Vienna', *Medicine Studies* 2 (2010), 71–92; 'Cultures of death and politics of corpse supply: anatomy in Vienna, 1848–1914', *Bulletin of the History of Medicine* 82 (2008), 570–607; Introduction to the special section on 'Science, medicine and nationalism in the Habsburg Empire from the 1840s to 1918', *Studies in History and Philosophy of Biological and Biomedical Sciences* 38 (2007), 679–86 (with E. Lafferton); 'Surgery and national identity in late nineteenth-century Vienna', *Studies in History and Philosophy of Biological and Biomedical Sciences* 38 (2007), 756–74.

Deborah R. Coen is Assistant Professor, Department of History, Barnard College, Columbia University, New York. Her current projects include 'The Earthquake Observers: Disaster Science, 1755–1935' and a history of Imperial Austria as a laboratory for studies of the relationship between nature and culture. She is the author of *Vienna in the Age of Uncertainty: Science, Liberalism, and Private Life* (2007) as well as 'Climate and circulation in Imperial Austria', *The Journal of Modern History* 82 (2010), 839–75, and co-editor, with James Fleming and Vladimir Janković, of *Intimate Universality: Local and Global Themes in the History of Weather and Climate* (2006).

Johannes Feichtinger is Senior Research Associate at the Commission for Cultural Studies and History of Theatre, Austrian Academy of Sciences, Vienna, Austria. His research interest is the history of Imperial and Republican Austria with emphasis on intellectual history, history of science and cultural studies. His publications include *Wissenschaft als reflexives Projekt. Von Bolzano über Freud zu Kelsen: Österreichische Wissenschaftsgeschichte 1848–1938* (2010); 'Positivismus und Machtpolitik: Ein wissenschaftliches Programm und dessen Transfer nach Österreich/Zentraleuropa. Zu einem Beispiel von Wissenstransfer', in Helga Mitterbauer and Katharina Scherke (eds), *Entgrenzte Räume: Kulturelle Transfers um 1900 und in der Gegenwart* (2005, 297–319); (ed. with Ursula Prutsch, Moritz Csáky) *Habsburg Postcolonial: Machtstrukturen und kollektives Gedächtnis* (2003).

Tibor Frank is Professor of History and Director of the School of English and American Studies at Eötvös Loránd University in Budapest, Hungary, where he has been head of the PhD program in American Studies since 2000. His research areas include transatlantic relations, international migration, the history of science, imagology and historiography. He was a Fulbright visiting professor at the University of California, Santa Barbara and at UCLA (1987–1990), an NEH visiting professor at the University of Nevada-Reno (1990–1991) and a recurrent visiting professor at Columbia University (2001, 2007, 2010). He is a recipient of the Humboldt Award (Bonn, Germany, 2002) and a corresponding fellow of the Royal Historical Society (London, 2006). His recent books include *Double Exile: Migrations of Jewish-Hungarian Professionals through Germany to the United States 1919–1945* (2009); and (ed. with Frank Hadler) *Disputed Territories and Shared Pasts: Overlapping National Histories in Modern Europe* (2011).

Marianne Klemun is Associate Professor and member of the Working Group in History of Science, Department of History, and since 2006 Vice Dean of the Faculty of Historical and Cultural Studies, University of Vienna. She is an elected member of the International Commission on the History of Geological Sciences (INHIGEO) and of the Commission for History of Science of the Austrian Academy of Sciences. Her research field is the history of natural history in its cultural and political dimensions. Her publications include (ed.) 'Wissenschaft und Kolonialismus', *Wiener Zeitschrift zur Geschichte der Neuzeit* 9/2 (2009) ed. with Veronika Hofer); 'Bildfunktionen in den Wissenschaften', *Wiener Zeitschrift zur Geschichte der Neuzeit* 7/1 (2007); 'Franz Unger and Sebastian Brunner on evolution and the visualization of Earth history: a debate between liberal and conservative Catholics', in *Geology and Religion: A History of Harmony and Hostility* (2009), 259–67.

Gábor Palló is Senior Research Fellow at the Institute for Research Organization of the Hungarian Academy of Sciences. His fields of research include history of chemistry and physics, twentieth-century history of natural sciences in Hungary, philosophy of science, history of migration of scientists and the relationship between science, politics and philosophy. His recent publications include *Zsenialitás és korszellem* [Genius and Zeitgeist] (2004); 'The advantage and disadvantage of peripheral ignorance: the gas adsorption controversy', *Ambix*, 57/2 (2010), 216–30.

Soňa Štrbáňová is Associate Professor, Centre for the History of Sciences and Humanities, Institute for Contemporary History, Academy of Sciences of the Czech Republic in Prague, President of the European Science Foundation and Effective Member of the International Academy of History of Science. She is the author or editor of the following works, among others: (with Jan Janko) *Science in Purkinje's Time* (1988); (with Ida Stamhuis and K. Mojsejová) *Women Scholars and Institutions, Studies in the History of Sciences and Humanities* (2004); (with Brigitte Hoppe and N. Robin) 'International networks, exchange and circulation of knowledge in life sciences 18th to 20th centuries', *Archives internationales d'histoire des sciences* (2006); (with Anita Kildebaek Nielsen) *Creating Networks in Chemistry: The Founding and Early History of Chemical Societies in Europe* (2008).

Jan Surman received his PhD from the University of Vienna in 2012 and was a research fellow at the Max-Planck-Institute for History of Science, Berlin in 2011. His main research interests are history of scientific transfer, history and philosophy of scientific language, and cultural history of Central Europe. His main publications include (ed. together with Gerald Angermann-Mozetič) *Dwa życia Ludwika Gumplowicza* (2010); 'Imperial, kolonial, hegemonial? die wissenschaften in der Habsburger-Monarchie', *Wiener Zeitschrift zur Geschichte der Neuzeit* 9/2 (2009), 119–33; 'Entre indifférence et transfert d'idées: la réception de la sociologie autrichienne en France avant la Première Guerre mondiale', *Austriaca. Cahiers universitaires d'information sur l'Autriche (Rouen)*, 63 (2007), 119–32.

Marius Turda is Reader in 20th-Century Eastern and Central European Biomedicine and Deputy Director, The Centre for Health, Medicine and Society, Oxford Brookes University. His main research interests are history of eugenics, racism and biopolitics from around 1800 to 1945, with a particular focus on Central and Southeastern Europe. His main publications include *Health, Hygiene and Eugenics in Southeastern Europe to 1945* (2011); *Modernism and Eugenics* (2010); and (ed. with Paul J. Weindling) *Blood and Homeland: Eugenics and Racial Nationalism in Central and Southeastern Europe, 1900–1940* (2007).

1

The Nationalization of Scientific Knowledge in Nineteenth-Century Central Europe: An Introduction

Mitchell G. Ash and Jan Surman

Interlinked at first by politics and the common use of the German language for scientific and scholarly communication, Central Europe became in the nineteenth century the site of a scientific system in which a free flow of ideas and to a certain degree of people enabled scientific relations to flourish. This Central European 'republic of letters' began to break apart in the second half of the century, as national disparities and nationalistic politics displaced allegiance to a common scientific community. The shift away from German as the symbolic language of imperial power, the adoption of national languages and the corresponding pressure towards single-language nationhood proved decisive in the end. This nationalization process was highly complex and contradictory.[1] Whether to remain affiliated with German-speaking *'Kultur'*, to create national sciences, to internationalize science beyond the German-speaking realm, or to do all of these things, was a lively topic of discussion throughout the post-1848 period. This volume examines interactions between emerging national cultures and cultural institutions, on the one hand, and cultures of science and scholarship, on the other hand, in this region. We ask two questions: how did the nationalization of the sciences work in this region during this period; and did this highly complex political, social and cultural process inevitably lead to a corruption of scientific objectivity, or rather to a transformation of the very definition of science and scholarship?

The volume extends recent discussions of epistemic cultures in the history of science to political entities and inquires into the role of politics in the production and circulation of scientific and scholarly knowledge. By doing so, the authors also hope to contribute to the general historical discussion of nation-building and the emergence of national and nationalistic cultures in nineteenth- and early twentieth-century (Eastern) Central Europe. The specific examples studied range from geology, seismology, physics and chemistry to eugenics and anatomy; humanistic disciplines such as philosophy, legal theory, language studies and ethnology are also discussed. The volume thus takes seriously the unified ideal of science and scholarship (*Wissenschaft*) shared by the actors themselves and questions stereotyped oppositions between supposedly objective (though corruptible) natural sciences and humanistic fields thought to be inherently nationalistic. The geographical focus on the late Habsburg Monarchy – which in this period was often explicitly described as a multi-ethnic empire (*Vielvölkerreich*) – makes possible the analysis of the emergence of nation and language as cultural values in the natural sciences as well as the humanities in a clearly delimited historical setting. Questions about scientific processes are intertwined here with political tensions, creating a situation in which scientific problems became political, with obvious implications for an understanding of historical as well as current politics and culture in this volatile region.

Nation-building and nationalism in the nineteenth century

The concepts of nation and nationalism have undergone many definitions and redefinitions since the classical works of Gellner, Hobsbawm and Anderson.[2] Though these works take different approaches, they share a critique of earlier work that had taken existing nation-states as givens and emphasized the cultural construction of entities called nations, each of which developed what Hobsbawm called 'invented traditions' in the course of modernization and industrialization. In such analyses the nation or national idea becomes central to identity formation in an era of secularization,[3] a marker of collective security and ideological stability in times of rapid social, economic and cultural change.[4] Yet at the same time these analyses make clear that nationalism was not only an outcome of this process, but itself a driving force and shaper of change.

Research on nation-formation and nationalism has moved in a number of directions during the past three decades. Without pretending to provide a complete survey, we focus here on four trends that have implications for the topic of this volume: sociocultural differentiation, regional and borderland perspectives, the 'centre–periphery' approach, and post-colonial perspectives.

In the first place, scholars have begun to ask how the idea of the nation as 'imagined community' was appropriated and enacted on different social levels, from elites to the common people. In the course of history the adjective 'national' has been applied to most spheres of human activity and cultural production, replacing local and regional foci through concentration on ethno-linguistically defined areas and spaces.[5] The process of enlarging or even creating national spaces, and thus imposing and rewriting identities, has proved to be more complex than the still widespread narratives representing nations as perennial and natural entities may suggest. Nationalization has been shown to be not a linear process, but a dialectical negotiation between elite and popular discourses, each of which is multifaceted and conflict-rich.[6] These interactions of imposed and negotiated identities have themselves undergone constant change. Polish nationalists, for example, aimed for a time to achieve unity within the geography of the ancient Commonwealth, which had been multicultural and multilingual. Once this project came to be seen as a failure – not only due to competing nationalisms but also due to a rethinking of the importance of language for national identity – the argumentation changed and became more exclusive.[7] One new focus of historical nationalism studies is thus on the interrelations of national and nationalist identity politics on different social levels and their effects.[8]

Regional studies have proved to be an effective way of opening up these complexities.[9] The role of conflicting nationalist discourses has been analysed in the south Bohemian town of Budweis (now České Budějovice).[10] In the case of children's education in Bohemia, Tara Zahra has shown how nationalized popular education systems replaced traditional bi- and multilingualism with monolingualism.[11] The process of 'making into' (e.g. Budweisers into Czechs) involves not only the assertion and passive acceptance of communities of belief, but also persuading or compelling people to accept one or the other nationalist discourse and thus to abandon their respective regional or mixed identities – a process that did not happen without conflicts, resistance or even failure for the nationalists.[12] Conflicts among ethnic groups in a common territory or region could open spaces for the preservation of non-national groups and even for the creation of new groupings.[13]

For such studies the issue of boundary regions has grown in importance. As 'privileged sites for articulation of national distinction',[14] frontiers offer the possibility of observing the creation of symbolic boundaries, persistent conflicts, cultural hybrids or interrelations between empire-formation and nationalism. Interesting in this respect is the collapse of the Habsburg Empire, the central government of which underestimated and later failed to control the centrifugal forces of nationalist movements.[15] The borderland idea can have multiple connotations, of course. The Polish term for the eastern borderlands of the interwar Polish state, '*Kresy*', denotes not only the relation of these territories to the earlier Commonwealth as a mythologized, paradisiacal space, but also the border between civilization and barbarism; '*kres*' literally means 'the end'.[16]

Thus, in recent years the nearly exclusive primacy of power politics that had once governed discussion of this topic has been decentred, if not displaced, by social and, more recently, cultural perspectives. The move to cultural approaches is eminently justifiable, given that the idea and ideal of a *Kulturnation* (cultural nation) has a history of its own, which in the eyes of the Habsburg Monarchy conflicted with that of the *Staatsnation* (nation-state), meaning in this case the idea of 'Austria' as coterminous with the lands ruled by the Kaiser, but in many respects was also formative for other *Staatsnationen* (nation-states; in this case Germany and Hungary as well as the Eastern European nation-states after 1918) throughout the region.[17] Of course, the centrality of language is obvious here, and this was emphasized by the actors at the time. It is not accidental that the Slavic vernaculars were codified during this period. However, a cultural perspective also requires focus on institutions – schools, universities, academies of arts and sciences, professional or other associations, and the media – which were sites of national (or nationalistic) cultural creation as well as cultural conflict. Seen in this light, the previously predominant role of the overseas colonial powers in the discussion of nationalism also becomes increasingly complicated, as it becomes clear that France and Germany (and, we might add, Italy) also found themselves enmeshed in processes of internal nationalization during this period, and that even the supposedly single-ethnic nation-states of the twentieth century always incorporated ethnic minorities.[18]

A third direction of research has focused on the issue of 'centre–periphery' relations. Originating in world-system theory, which referred to political–economic power relations and social structures, this conception worked at first from a simple dichotomy of development versus underdevelopment, with 'development' limited only to the supposed 'centre'; at the same time, such dualisms were modified to some extent by pointing to the local dominance of creole elites in the capital cities and provincial capitals of the supposed 'periphery'.[19] While this model can be justly criticized for being (West) Eurocentric,[20] its development from a 'generalizing' to a 'relativizing' project[21] shows productive potential. The so-called periphery is no longer seen merely as a region of exploitation, but also as a space with a structure of its own, in interplay with other regions as well as with the 'centre'. One might agree with the Galician-born Jewish writer Joseph Roth – albeit without adopting his nostalgic tone – that 'The essence of Austria [meaning: the Habsburg Empire] is the periphery.'[22] Consistent with the emphasis on bottom-up rather than top-down political perspectives, and thus on the role of regions and borderlands, just described, the 'centre–periphery' idea has also been modified in studies of the Habsburg Monarchy. Alongside the long established opposition of Vienna and also Budapest as imperial centres and the other imperial territories, attention has turned to 'centres' within the alleged 'periphery' – for example, regions within Bohemia, Bohemia versus Moravia, or Galicia as a multi-ethnic entity unto itself – as well as migration issues.[23]

Originally the 'centre–periphery' dichotomy applied mainly to colonial empires and to the successor states of former colonies in relation to their European metropoles. The political situation of Central Europe does not correspond to such a classical colonial setting, but recent research has suggested that models from post-colonial studies might be applied to cultural processes in the region. Ursula Prutsch, for example, calls the Habsburg Empire 'soft colonial' and emphasizes its declining influence in the nineteenth century vis-a-vis territorial overseas empires.[24] Combining a post-colonial perspective with the regional focus described above, Robert Donia speaks of Bosnia-Herzegovina as a Habsburg 'proximate colony'.[25] Hans-Christian Maner traces the annexation of Galicia to a colonialist project hidden behind a discourse of cultural transfer and civilizing mission,[26] and Philipp Ther proposes a transnational approach to the creation of a German 'continental empire'.[27] The many differences between overseas colonial empires and (East-)Central Europe – including the lack of racial codification (excepting that of the Jews, then widely identified as a 'race'), the multi-centric rather than exclusively metropolitan administration of the Habsburg Monarchy, and increasing national autonomy within this multi-centric framework – have been cited in support of the rejection of the relevance of post-colonial theory for this region.[28] However, the continued use of German as a common language of administration, at least in 'Cisleithania' after the establishment of the Dual Monarchy in 1867, and the prominence of colonialist discourse in varied forms even in an empire without overseas colonies might suggest a different view. From a wider European perspective, the view of respective 'others' by national elites appears to have differed largely according to a geographical codification: while the 'West', following imperialist discourse, was perceived as a place of culture and especially of civilization, the 'East' was represented as a space of barbarism and chaos.[29] Central Europe, on the other hand, was presented as a transmitter of ideas or as a bulwark defending Western civilization from barbarism.[30] As some of the chapters in this volume will show, the idea of a 'civilizing mission' of the German-speaking centre in Vienna for the rest of the empire was clearly important for the organization of scientific work (see below).

Such post-colonial perspectives can be applied at the regional level also. Though it was long contested, the idea that Central Europe was influenced by regimes with (partially) colonial characteristics has been taken up especially by Polish historians,[31] and has recently been introduced also to analyse the specific features of cultural productivity under imperial circumstances, and partially extended into the Soviet period.[32] This has brought with it a re-evaluation of the dual role of Poland as 'colonized and colonizer',[33] not only in the time of statehood and the Polish-Lithuanian Commonwealth, but also in the period of the partitioned state, during which an assumption of cultural hegemony guided a Polish 'civilizing mission' against Ruthenians or Lithuanians. The process of 'othering' stabilized not only imperial but also local identities, among them national ones, separating Germans from Slavs or Poles from Ruthenians.[34] Studies of self-identification as colonized peoples can be placed in this context, ranging from Ludwik Powidaj's 'subaltern' comparisons of Poles and Indians in 1864[35] to Slovak literature of the interwar period.[36]

The idea of 'micro-colonialisms'[37] suggests that imperial and nationalistic 'nation-building' strategies in the nineteenth century were not opposed, but may have been interdependent, both on the discursive level and in *Realpolitik*. Examples of such interdependence are the use of images from imperial projects like the *Kronprinzenwerk* (a vast compilation of physiognomies, clothing styles and customs of the empire's ethnic groups initiated by Crown Prince Rudolph and published in eighteen volumes from 1884 to 1902) in nationalist discourse,[38] or the intentional undermining of Habsburg imperial history via the delimitation of a national and an imperial past by national scholars.[39] The 'colonial gaze' includes, for example, processes of orientalization,[40] inscriptions of cultural hegemony or translations of racial concepts into ethnic[41] or linguistic[42] ones. As is well known, Jews were regarded as an internal 'other' from the beginning of the nation-building project, even – perhaps especially – after they were granted citizenship rights in Cisleithania in 1867, and despite – or perhaps because of – the fact that Jewish intellectuals inscribed themselves into the process of nationalization through cultural assimilation.[43]

It is not our purpose to choose among, let alone to verify or falsify any of, these recent research trends. Rather, we seek here to explore the varied roles of the sciences as cultural enterprises in processes of nationalization. In doing so, we treat scientists and the sciences not only as reactions to or additional exemplars of, but also as co-creators of such processes.

Nationalization, internationalization and the sciences in Central Europe

The role of the sciences and scholarship in nationalization processes has, of course, not been entirely ignored by historians.[44] General historians, however, when they mention scientists or scholars at all, often tend to focus selectively upon particular groups of academics, whom we might call the usual suspects – literati, philosophers, linguists, historians and, more recently, ethnologists.[45] In our opinion, the assumption that humanists were the leading actors in the co-construction of science and national identity, though not entirely false, requires a new look and certainly supplementation, perhaps even modification. The now famous (and overly simple) 'two cultures' divide between humanistic disciplines and the natural sciences was only beginning to form at the time; it is therefore inappropriate to presuppose its existence and political or cultural effectiveness in this period. Moreover, both natural scientists and humanists in mid- and late nineteenth-century Central Europe were educated largely in the same elite secondary schools and should thus be regarded as members of a common culture. Thus it would be surprising if natural scientists, as members of the educated elite, had not also participated in the nationalization process.

Seen in this light, historians' perhaps understandable biases towards word-centred fields and their corresponding aversion to engaging with technical questions of natural scientific knowledge may support a tendency to ignore the potentially significant roles and specific problems of the natural sciences and scientists that are at the centre of this volume. An additional factor might be an uncritical acceptance of the assumption that the natural sciences are per se international by virtue of their 'objectivity'.[46] However, as recent work in history of science has shown, and as will be shown here as well, the very internationality presupposed to characterize (natural) science was itself a historical invention.[47] This is also true of the idea of impersonal (or supra-personal) 'objectivity'.[48] Both constructs emerged, along with the institutional foundations that supported them, in the very nineteenth century that is generally characterized as the age of nationalism. Indeed, as the chapters in this volume show, the claim that genuine science is per se international was itself central to intramural debates and struggles for institutional power, both within and among nationalizing scientific communities. This tension-filled, dynamic interaction of cultural nationalism and scientific objectivity is, or should be, central to what is meant by the often stated claim that the nation-state and cultural modernity emerged together.

In a volume of essays on the subject, historians Ralph Jessen and Jakob Vogel open up five related issues for discussion: the institutionalization of the (natural) sciences and the humanistic disciplines in the context of the emerging nation-state; the role of the 'nation' as an argument (or trope) both within the sciences and in their relations with state sponsors; the function of natural sciences (and, we might add, technological achievements) as symbolic resources for nationalistic discourse; contributions of the sciences and scholarship to the 'scientific' construction of nations and national cultures; and the already mentioned dynamic tension between nationalistic/patriotic impulses and transnational or universalist orientations in the sciences.[49] As Jessen and Vogel make clear, none of these issues can or should be limited to the natural sciences or the humanities alone; rather, all of them are present across the spectrum of disciplines, albeit in different ways.

Historians of science and medicine have themselves taken up this subject only recently, after taking a cultural turn of their own.[50] Of particular relevance to this volume is an issue of the journal *Studies in History and Philosophy of Biological and Biomedical Sciences* entitled 'Science, medicine and nationalism in the Habsburg Empire from the 1840s to 1918', published in 2007.[51] Drawing upon the vast literature on science and imperialism (discussed further below), the authors provide examples in support of the view that 'it was in the interplay between nationalism and imperial aspirations, regional concerns and "central" impulses as well as international rivalries and internal interests that new forms of disciplinary knowledge and practices were created, to make sense of the empire.'[52] As a case in point,

Emese Lafferton argues that ethnologists and physical anthropologists scientifically constructed the idea of a Hungarian nation and supported political liberalism by delineating the ethnic diversity of nineteenth-century 'Transleithania' in ways that do not fit the models derived from the British, French or German cases.[53] Recent studies have detailed the reception of Darwinism in this region from a multi-centric perspective,[54] and numerous writings by Marius Turda and others have explored the relationships of eugenics, racism and nationalism in (East-)Central Europe from comparable perspectives.[55]

Given the now accepted fact that natural scientists, along with humanists and social scientists, were also involved in, and indeed were co-creators of, cultural processes of nation-formation in this period, the question arises what, if anything, is specific to their involvement. Reducing somewhat the list of issues proposed by Jessen and Vogel, mentioned above, we discuss this question in an exploratory way with respect to three issues: language and symbolic constructions of nationhood; community loyalties and 'centre–periphery' issues; and imperialist science in an empire without overseas possessions.

The importance of language in the creation of national identities has long been clear. Philologists such as Josef Dobrovský for Czech or Onufry Kopczyński for Polish established guidelines for the (re)invention of national languages; these included their historization and vernacularization, which were also intended to eliminate or lessen foreign influences on vocabulary and syntax. Their projects were both political and philosophical, and were closely linked to discussions among French and German linguists of the time.[56] This is a clear example of what Jessen and Vogel call the 'scientific' construction of national cultures. As Jan Surman shows in this volume, however, some of the same linguists were also natural scientists who were engaged in efforts to develop standard scientific terminologies in their several languages.[57] The tensions and debates between national-linguistic 'purism' and the requirements of scientific communication become especially evident in such cases.

The language issue was complicated in the case of the natural sciences by the codification of scientific terminologies and their expression in the very national languages that were themselves being newly codified in the nineteenth century. Given that these scientific terminologies were often created during the period under discussion, it should be clear that we cannot speak here of fixed, previously established scientific terms that only needed to be translated into (East-)Central European vernaculars. Rather, in many natural sciences, most prominently in chemistry, the international terminologies of the disciplines themselves were being established at the same time that efforts were being made to develop such terminologies in national languages.[58]

Further examples of the impact of language-centred cultural hegemony in the sciences are cases of unequal scientific transfer and priority disputes.

As William H. Brock has shown, the insertion of nationalistic rhetoric into questions of scientific validity was popular already in the eighteenth century; relations between 'German' and 'French' chemistry were complicated by nationalism on both sides. Here the rejection or acceptance of knowledge claims was not an issue of science alone, but rather of community membership and nationalistic alignment.[59] In the nationalized spaces of Central Europe the linguistic issues just mentioned reinforced this process; as contemporaries realized, scientific papers published in regional languages were not likely to be noticed at all outside the local community. Dual publication of the same results in a national and a widely read language was often seen as a solution,[60] but this strategy, too, became caught up in discourses of cultural dependency. For example, claims of priority for the discovery of the condensation of oxygen or the creation of the periodic table of chemical elements were often rejected or ignored by scholars who found it difficult to imagine that 'peripheral' scholars were not emulating the work of their 'betters' in the metropoles, but presenting original results.[61]

This example points to the evident linkage between cultural creation and 'centre–periphery' issues in the sciences. The recent emphasis on the multiplicity of 'centres' and 'peripheries' in (East-)Central Europe has resulted in a new emphasis on space that is particularly relevant to historical studies of science. The 'centre on the periphery' idea[62] denotes a particular regional scientific development that was enabled through particular circumstances at the periphery, but was also important for the metropolitan centre, in this case for general science. While the social component of knowledge production is reasserted here, this approach can also provide ironic reassurance that regional innovation on the periphery is only an occasional occurrence. The term 'centre' denotes here a resource-rich spatial congestion, which defines the shape of science. This idea of a normative cultural dichotomy with the metropoles of the 'West' defining what counts as 'science' and other places adapting to such definitions raises questions about the formation of scientific spaces, in which local knowledge production appears condemned to fall behind, whereas the idea of normative and generalized 'science' is ascribed almost entirely to the best work done in the 'West'.

One can see such schemata at work also in studies of the transmission and reception of science, most of which ask how certain ideas originating in the metropoles became globalized, locally received and appropriated elsewhere.[63] Current historiography, revolving largely around English, French and German research centres, thus overshadows local knowledge production and circulation systems. While this relative imbalance is partially caused by language and communication constraints,[64] concentration on 'global' developments so defined certainly limits the visibility of cultural variability and thus the broader base of science.

The idea of 'periphery' has also grown in popularity and has acquired more positive connotations in the process. The working group 'Science and

Technology in the European Periphery', for example, aims to encourage research on spaces that are not part of a historiographical canon of science 'still shaped by a central focus on French, British, German, and increasingly US national narratives ("the big four")'.[65] Clearly, the authors of this website believe that being seen on the 'periphery' of the 'World System' of science is better than not being seen at all; but by writing in this way, they reinforce the classic dualism of 'centre' and 'periphery' that they claim to oppose. This development, although it certainly multiplies perspectives on science, thus follows the predominant inscriptions of global science and helps to secure their hegemonic position.

As stated above, such cultural perspectives cannot and should not be separated from institutionalization issues. An oft cited case of radical conflict is the division of Prague University in 1882 into the German-speaking Charles University and a Czech counterpart.[66] Though plainly spectacular, this case may not be typical; more frequent were conflicts over the use of national languages in teaching and research within existing institutions, which led to serious divisions and often to the migration of affected individual scholars and scientists to other places, but did not necessarily lead to institutional break-up. Equally important as 'centres on the periphery' are the national academies of sciences founded in this period. The Hungarian Academy of Sciences (established in 1845), the Academy of Sciences and Arts in Cracow (established in 1872 and renamed 'Polska Akademia Umiejętności' in 1918), the Shevchenko Scientific Society (established in 1873 and regarded by its members as a Ukrainian academy of sciences even before achieving official status as such in 1893), and the Czech Academy of Sciences and Arts (founded in 1890) can all be regarded as cultural symbols and active constituents of national identity in (East-)Central Europe, and also as 'national' institutions established long before there were nation-states in these places.[67] Since they were also sites of philological research on the history, semantics and general structure of vernacular languages, they are clearly examples of the interaction of cultural history and *'Gelehrtenpolitik'*.

Central here is the tension between national (or 'patriotic') loyalties and membership in international scientific communities.[68] Of course, this tension was by no means limited to Central Europe. Yet the predominant version of this discourse, which not coincidentally became established in the nineteenth century, acknowledged no such tension or contradiction. Rather, scientific and technological achievements, suitably certified by international acceptance or economic success, were regarded as cultural resources in a competition for prestige, respect and cultural power. As the great German physiologist and physicist Hermann Helmholtz put it in 1862,

> every nation is interested in the progress of knowledge on the simple ground of self-preservation, even were there no higher desires of an ideal

character to be satisfied; and [this is true] not merely in the development of the physical sciences, and their technical application, but also
in the progress of legal, political, and moral sciences, and of the necessary historical and philological studies. No nation which would be independent and influential can afford to be left behind in the race.[69]

Such statements clearly exemplify what Ralph Jessen and Jakob Vogel
call the idea of 'nation' as argument. Notable in the text by Helmholtz just
cited – and widely noticed at the time – are the statement that 'knowledge
is power' and his use of military metaphors in this context.[70] Not until the
breakdown of international scientific and scholarly communication and
the self-mobilization of scientists, scholars and other intellectuals during
the First World War did this internal contradiction within the cultures of
science become too obvious to ignore.[71] Nonetheless, as many of the chapters in this volume show, such tensions became visible much earlier precisely
in the emerging national cultures of (East-)Central Europe. Institutional
affiliations, political commitments and knowledge claims were all involved;
we are speaking here of a subtle blending of community membership and
epistemic commitments.

It cannot be ignored, however, that while the use of 'nation' as argument grew more visible in the nineteenth century, this did not hinder
the practice of cultural exchange and contacts. Quite the contrary: international experience and contacts remained crucial for academic careers,
and became more vital as the century went on.[72] Cultural entanglement
was facilitated by imperial circumstances but increasingly went beyond
the boundaries of the Habsburg Monarchy towards centres in Berlin,
Naples, Paris and, increasingly, Great Britain. In this regard 'German
culture' remained both appealing on account of its scientific connotations
and distressing through its imperial connotations; Julian Dybiec called the
Germans in Polish culture of the late nineteenth century at once 'oppres-
sors and teachers'.[73]

As a major European power without colonial possessions overseas, the
Habsburg Monarchy clearly occupies an unusual, and often ignored,
position within the much studied complex 'science and empire'.[74] Austrian
scientists and scholars participated in the far-flung expansion of the
'empire of science' in the nineteenth century, and thus in the attempts at
symbolic capital acquisition that these entailed. Studies of colonialism and
the sciences in (East-)Central Europe have focused on Bosnia[75] and overseas exploration.[76] Examples of the latter from the period discussed in this
volume include the round-the-world voyage of the frigate 'Novara' (1857–59)
and the polar expedition of the 'Admiral Tegetthof' (1872–74).[77] These spectacular and well publicized events may have compensated in some ways
for the empire's lack of overseas colonial territories. At the same time they
focused public discussion of and political credit for these enterprises on the

Habsburg Monarchy and the Austrian Academy of Sciences, in effect identifying imperial institutions with the Austrian 'nation' per se.

Of still greater importance for the topic of nationalization are cases of
'internal' colonization in a largely landlocked empire.[78] As earlier literature
and chapters in this volume all show, organized, large-scale cartographic,
survey and data-gathering projects in the sciences played central roles in
the neo-absolutist project of establishing unity in the monarchy after the
revolutions of 1848. Examples include the foundation of the central agency
for meteorology, climate research, and seismology in Vienna in 1851, and
the Imperial Geological Survey in 1849.[79] The extended networks of actors
in such projects were all carefully coordinated and staffed from Vienna;
the aim was to achieve nothing less than the science-based construction
of 'Austria' as a naturalized empire, despite the fact that it largely lacked
natural borders. Similarly, the departments of the Imperial Museum of
Natural History in Vienna, housed in an elaborate historicist building
constructed on the Ringstrasse in precisely this period, served as gathering points for research collections in botany, mineralogy and many
other sciences, while the museum's public exhibition rooms were (and to
some extent still are) decorated with landscapes of the Kaiser's territories
from which the exhibited objects had come.[80] At the same time, 'centres'
in the 'periphery', such as regional and so-called national museums,
served as forums for 'national' cultural display and self-creation outside
the Viennese 'centre'.[81]

Such examples show that the special case of Central Europe is a highly
relevant focal point both for general history and the history of science in
relation with one another. Of course, 'Central Europe' has itself been a
contested concept since the early twentieth century. We focus here on the
Habsburg Monarchy but include German developments in so far as they
impinged on the region, and also discuss trends in Polish-speaking regions
not incorporated into the monarchy. It was precisely during this period that
a self-defined multi-ethnic entity (*Vielvölkerstaat*) increasingly became a
multinational one. In this vast territory nationalization was not pursued
in a top-down manner, as in Russia or the US, but in opposition to central
authority. Nonetheless, German remained the language of central administration, and a major language in the sciences, even after the establishment
of the Dual Monarchy and, as chapters in this volume show, aspects of what
might be called a colonial perspective, including the rhetoric of science in
service to a 'civilizing mission', are visible in the attitudes of state officials
and Viennese scientists.

Whatever viewpoint is chosen, all of the volume's authors maintain a
strict historical stance. The monarchy ultimately collapsed – and many
thought with some justification that it had already become ungovernable
by the early 1900s – but this outcome should not be seen as inescapable, nor
distract us from recognizing the positive potential that many scholars and

scientists saw in national self-assertion in earlier periods. We are well aware of the dangers of positing a negative teleology here.

The chapters in outline[82]

The first two chapters of this volume pursue some of the wider issues raised in this introduction from parallel viewpoints. In a chapter entitled 'Science and its Publics: Internationality and National Languages in Central Europe', Jan Surman focuses on the factor generally considered central to the process of nationalization – the emergence of and insistence upon vernacular languages. If 'imagining community' means creating particular interpretations of regional identities and imposing cultural definitions that create boundaries with other communities, its scope includes all spheres of public life. This highly disputed process of impoverished emancipation included the scientific landscape of the Habsburg monarchy; universities, which became important scientific research institutions at just this time, and newly founded scientific societies were favoured battlefields. This chapter examines the nationalization of the sciences and scholarship across a wide range of disciplines through the long nineteenth century. Many of the conflicts described here can also be seen as conflicts among 'imagined communities' – of nationalistic language 'purists', for example – trying to mark their territories, but Surman shows that they also involved struggles over the articulation of scientific knowledge. Language worked at the meta-level, defining by whom and how knowledge was to be presented, and to a certain extent influencing what was researched as well. Science and its practices thus played a not inconsiderable role in processes of cultural boundary creation and maintenance; the importance of language for styles of research became an acknowledged argument in broader cultural debates, and in this way the sciences entered public discourse.

While the nineteenth century is usually described as a time when nationalism and internationalism fell apart, Surman's analysis of the language issue in the scientific communities of the Habsburg Monarchy presents them as highly interdependent. In both the creation and the use of scientific languages, a turn from nationalism to internationalism can be observed by the 1890s, but this internationalism was quite different from the structure of the 'republic of letters' in pre-national times. While both science and scholarship became increasingly international, scientists and scholars also represented the national communities into which they inscribed themselves. At the very same time that impersonal objectivity was coming to be accepted as a core value of science, alleged peculiarities of 'national styles' grew in importance. From this perspective language played a pivotal role, linking nationalistic ideology with the allegedly international character of the scientific community. Surman argues that a process of co-creation occurred: academic communities supported nationalization processes, creating and at the same being formed by nationalistic discourse, and yet precisely this nationalization, once achieved, became the basis for later moves towards international standing.

In '"Staatsnation", "Kulturnation", "Nationalstaat"': The role of National Politics in the Advancement of Science and Scholarship in Austria from 1848 to 1938', Johannes Feichtinger argues that when historians consider national styles of science, it is appropriate to treat the Habsburg Monarchy as a special case. In Austria modern scholarship has confronted both the concept of the nation-state (*Staatsnation*) and that of a culturally defined nation (*Kulturnation*). If science and scholarship used political resources to establish themselves scientifically, they ran the risk of becoming engaged in diametrically opposed political projects. For advocates of '*Staatsnation*', identity and commitment were primarily based on the principle of dynastic rule over the Habsburg territory. Proponents of the '*Kulturnation*' constructed their specific national understanding by using cultural difference as a means of demarcation. Those subscribing to the '*Gesamtstaat*' (meaning loyalty to the Monarchy as a whole and especially to the person of the Emperor) usually attacked the adherents of the '*Kulturnation*' as nationalists, or language nationalists, but in fact both positions strove for nationhood, though conceived in opposite terms. The chapter uses numerous examples from the humanities, especially philosophy and legal theory, and also alludes to cases from the natural sciences to show how the Austrian academic community of the nineteenth and early twentieth centuries tried to meet this challenge, while at the same time dealing in various ways with the need to establish and maintain some form of scientific autonomy.

The following chapters focus on interactions of institutional and epistemic perspectives. In her chapter, 'National Agreement as Style of Thinking? The Geological Survey of the Habsburg Empire (1849–67)', Marianne Klemun discusses the way in which geology contributed to the symbolic construction and practical maintenance of a multinational Habsburg Monarchy and formed the basis of the project of political harmonization after the failed revolution of 1848. The quintessential task of the Imperial Geological Survey in Vienna, founded in 1849, was clearly defined as the comprehensive incorporation of geological knowledge from all of the nations under the Habsburg crown. Within fourteen years this project produced the desired consistent geological mapping covering an enormous, geologically diverse territory extending from Lombardy to Bukovina and from Dalmatia to the gorge of the Elbe. The result, the *Geological Survey Map of the Austrian-Hungarian Monarchy*, published in 1867, was achieved by means of an elaborate set of negotiated relationships for the fieldwork in which many geologists took part; this involved a practical culture of 'mixing' or 'agreeing'. Consensual relations of unity gained strength in this context; the goal was a uniform transformation of these countries into a geologically coherent supranational territory, modelled as a unified entity and also scientifically defined.

Stratigraphy gave the map an abstract temporal dimension that was both naturally determined and at the same time profoundly political.

In his essay, 'Scientific Nationalism: A Historical Approach to Nature in Late Nineteenth-Century Hungary', Gábor Palló reinforces the volume's central claim that the cultural fertility of the Austro-Hungarian Dual Monarchy extended to the natural sciences. Following Ernest Gellner's three criteria typology of nationalism – the existence of a centralized power, education, and shared (high) culture – he argues that it is sensible to look for nationalist features in Hungarian science in the same way as it is sensible to look for nationalist features in Hungarian literature or dance. Since nationalism is a political principle, Palló argues, nationalist science should be considered to be a political actor in realizing the goal of constructing a (linguistically) homogenous high culture. In both parts of the Dual Monarchy, a number of important results were achieved and scientists achieved influence within and beyond their borders. However, according to Palló, the political power positions of the two parts, Cisleithanien and Transleithanien, 'Austria' and Hungary, were not symmetrical. This asymmetry was reflected in the continuation of Hungarian nationalism, born in the late eighteenth century. Nationalism was a characteristic feature of Hungarian culture, literature, music and science. Compared with Austrian universalism, however, Palló maintains, Hungarian scientific thinking was local, practical and historical. The scientific controversies at the Hungarian Academy of Sciences argued for the importance of national science. In addition to sociological, political and linguistic endeavours, nationalism influenced the content of scientific research through its traditional natural historical approach. The chapter details the peculiar natural historical approach and its manifestations in chemistry, biology and physics in Hungary in the late nineteenth and early twentieth centuries.

In his chapter, 'Acts of Creation: The Eötvös Family and the Rise of Science Education in Hungary', Tibor Frank approaches the case of Hungary by examining the political biographies of and the science education policies advocated and instituted by Baron József Eötvös and his physicist son Baron Loránd Eötvös; also discussed is Ágoston Trefort, brother of the elder and uncle of the younger Eötvös. Both father and son were ministers of education and religious affairs in Hungary at least briefly – the elder Eötvös held that post twice – while Trefort served in the position from 1872 to 1888. In addition, both Trefort and the younger Eötvös were elected President of the Hungarian Academy of Sciences. This family network was thus in a position to articulate its ideas on nationality and science and put them into practice. Frank's discussion is set against the panorama of Hungarian history from 1848 through the establishment of the Dual Monarchy in 1867 to the period of unparalleled prosperity and cultural-intellectual creativity that followed a generation later. He notes that Mór Kármán (father of the world-famous aerodynamicist Theodor Kármán) introduced a secondary education reform inspired by the German Gymnasium in the 1870s at the instigation of the elder Eötvös, and presents the thesis that for the elder Eötvös, his German-trained scientist son, and many others at that time German culture remained paradigmatic for their thinking. For these three men there was apparently no contradiction in principle between loyalty to German *Kultur* and Hungarian patriotism.[83]

At the same time, however, Hungarian became the language of secondary-school instruction after 1848 and gradually became pre-eminent also in university instruction after the 'Compromise' of 1867. Loránd Eötvös appears to have seen no difficulty in establishing circles for elite mathematical discussion and supporting a mathematics achievement contest for high-school students that was eventually named after him, all conducted in Hungarian, while publishing his most important physics research in German. Frank acknowledges that this ideal of bi- or even multilingual nationality was not universal at the time. Nonetheless, for him the Eötvös family personifies a model of Hungarian creativity combined with political liberalism that was exemplified to a greater or lesser degree by many other outstanding scholars, scientists and cultural leaders of this period and afterwards.

Soňa Štrbáňová discusses epistemic and institutional dimensions in her chapter, 'Patriotism, Nationalism and Internationalism in Czech Science: Chemists in the Czech National Revival'. She describes the dilemma faced by her subjects in stark terms: 'To be a good son of one's nation or to become involved in supranational scientific networks?'. These choices were only apparently in opposition to one another. The notion of 'national style' was widespread in European science of the second half of the nineteenth century, and some of its characteristic features acquired distinct form in the multinational and multi-ethnic Habsburg Monarchy. With respect to the Czech Lands, science tended to be both 'patriotic science' (with nationalistic attributes) and 'internationalist science'; however, these directions were not inevitably divergent. The 'provincial patriotism' (*Landespatriotismus*) of the first half of the nineteenth century turned into fierce national or ethnic patriotism in the second half. Peculiar to the Czech Lands was the parallel manifestation of such attitudes in Czech and German linguistic environments. At the same time, the rapid development of European science made international cooperation attractive for Czech scientists. All these tendencies were reflected in the attitudes and actions of individual scientists or scientific institutions in the Czech Lands.

What did it mean to be a patriotic Czech scientist in the Czech Lands, and how did patriotism shape the involvement of Czech scientists in European science? As Štrbáňová shows (as does Surman more broadly in his chapter), patriotism in science played a positive role in promoting the formation of a Czech scientific terminology, as well as Czech research and educational institutions and communication bases. It also enabled Czech scientists to participate in international cooperation among Slavic scientists. However, in the 1880s and 1890s patriotism in science started to shift into nationalism and chauvinism, marked by a sharp demarcation from German science. These tendencies, along with an effort to abuse science for political ends, created barriers obstructing participation by Czech scientists in international networks. Chemists, however, soon became aware of this threat; in response they and other natural scientists nurtured international ties and attempted to counteract nationalism. Developments in the humanities were different, as demonstrated by the example of history.[84] However, both scientists and humanists found appropriate ways of communicating with international scholarship, and this led to methodologically and thematically fruitful dialogue.

In her chapter, 'Fault Lines and Borderlands: Earthquake Science in Imperial Austria', Deborah Coen returns to institutional and epistemic issues raised by Marianne Klemun while focusing on a later period. The catastrophic Ljubljana earthquake of 1895 spurred the Imperial Academy of Sciences in Vienna to initiate a macroseismological observation network of lay volunteers, on the Swiss model. The question was how to organize such a network for the multinational empire: how to coordinate expert–lay communication in a state driven by linguistic and cultural divides, and how to impose a uniform observation system in a territory ranging from the seismically active lands of Southern Europe to the steppes of Galicia and Bukovina, where earthquakes were virtually unknown. The chapter examines the network's decentralized structure. It draws on the reflections of both Austrian and foreign scientists on the advantages and disadvantages of decentralization, as well as on archival letters, telegrams and questionnaires that reveal the texture of communication within the network. Coen is interested in the ways that the perceived geography of a continental empire – non-natural borders, 'hybrid' frontiers, ambiguities of centre and periphery – affected the construction and mapping of seismological knowledge.

In his chapter on eugenics in Hungary, Marius Turda focuses tightly on the public debate on eugenics that took place between 1910 and 1911. As he shows, the debate is important, first of all, because it created an auspicious environment for the nationalization of eugenic knowledge that was to occur in early twentieth-century Hungary. Although they used the internationally recognizable language of evolutionary science, Hungarian eugenicists expressed specifically local imperatives. In doing so, they became co-creators of the national context, enabling Hungarian academic knowledge about evolution and heredity to be expressed publicly. Eugenics, in this context, was seen as a mechanism capable of decoding particular social and national predicaments, an expression of the ideal of a healthy nation in the face of dramatic demographic and social changes. In addition, however, Turda shows that the debate had an international dimension, illustrating the level of scientific sophistication achieved by Hungarian eugenicists at the time; in other words, their scholarly engagement with emerging European trends in heredity. Finally, Turda suggests that the Hungarian debate also had regional importance, because it was the first public debate on eugenics in Austria-Hungary. Its particulars may thus help us to understand other national eugenic movements in this region.

In the volume's final chapter, 'The Politics of Anatomy in Late Nineteenth-century Vienna', Tatjana Buklijas returns to the Habsburg capital. Around 1900 Viennese anatomy was internationally famous as a result of the elegant specimens displayed at world exhibitions, widely read textbooks and innovative atlases, as well as the easy access to dissectible bodies that attracted students from countries as far away as the US. But it was also a discipline divided among politically, educationally and scientifically opposed professorships . In her chapter, Buklijas details the lives and careers of anatomists who held the two normal anatomy chairs at the University of Vienna at that time, Emil Zuckerkandl and Carl Toldt, demonstrating the differences between the two anatomical disciplinary orientations practised in fin-de-siècle Vienna and their close links with the political views and social networks with which the two anatomists were allied. In doing so she shows how anatomical divergences can be understood only if we put them back into the contemporary Austrian political and social context, in terms of the growing middle-class rift along ethnic and religious lines. Thus, Buklijas presents a fine example of the nationalization of scientific knowledge at the centre of the Monarchy itself.

Science and historical memory – nationalism unfinished?

The idea that the sciences ceased to be an issue in nationalisation – or re-nationalization – processes after the fall of Communism is obviously wrong. Intensive sponsorship programs by national research organizations, by aiming to strengthen national representation in 'international' science, reinforce long-standing policies of the kind described in this volume. Among the most generously endowed scholarships are homecoming/reintegration scholarships, offered both at the level of the European Union[85] and at national level.[86] Their aim is either to enable research stays in participating countries with a compulsory return phase, or to reintegrate internationally active scholars into the sciences in their countries of origin (in some cases in countries were they have acquired residency status). It will be interesting to follow the impact of such policies (and many other EU-financed networking programmes) at a time when it is finally being acknowledged throughout Europe and elsewhere that more than a century of enforced monolingualism has been severely damaging to science and scholarship, particularly in (East-)Central Europe but also in former scientific power centres like France and Germany.

At least as important as such programmes is the use of the scientific past in the memory politics of (East-)Central Europe. Already in the nineteenth century, 'great men of science' – scientists then always being male – were mobilized as resources for national commemoration and in this way contributed to stabilizing the historical identity of their nations.[87] From then until now, scientists and scholars have continued to be emblematic figures in this sense. National, or rather nationalistic, attributions assigned to Renaissance figures such as the astronomer Yuriy Drohobych in Ukraine, Copernicus in Poland or Comenius in the Czech Republic help to locate the nation in a historical continuum, and at the same time to place it favourably in the context of international scientific development. In recent versions of this cultural game, in a nod to multiculturalism, the peripatetic scholarly careers and varied education of these famous scholars is acknowledged. Yet in the end language remains decisive as a demarcation criterion of nationality attributions, despite long-standing debates on whether the scholar in question ever actually spoke the national language in question (e.g. whether Copernicus spoke Polish) or which version of that language the scholar employed (Drohobych).

The politics of historical memory is by no means limited to such issues, of course, but enters everyday cultural life in many forms. Commemorative celebrations strengthen (national) identity communities by providing support for cultural self-esteem. In this regard university jubilees, or commemorations of famous scholars' or scientists' birth or death dates, continue to take pivotal representational roles.[88] In contemporary (East-)Central Europe, such cultural strategies appear to provide symbolic compensation for these countries' relatively weak international scientific standing – an ironically inverted version of the invocations of science-based technology in support of national power politics that became common in the nineteenth century.

We cannot go into detail about this here, but surely it is not out of place to mention examples of the visibility of science and scientists as cultural symbols in prominent and visible places – even as mundane as banknotes or brand names. Copernicus, for example, was on the 1000 zloty note, and his name once also graced a well-known gingerbread factory in Toruń. In the summer of 2010, the first modern science museum in Poland – called, of course, the Copernicus Science Centre – opened with an extraordinary outdoor multimedia show, 'Big Bang', by Peter Greenaway and Saskia Boddeke, which gained extensive media coverage.[89] Comparable examples exist throughout (East-)Central Europe, wherever the euro – with architectural rather than portrait imagery on its banknotes – has not yet been introduced.[90] The pictorial invocations of glorious national scientific pasts in everyday currency instruments has much the same functions as already described; this too creates an illusion of timeless national continuity (invented tradition) and assigns national identities to scientists and scholars (at least in local eyes), even though Copernicus (as mentioned above),

Marie Skłodowska Curie (= Curie Skłodowska), Grigorij Savvich Skorovoda/ Hryhorii Savych Skovoroda and Johann Weichard Freiherr von Valvasor/ Janez Vajkard Valvasor were (and still are) nationally contested figures. Of course, it is not possible to survey all aspects of this vast and complex topic in a single short volume. Our purpose is, rather, to open up lines of inquiry, and in doing so to cross academic boundaries between general history and the history of science on the one hand, and between the history of natural science or medicine and that of the humanities on the other. The many specific differences among the national histories and fields of knowledge discussed here are obvious; and yet it appears clear to us that there are overriding common patterns – not least the ambivalence between the drive to establish national identities and the equally powerful need to gain standing in transnational scientific cultures – that become visible when studies from varied locations are brought together.

Notes

1. For discussion of the shift from multi- to monolingualism see Jan Fellerer (2005) *Mehrsprachigkeit im galizischen Verwaltungswesen (1772–1914). Eine historisch-soziolinguistische Studie zum Polnischen und Ruthenischen (Ukrainischen)* (Cologne, Weimar: Böhlau), especially 279–80; Tomasz Kamusella (2001) 'Language as an instrument of nationalism in Central Europe', *Nations and Nationalism*, 7/2, 235–51; idem (2009) *The Politics of Language and Nationalism in Modern Central Europe* (Basingstoke: Palgrave Macmillan).
2. Ernest Gellner (1983) *Nations and Nationalism* (Ithaca, NY: Cornell University Press); Eric J. Hobsbawm (1990) *Nations and Nationalism since 1780: Programme, Myth, Reality* (Cambridge: Cambridge University Press); Benedict Anderson (1991) *Imagined Communities: Reflections on the Origin and Spread of Nationalism*, rev. ed. (London: Verso). For a more traditional approach, see also John Breuilly (1982) *Nationalism and the State* (Manchester: Manchester University Press).
3. For studies of nationalism as secular religion, see, for example, George L. Mosse (1975) *The Nationalisation of the Masses: Political Symbolism and Mass Movements from the Napoleonic Wars through the Third Reich* (New York: Howard Fertig); Norbert Elias (1996) *The Germans: Power Struggles and the Development of Habitus in the Nineteenth and Twentieth Centuries* (Cambridge: Polity Press); Anthony D. Smith (2010 [2001]) *Nationalism: Theory, Ideology, History*, 2nd ed. (Cambridge: Polity Press).
4. Miroslav Hroch (2007) 'National Romanticism', in *Discourses of Collective Identity in Central and Southeast Europe 1770–1945 Vol. 2. National Romanticism: The formation of National Movements*, eds. Balázs Trencsényi and Michal Kopeček (Budapest, New York: Central European University Press), 4–18.
5. On the spatial dimension of nationalism in central Europe see, for example, Patrice M. Dabrowski (2008) 'Constructing a Polish landscape: The example of the Carpathian frontier', *Austrian History Yearbook*, 39, 46–65.
6. Partha Chatterjee (1993) *The Nation and its Fragments: Colonial and Postcolonial Histories* (Princeton, NJ: Princeton University Press), 159.
7. This process was analysed in the nineteenth century, for example, by sociologist Ludwik Gumplowicz. For a recent analysis see Brian Porter (2000)

When Nationalism Began to Hate: Imagining Modern Politics in Nineteenth-century Poland (New York: Oxford University Press).

8. Of course, none of this was limited to (East-)Central Europe. See, for example, the classic study by Eugen Weber (1979) *Peasants into Frenchmen: The Modernisation of Rural France* (London: Chatto and Windus).

9. For the German case, see Siegfried Weichlein (2004) *Nation und Region: Integrationsprozesse im Bismarck-Reich* (Düsseldorf: Droste).

10. Jeremy King (2005) *Budweisers into Czechs and Germans: A Local History of Bohemian Politics 1848–1948* (Princeton, NJ: Princeton University Press).

11. Tara Zahra (2008) *Kidnapped Souls: National Indifference and the Battle for Children in the Bohemian Lands, 1900–1948* (Ithaca, NY: Cornell University Press).

12. See, for example, James E. Bjork (2008) *Neither German nor Pole: Catholicism and National Indifference in a Central European Borderland* (Ann Arbor: University of Michigan Press).

13. Tomasz Kamusella (2007) *Silesia and Central European Nationalism: The Emergence of National and Ethnic Groups in Prussian Silesia and Austrian Silesia, 1848–1918* (West Lafayette, IN: Purdue University Press).

14. Peter Sahlins (1989) *Boundaries: The Making of France and Spain in the Pyrenees* (Berkeley, 1989), 271.

15. On this issue see Hans-Christian Maner (ed.) (2005) *Grenzregionen der Habsburgermonarchie im 18. und 19. Jahrhundert: Ihre Bedeutung und Funktion aus der Perspektive Wiens* (Münster: LIT Verlag).

16. For an earlier critique of the concept, which preceded the intensification of the debate by at least a decade, see Daniel Beauvois (1994) 'Mit "kresów wschodnich" czyli jak mu położyć kres' [The Myth of "Eastern Borderlands" and how to end it], in Wojciech Wrzesiński (ed.), *Polskie mity polityczne XIX i XX wieku* [Polish political myths of the nineteenth and twentieth centuries] (Wrocław: Wydawnictwo Uniwersytetu Wrocławskiego), 93–105. Recently, critiques of this terminology have brought about a post-colonial turn in Polish self-reflection on its political position and cultural identity politics. See, for example, Bogusław Bakuła (2006) 'Kolonialne i postkolonialne aspekty polskiego dyskursu kresoznawczego (zarys problematyki)' [Colonial and postcolonial aspects of Polish borderland-science discourse: an outline], *Teksty Drugie*, 6, 11–33.

17. For an extended discussion of these complex relationships, see the chapter by Johannes Feichtinger in this volume.

18. See, for example, Holm Sundermann (2007) 'Die Ethnisierung von Staat, Nation and Gerechtigkeit. Zu den Anfängen nationaler "Homogenisierung" im Balkanraum', in Matthias Beer (ed.), *Auf dem Weg zum ethnisch reinen Nationalstaat? Europa in Geschichte und Gegenwart*, 2nd ed. (Tübingen: atempto), 69–90.

19. Immanuel M. Wallerstein (2011) *The Modern World System*, 4 vols. (Berkeley, CA: University of California Press). First published 3 vols., 1976–1988; André Gunder Frank (1970) *Latin America: Underdevelopment or Revolution. Essays on the Development of Underdevelopment and the Immediate Enemy* (New York: Monthly Review Press). See also Immanuel M. Wallerstein (1998) 'The Construction of peoples: Racism, nationalism, ethnicity', Chap. 4, in Etienne Balibar, Immanuel M. Wallerstein (eds.) *Race, Nation, Class: Ambiguous Identities* (London: Verso).

20. For an indirect critique of this model, see Dipesh Chakrabarty (2000) *Provincializing Europe: Postcolonial Thought and Historical Difference* (Princeton: Princeton University Press).

21. For this distinction see Richard Handler (ed.) (2006) *Central Sites, Peripheral Visions: Cultural and Institutional Crossings in the History of Anthropology* (*History of Anthropology, Volume 11*) (Madison, WI: University of Wisconsin Press).

22. 'Das Wesen Österreichs ist die Peripherie'. Joseph Roth (1987 [1938]) *Die Kapuzinergruft* (Cologne: Kiepenheuer & Witsch), 17.

23. Pieter M. Judson and Marsha L. Rozenblit (eds.) (2005) *Constructing Nationalities in East Central Europe* (Austrian and Habsburg Studies, 6) (Oxford and New York: Berghahn Books); Pieter M. Judson (2006) *Guardians of the Nation: Activists on the Language Frontiers in Imperial Austria* (Cambridge, MA: Harvard University Press); Marius Turda (2005) *The Idea of National Superiority in Central Europe, 1880–1918* (New York: Edwin Mellen Press).

24. Ursula Prutsch (2003) 'Habsburg postcolonial', in Ursula Prutsch, Moritz Csáky and Johannes Feichtinger (eds.) *Habsburg Postcolonial: Machtstrukturen und kollektives Gedächtnis* (Innsbruck: Studien-Verlag), 33–43, here 36.

25. Robert Donia (2007) 'The Proximate Colony. Bosnia-Herzegovina under Austro-Hungarian Rule', http://www.kakanien.ac.at/beitr/fallstudie/RDonia1.pdf.

26. Hans-Christian Maner (2007) *Galizien. Eine Grenzregion im Kalkul der Donaumonarchie im 18. und 19. Jahrhundert* (Munich: IKGS-Verlag), 49.

27. Philipp Ther (2004), 'Deutsche Geschichte als transnationale Geschichte. Polen, slawophone Minderheiten und das Kaiserreich als kontinentales Empire', in Sebastian Conrad (ed.) *Das Kaiserreich transnational. Deutschland in der Welt 1871–1914* (Göttingen: Vandenhoeck & Ruprecht), 129–48.

28. See, for example, Alexei Miller (2003) *The Ukrainian Question: The Russian Empire and Nationalism in the Nineteenth Century* (Budapest, New York: Central European University Press); Veronika Wendland (2010), 'Imperiale, koloniale und postkoloniale Blicke auf die Peripherien des Habsburgerreiches', in Claudia Kraft and Alf Lüdtke (eds.), *Kolonialgeschichten: Regionale Perspektiven auf ein globales Phänomen* (Frankfurt am Main: Campus Verlag), 215–35.

29. From the vast literature on this topic, see especially Maria Janion (2006) *Niesamowita Słowiańszczyzna: fantazmaty literatury* [Amazing Slavdom: The Literary Imagination] (Kraków: Wydawnictwo Literackie) and Jaroslav Hrycak (2004) *Strasti za nacionalizmom* [Passion for Nationalism] (Kyjiv: Krytyka), especially the chapters 'I my v Jevropi?' [We in Europe?] (pp. 24–36) and 'Istorija vid Pjatnyci' [History on Friday] (pp. 309–24). For a general overview see Alexander Maxwell (ed.) (2011) *The East–West Discourse: Symbolic Geography and Its Consequences* (Oxford, Frankfurt am Main: Peter Lang).

30. The term *antemurale* is mostly linked to Poland but can be found as a self-identification throughout Central Europe, from Estonia through Ukraine to Kosovo. See, for example, Chantal Delsol, Michel Masłowski and Joanna Nowicki (eds.) (2002) *Mythes et symboles politiques en Europe centrale* (Paris: Presses Universitaires de France); for Ukraine see the project of Liliya Berezhnaya, *'Die ukrainische Bastion'* – *Vormauer Europas und antemurale christianitatis. Nationalisierung eines Mythos*, http://www.uni-muenster.de/Religion-und-Politik /forschung/projekte/b15.html.

31. Paradigmatic here is Jan Kieniewicz (2008) 'Polski los w Imperium Rosyjskim jako sytuacja kolonialna' [Polish fate in the Russian Empire as a colonial situation], in idem (ed.) *Ekspansja, kolonializm, cywilizacja* [Expansion, colonialism, civilization] (Warszawa: DiG), 244–262, or the collection of articles in idem (ed.) (2009) *Silent Intelligentsia. A Study of Civilisational Oppression* (Warsaw: Institute of Interdisciplinary Studies 'Artes Liberales,' University of Warsaw).

32. Janusz Korek (ed.) (2007) *From Sovietology to Postcoloniality: Poland and Ukraine from a Postcolonial Perspective* (Stockholm: Södertörn Academic Studies 32). For critical discussion of differences between postcolonialism and the particularities of post-partition (1793–1918) and post-1945 Poland (1945–89), see Hanna Gosk (2008) 'Polskie opowieści w dyskurs postkolonialny ujęte' [Polish stories framed in postcolonial discourse], in idem and Bożena Karwowska (eds.) *(Nie) obecność: Pominięcia i przemilczenia w narracjach XX wieku* [Presence/Absence: Omissions and Concealments in narrations of the twentieth century] (Warszawa: Elipsa), 75–88.

33. Maria Janion (2004) 'Rozstać się z Polską?' [To part with Poland?], *Gazeta Wyborcza*, 02./03.10.2004, 14–16, here 16.

34. See, for example, Pieter M. Judson (1993) 'Inventing Germans: class, nationality and colonial fantasy at the margins of the Habsburg Monarchy', *Social Analysis*, 33, 47–67; Danuta Sosnowska (2008) *Inna Galicja* [The other Galicia] (Warszawa: Elipsa).

35. Izabela Surynt (2008) 'Postcolonial studies and the "Second World": twentieth-century German national-colonial constructs', *Werkwinkel*, 3/1, 61–87.

36. Jozef Špetko (1986) 'Ubližovanie – mýtus a syndrom' [Rapprochement – myth and syndrome], *Premeny*, 3, 3–13.

37. Johannes Feichtinger, Ursula Prutsch and Moritz Csáky (2003), 'Vorwort', in idem (eds.) *Habsburg Postcolonial: Machtstrukturen und kollektives Gedächtnis* (Innsbruck: Studien Verlag), 11.

38. Christiane Zintzen (ed.) (1999) *Die österreichisch-ungarische Monarchie in Wort und Bild. Aus dem Kronprinzenwerk des Erzherzog Rudolf* (Cologne, Weimar, Vienna: Böhlau-Verlag); Regina Bendix (2003) 'Ethnology, cultural reification, and the dynamics of difference in the Kronprinzenwerk', in Nancy M. Wingfield (ed.) (2003) *Creating the Other: Ethnic Conflict and Nationalism in Habsburg Central Europe* (New York/Oxford: Berghahn Books), 149–66.

39. Serhii Plokhy (2005) *Unmaking Imperial Russia: Mykhailo Hrushevsky and the Writing of Ukrainian History* (Buffalo, NY: University of Toronto Press).

40. Izabela Surynt (2004) *Das 'ferne', 'unheimliche' Land: Gustav Freytags Polen* (Dresden: Thelem Verlag).

41. Brigitte Fuchs (2003) *'Rasse', 'Volk', 'Geschlecht': Anthropologische Diskurse in Österreich, 1850–1960* (Frankfurt am Main: Campus Verlag).

42. Mykola Riabchuk (2000) *Vid Malorosiji do Ukrajiny: paradoksy zapizniloho nacije tvorennja* [From 'Little Russia' to Ukraine: Paradoxes of Delayed Nation Formation] (Kyjiv: Krytyka).

43. See, for example, Alina Cała (1989) *Asymilacja Żydów w Królestwie Polskim 1864–1897: postawy, konflikty, stereotypy* [Assimilation of Jews in the Kingdom of Poland 1864–1897: attitudes, conflicts, stereotypes] (Warsaw: Państwowy Instytut Wydawniczy); Michal Frankl (2007) *Emancipace od Židů. Český antisemitismus na konci 19. Století* [Emancipation from the Jews. Czech anti-Semitism at the end of the nineteenth century] (Praha: Paseka); Steven Beller (1989) *Vienna and the Jews 1867–1938: A Cultural History* (Cambridge: Cambridge University Press); Klaus Hödl (2006) *Wiener Juden, Jüdische Wiener: Identität, Gedächtnis und Performanz im 19. Jahrhundert* (Innsbruck, Vienna: Studien-Verlag). For examples of the impact of Jewish assimilation on science, see John Efron (1994) *Defenders of the Race: Jewish Doctors in Fin-de-Siècle Europe* (New Haven and London: Yale University Press), 141–53; Veronika Lipphardt (2008) *Biologie der Juden: Jüdische Wissenschaftler über 'Rasse' und Vererbung 1900–1935* (Göttingen: Vandenhoeck & Ruprecht).

44. The stronger claim by Carol Harrison and Ann Johnson that 'Research on nationalism has largely ignored the nexus between science and national identity' applies, if at all, only to the literature in English. Carol E. Harrison and Ann Johnson (2009) 'Introduction: science and national identity', in idem (eds.) *National Identity: The Role of Science and Technology* (Osiris, new series, vol. 24) (Chicago: University of Chicago Press), 4.

45. For references to literature on linguists and language studies, see the chapter by Jan Surman in this volume. On ethnology, see Andrew Zimmermann (2001) *Anthropology and Anti-Humanism in Imperial Germany* (Chicago: University of Chicago Press); H. Glenn Penny and Matti Bunzl (eds.) (2003) *Worldly Provincialism: German Anthropology in the Age of Empire* (Ann Arbor: University of Michigan Press); Karl Pusman (2008) *Die 'Wissenschaften vom Menschen' auf Wiener Boden (1870–1959): Die Anthropologische Gesellschaft und die anthropologischen Disziplinen im Fokus von Wissenschaftsgeschichte, Wissenschafts- und Verdrängungspolitik* (Vienna: Böhlau); Irene Ranzmaier (forthcoming) *Die Anthropologische Gesellschaft in Wien und die akademische Etablierung anthropologischer Disziplinen an der Universität Wien, 1870–1930* (Vienna: Böhlau). On archaeology and prehistory, see Paul Graves-Brown (ed.) (1996) *Cultural Identity and Archaeology: The Construction of European Communities* (London: Routledge); Heiko Steuer (2001) *Eine hervorragend nationale Wissenschaft: Deutsche Prähistoriker zwischen 1900 und 1995* (Berlin: De Gruyter). On history, see, for example, Stefan Berger (1997) *The Search for Normality: National Identity and Historical Consciousness in Germany since 1800* (Providence, RI: Brown University Press); Eckhardt Fuchs and Benedikt Stuchtey (2002) *Across Cultural Borders: Historiography in Global Perspective* (London: Rowman and Littlefield); Christoph Conrad and Sebastian Conrad (eds.) (2002) *Die Nation schreiben: Geschichtswissenschaft im internationalen Vergleich* (Göttingen: Vandenhoeck & Ruprecht); Hans-Peter Hye, Brigitte Mazohl and Jan Paul Niederkorn (eds.) (2009), *Nationalgeschichte als Artefakt: Zum Paradigma Nationalstaat in den Historiographien Deutschlands, Italiens und Österreichs* (Vienna: Böhlau). For recent studies on history and historians in Eastern Central Europe, see Pavel Kolár (2008) *Geschichtswissenschaft in Zentraleuropa: Die Universitäten Prag, Wien und Berlin um 1900*, 2 Halbbände (Leipzig: Leipzig University Press); Monika Baár (2010), *Historians and Nationalism in East-Central Europe in the Nineteenth Century* (New York: Oxford University Press).

46. This traditional view, which uncritically reproduces hopes of the era in question and tells only one side of the story, is nicely reflected in the statement that in the age of nationalism the sciences 'functioned as some kind of universal language – a bond or bridge between nations and not a bar', Hans Hauge (1996) 'Nationalizing science', in Roger Chartier and Pietro Corsi (eds.) *Sciences et langages en Europe* (Paris: Éditions de l'École des Hautes études en sciences sociales), 159–68, here 160.

47. For a brief general discussion, see Mitchell G. Ash (2000) 'Internationalisierung und Entinternationalisierung der Wissenschaften im 19. und 20. Jahrhundert – Thesen', in Manfred Lechner and Dietmar Seiler (eds.) *zeitgeschichte.at. Österreichischer Zeithistorikertag 1999* (Innsbruck: Studien-Verlag), 4–12. For earlier literature on scientific internationalism, see Brigitte Schroeder-Gudehus (1990) 'Nationalism and Internationalism', in Robert C. Olby, George N. Cantor, J.R.R. Christie and M.J.S. Hodge (eds.) *Companion to the History of Modern Science* (London: Routledge), 909–19; Elisabeth Crawford (1992) *Nationalism and Internationalism in Science, 1880–1939 – Four Studies of the Nobel Population* (Cambridge: Cambridge University Press).

48. Lorraine Daston and Peter Galison (2007) *Objectivity* (Cambridge, MA: Zone Books).
49. Ralph Jessen and Jakob Vogel (2002) 'Einleitung. Die Naturwissenschaften und die Nation', in idem (eds.) *Wissenschaft und Nation in der europäischen Geschichte* (Frankfurt am Main: Campus Verlag), 7–37, here 18.
50. Ludmilla Jordanova (1998) 'Science and nationhood: cultures of imagined communities' in Geoffrey Cubitt (ed.) *Imagining Nations* (Manchester: University of Manchester Press), 192–211; idem (1996) 'Science and national identity', in Roger Chartier and Pietro Corsi (eds.) *Sciences et langages en Europe*, 221–31; David Edgerton (2003) 'Science in the United Kingdom: a study in the nationalisation of science', in Dominique Pestre and John Krige (eds.) *Companion to Science in the Twentieth Century* (London: Routledge), 759–75; Harrison and Johnson (eds.) *National Identity: The Role of Science and Technology*, raises some of the relevant issues, but is devoted mainly to developments in the twentieth century and especially to post-colonial science.
51. Tatjana Buklijas and Emese Lafferton (2007) 'Science, medicine and nationalism in the Habsburg Empire from the 1840s to 1918', *Studies in History and Philosophy of Biological and Biomedical Sciences*, 38/4, 679–86.
52. Ibid., 685.
53. Emese Lafferton (2007) 'The Magyar moustache: the faces of Hungarian state formation, 1867–1918', *Studies in History and Philosophy of Biological and Biomedical Science*, 38/4, 706–32.
54. On the reception of Darwinism in Hungary, see also Sándor Sóos (2008) 'The scientific reception of Darwin's work in nineteenth-century Hungary' and Katalin Mund (2008), 'The reception of Darwin in nineteenth-century Hungarian society', in Eva-Marie Engels and Thomas F. Glick (eds.) *The Reception of Charles Darwin in Europe*, vol. 2 (London: Continuum), 430–40 and 441–62, respectively.
55. Marius Turda (2007) 'Race, politics and nationalist Darwinism in Hungary, 1880–1918', *Ab Imperio*, 139–64; Marius Turda and Paul J. Weindling (eds.) *Blood and Homeland: Eugenics and Racial Nationalism in Central and Southeastern Europe, 1900–1940* (Budapest: CEU Press); Marius Turda (2010) *Modernism and Eugenics* (Basingstoke: Palgrave Macmillan); idem (2011) *Health, Hygiene and Eugenics in Southeastern Europe to 1945* (Budapest: CEU Press).
56. For literature on this topic see the chapter by Jan Surman in this volume.
57. Such multiple linkages are not as surprising as they may seem from a current perspective. The vast array of disciplines and specialities within disciplines now taken for granted was not yet in place anywhere in Europe in the early nineteenth century. See Rudolf Stichweh (1984) *Zur Entstehung des Systems moderner Disziplinen: Physik in Deutschland 1740–1890* (Frankfurt am Main: Suhrkamp); idem (1994) *Wissenschaft, Universität Profession: Soziologische Analysen* (Frankfurt am Main: Suhrkamp); Mitchell G. Ash (1999) 'Die Wissenschaften in der Geschichte der Moderne (Antrittsvorlesung am Institut für Geschichte der Universität Wien, 2. April 1998)', *Österreichische Zeitschrift für Geschichtswissenschaften*, 10, 105–29, English abstract on p. 131.
58. See Olga A. Valkova (2002) 'Wissenschaftssprache und Nationalsprache. Konflikte unter russischen Naturwissenschaftlern in der Mitte des 19. Jahrhunderts', in Jessen and Vogel (eds.) *Wissenschaft und Nation*, 59–79; see also the papers by Jan Surman and Soňa Štrbáňová in this volume.
59. William Brock (1992) *The Chemical Tree: A History of Chemistry* (New York: W.W. Norton), 87.

60. See the chapter by Jan Surman in this volume.
61. On the priority dispute over the liquefaction of oxygen, see Zdzisław Wojtaszek et al. (1990) *Karol Olszewski* (Zeszyty Naukowe Uniwersytetu Jagiellońskiego, 899; Universitatis Iagellonicae Acta Chimica fasc. 33) (Warszawa, Kraków: Państwowe Wydawnictwo Naukowe), 86–98. On Mendeleev's periodic table see Michael D. Gordin (forthcoming) 'The textbook case of a priority dispute: D.I. Mendeleev, Lothar Meyer, and the periodic system', in Jessica Riskin and Mario Biagioli (eds.) *Nature Engaged: Science in Practice from the Renaissance to the Present* (New York: Palgrave Macmillan).
62. See, for example, Svante Lindquist (ed.) (1993) *Center on the Periphery, Historical Aspects of 20th-Century Swedish Physics* (Canton, MA: Science History Publications); Louise Hecht (2005) 'The beginning of modern Jewish historiography: Prague – A center on the periphery', *Jewish History*, 19, 347–73.
63. David N. Livingstone (2003) *Putting Science in its Place: Geographies of Scientific Knowledge* (Chicago: University of Chicago Press). See also the studies of the reception of Darwinism in Hungary cited above (note 53).
64. See A. Suresh Canagarajah (2002) *A Geopolitics of Academic Writing* (Pittsburgh: University of Pittsburgh Press).
65. See http://147.156.155.104/?q=node/3 (last accessed 21.12.2010). In the descriptions of participating projects the word 'periphery' is sometimes used with inverted commas and sometimes not.
66. Crawford, *Nationalism*, pp. 37, 87; Gary B. Cohen (1996) *Education and Middle-class Society in Imperial Austria 1848–1918* (West Lafayette, IN: Purdue University Press). For an example of the impact see Lenka Vodrážková-Pokorná (2006) *Die Prager Germanistik nach 1882: Mit besonderer Berücksichtigung der bis 1900 an die Universität berufenen Persönlichkeiten* (Frankfurt am Main: Peter Lang), esp. 67–73.
67. On the Cracow Academy see Renato Mazzolini (1995) 'Nationale Wissenschaftsakademien im Europa des 19. Jahrhunderts', in Lothar Jordan and Bernd Kortländer (eds.) *Nationale Grenzen und internationaler Austausch: Studien zum Kultur- und Wissenschaftstransfer in Europa* (Tübingen: Mohr), 245–60. For literature on the Academies of Sciences in Budapest and Prague, see the chapters by Gábor Palló and Soňa Štrbáňová in this volume. In contrast, the Imperial Academy of Sciences in Vienna (founded in 1847 and only later called 'Austrian') can be regarded as part of an effort towards the construction of an imperial 'nation' that was redoubled in the neo-absolutist era (see below).
68. A possible confusion of concepts should be mentioned here. In Czech, Polish and Ukrainian the concepts nationalism and patriotism have different connotations; while the first is negative and conceptually near to chauvinism, patriotism is positive. On this point see Porter, *When Nationalism Began to Hate* (cit. note 7). For recent discussion of this terminology in Polish, see Krzysztof Jaskułowski (2009) *Nacjonalizm bez narodów: Nacjonalizm w koncepcjach anglosaskich nauk społecznych* [Nationalism Without Nations. Nationalism in anglophone social sciences] (Wrocław: Wydawnictwo Uniwersytetu Wrocławskiego).
69. Hermann Helmholtz (1862) 'On the relations of natural science to science in general', in idem (1995) *Science and Culture: Popular and Philosophical Essays*, ed. David Cahan (Chicago: University of Chicago Press), 76–95, here 92.
70. Ibid. See also Ash 'Die Wissenschaften'; Mitchell G. Ash (2002), 'Wissenschaft und Politik als Ressourcen für einander', in Rüdiger vom Bruch and Brigitte Kaderas (eds.) *Wissenschaften und Wissenschaftspolitik: Bestandsaufnahmen zu Formationen, Brüchen und Kontinuitäten im Deutschland des 20. Jahrhunderts* (Stuttgart: Steiner), 32–51.

71. Nationalistic divisions in the sciences and the mobilization of scientists for war and propaganda during the First World War are well studied topics. For literature in German see, for example, Jürgen and Wolfgang Ungern-Sternberg (1996) *Der Aufruf 'An die Kulturwelt!' Das Manifest der 93 und die Anfänge der Kriegspropaganda im Ersten Weltkrieg* (Stuttgart: Steiner); Stefan L. Wolff (2001) 'Physiker im "Krieg der Geister"' (Arbeitspapiere des Münchener Zentrums für Wissenschafts- und Technikgeschichte, Munich).

72. See, for example, Jana Mandlerová (1969) 'K zahraničním cestám učitelů vysokých škol v českých zemích (1888–1918)' [Travels abroad by university teachers in the Czech Lands (1888–1918)], *Dějiny věd a techniky*, 4, 232–46; Maria Julita Nedza (1973) *Polityka stypendialna Akademii Umiejętności w latach 1878–1920: Fundacje Gałęzowskiego, Pileckiego i Osławskiego*. [The stipends policy of the Academy of Sciences and Arts 1878–1920: The endowments of Gałęzowski, Pilecki and Osławski] (Wrocław: Zakład Narodowy im. Ossolińskich).

73. Julian Dybiec (2005) 'Prześladowca i nauczyciel. Niemcy w nauce i kulturze polskiej 1795–1918' [Oppressor and Teacher. Germans in Polish science and culture 1795–1918], in Bogusław Dopart, Jacek Popiel and Marian Stala (eds.) *Literatura, kulturoznawstwo, uniwersytet. Księga ofiarowana Franciszkowi Ziejce w 65. rocznicę urodzin* [Literature, cultural sciences, university. Festschrift for Franciszek Ziejka in honor of his 65th birthday] (Kraków: Universitas), 455–68; see also the chapter by Tibor Frank in this volume.

74. Lewis Peyenson (1993) *Civilizing Mission: Exact Sciences and Colonial Expansion 1830–1940* (Baltimore: Johns Hopkins University Press); Richard Drayton (2000) *Nature's Government: Science, Imperial Britain and the 'Improvement' of the World* (New Haven, CT: Yale University Press); Roy MacLeod (ed.) (2000) *Nature and Empire: Science and the Colonial Enterprise* (Osiris, vol. 15. Chicago: University of Chicago Press); Kapil Raj (2006) *Relocating Modern Science; Circulation and the Construction of Knowledge in South Asia and Europe, 1650–1900* (New York: Palgrave Macmillan).

75. See, for example, Christian Marchetti (2007) 'Scientists with guns: on the ethnographic exploration of the Balkans by Austria-Hungarian scientists before and during World War I', *Ab Imperio*, 1, 165–90.

76. Marianne Klemun (ed.) (2009) 'Wissenschaft und Kolonialismus', *Wiener Zeitschrift zur Geschichte der Neuzeit*, 9/1.

77. Walter Sauer (ed.) (2002) *K. und k. kolonial. Habsburgermonarchie und europäische Herrschaft in Afrika* (Vienna: Böhlau Verlag); Christa Riedl-Dorn (ed.) (2010) *Novara – das Vermächtnis* (Vienna: Böhlau Verlag); Ursula Rack (2010) *Sozialhistorische Studie zur Polarforschung anhand von deutschen und österreich-ungarischen Polarexpeditionen zwischen 1868 1939* (Berichte zur Polar- und Meeresforschung, 618. Bremerhaven).

78. Jan Surman (2009) 'Imperial knowledge? Die Wissenschaften in der späten Habsburg-Monarchie zwischen Kolonialismus, Nationalismus und Imperialismus', *Wiener Zeitschrift zur Geschichte der Neuzeit*, 9/2, 119–33.

79. On the role of mapping see, for example, Pieter M. Judson (1996) 'Frontiers, islands, forests, stones: mapping the geography of a German identity in the Habsburg Monarchy, 1848–1900', in Patricia Yeager (ed.) *The Geography of Identity* (Ann Arbor: University of Michigan Press), 382–406; Irina Popova (2003) 'Representing national territory: cartography and nationalism in Hungary 1700–1848', in Nancy M. Wingfield (ed.) *Creating the Other: Ethnic Conflict and Nationalism in Habsburg Central Europe* (Oxford and New York: Berghahn Books), 20–38. For more

general perspectives see Robert Kaiser (2001) 'Geography', in *The Encyclopedia of Nationalism*, volume 1 (San Diego: Academic Press), 315–33; David Guggerli and Daniel Speich (2002) *Topographien der Nation: Politik, topographische Ordnung und Landschaft im 19. Jahrhundert* (Zurich: Chronos); Janes R. Akerman (ed.) (2009) *Imperial Map: Cartography and the Mastery of Empire* (Chicago: University of Chicago Press). On the role of data-gathering networks in Habsburg-era geology see the chapter by Marianne Klemun in this volume. On the Empire's central institute for meteorology and seismology see Christa Hammerl, Wolfgang Lenhardt, Reinhold Steinacker and Peter Steinhauser (eds.) (2001) *Die Zentralanstalt für Meteorologie und Geodynamik 1851–2001: 150 Jahre Meteorologie und Geophysik in Österreich* (Graz: Leykam). On the creation of data-gathering and reporting networks and the circulation of knowledge in climatology see Deborah R. Coen (2006) 'Scaling down: The "Austrian" climate between Empire and Republic', in James Rodger Fleming, Vladimir Jankovic and Deborah R. Coen (eds.) *Intimate Universality: Local and Global Themes in the History of Weather and Climate* (Sagamore Beach: Science History Publications), 115–40; Deborah R. Coen (2010) 'Climate and circulation in Imperial Austria', *The Journal of Modern History*, 82, 839–75. In certain respects these correspondence networks and survey projects are comparable with those in the US – also a land-based empire – at roughly the same time. See Daniel Goldstein (1985) '"Yours for science": The Smithsonian Institutions's correspondents and the shape of the scientific community in nineteenth-century America', *Isis*, 85, 573–99; Robert V. Bruce (1987) *The Launching of American Science 1846–1876* (Ithaca, NY: Cornell University Press).

80. On the Imperial and Royal Natural History Museum in Vienna see Christa Riedl-Dorn (1998) *Das Haus der Wunder: Zur Geschichte des Naturhistorischen Museums in Wien* (Vienna: Holzhausen).

81. Marlies Raffler (2007) *Museum – Spiegel der Nation? Zugänge zur historischen Museologie am Beispiel der Genese von Landes- und Nationalmuseen in der Habsburgermonarchie* (Vienna: Böhlau).

82. Earlier versions of most of these chapters were presented at the the third international conference of the European Society for the History of Science in Vienna, 10–12 September 2008, and at the XXIII International Congress for History of Science and Technology in Budapest, 28 July – 2 August, 2009.

83. Gábor Palló also discusses this issue with respect to the younger Eötvös in his chapter.

84. See the works by Kolár and Baár (cited above, note 45).

85. Marie Curie European Reintegration Grants.

86. See, for example, the scholarship Powroty/homing (now Homing plus) offered by the Foundation for Polish Science/ Fundacja na Rzecz Nauki Polskiej, http://www.fnp.org.pl/programmes/overview_of_programmes/grants_and_scholarships/homing_programme; http://www.fnp.org.pl/programmes/overview_of_programmes/grants_and_scholarships/homing_plus_programme. The Austrian Science Foundation also had a Schrödinger Rückkehrprogramm, which was abolished in 2003. See Katharina Warta (2006) *Evaluation of the FWF Mobility Programs Erwin Schrödinger and Lise Meitner* (Vienna: Technopolis Forschungs- und Beratungsgesellschaft mbH), http://www.fwf.ac.at/de/downloads/pdf/fwf_mobility_report.pdf.

87. For studies of such practices see for example, Pnina Abir-Am and Clark Elliot (eds.) (1999) *Commemorative Practices in Science: Historical Perspectives on the Politics of Collective Memory* (Osiris, vol. 14. Chicago: University of Chicago Press).

88. The history of science is hardly immune from this trend. Posters accompanying plenary lectures at the International Congress of History of Science and Technology in Budapest in the summer of 2009 followed much the same pattern, depicting important contributions of Hungarian scientists and technicians and thus continuing a long-standing pattern of commemoration-oriented historiography, with no analysis or contextualization whatever.

89. *Big Bang will open the Copernicus Science Centre*, online: http://www.naukawpolsce.pap.pl/palio/html.run?_Instance=cms_naukapl.pap.pl&_PageID=1&s=szablon.depesza&dz=szablon.depesza&dep=376783&data=&lang=&_CheckSum=1312044493 .

90. One can find Comenius on the Czech 200 krona banknote, as well as Tomáš Masaryk and historian František Palacký on other notes. Ukrainian banknotes are emblazoned with portraits of the eighteenth-century philosopher and poet Grigorij Savvich Skorovoda/ Hryhorii Savych Skovoroda (500 hryven', the highest banknote, introduced in 2006) and Mykhailo Hrushevskyi (50 hryven'), both members of the Kiev-Mohyla Academy. Polish zloty notes used to depict, apart from Copernicus, the chemist Maria Skłodowska Curie (= Marie Curie Skłodowska) and the eighteenth-century philosopher and geologist Stanisław Staszic. On Slovenia's banknotes (before the introduction of the euro) Carniola-born Janez Vajkard Valvasor, Fellow of The Royal Society (20 tolarjev) and astronomer Jurij Bartolomej Vega/Georg Freiherr von Vega (50 tolarjev) were commemorated. In Slovakia the linguist Anton Bernolák was honoured in this manner. Also before the introduction of the euro, Austria depicted the physicist Erwin Schrödinger (1000 Schilling) and economist Eugen Ritter von Böhm-Bawerk (100 Schilling), as well as sociologist Rosa Mayreder and psychoanalyst Sigmund Freud, who were excluded from the professional scientific community in their lifetimes.

2
Science and Its Publics: Internationality and National Languages in Central Europe

Jan Surman

Shortly after the collapse of the Habsburg Monarchy, the newly emerged successor states faced fundamental changes in the scientific infrastructure of Central Europe. The end of the multinational empire meant a gain of autonomy in science policy, and political sovereignty was perceived as a new chance to pursue what seemed to have been restricted in the past, the cultivation of science independently of foreign influences. Contemporary discussions as to what form the new science should take, however, showed that the nationalization of science that had been pursued during the nineteenth century had lost its attractiveness and was being gradually replaced by pronounced internationalism.[1] With the nation seen as the primary point of reference, science was now consciously transgressing cultural boundaries. For example, linguist Andrzej Gawroński stated during a debate on the reform of scientific infrastructure in Poland after 1918 that only when scholars representing different styles of research met and communicated could objective and unbiased truth be obtained.[2] In this regard internationalism and interculturality became the same term. Whereas nationality represented unified and standardized culture, international science was to be a mixture of psychologically and linguistically defined styles of research, which would now be called 'national sciences'.[3]

Science and scholarship had not lost their pivotal role in nation-building policy – philosopher Tomáš Garrigue Masaryk (1850–1937) became the first president of Czechoslovakia; Gabriel Narutowicz (1865–1920), professor of hydroelectric engineering at Zurich Polytechnic, became the first president of the Polish Republic; and historian Mykhailo Hrushevskyi (1866–1934) was elected chair of the Ukrainian revolutionary parliament Tsentral'na Rada and was thus de facto president of Ukraine (see below). The pronounced role of academics as public intellectuals joined here with a scientific ethos of impartiality, but by turning to Masaryk, Narutowicz and Hrushevskyi a conscious choice was made for internationally renowned scholars in preference to professional politicians. The idea of enacting nationality through the implicit addition of an international component was ironic, but at the same time iconic for the situation after 1918, since such moves were representative of changes taking place between the 'Springtime of Nations' and the 'League of Nations'.

As Phillipp Ther has argued for the opera, nationalization and internationalization are not exclusive categories; processes of nationalization in Central Europe resulted in new spaces of influence. The 'Europe of Opera' that developed at the end of the nineteenth century was characterized by the redefinition of music as a universal, and thus border-transcending, activity – in contrast to the decidedly national inscription it had had during the nation-building process. Internationalization was also achieved through pluralization of transfer channels, from a few centres at the beginning of the nineteenth century to a variety of local (here Slavic) subspaces on the eve of the First World War.[4] A similar turn to internationalism can be observed in the sciences, both in orientation and in rhetoric. Still, as I will argue, this was made possible only through the solidification of science as a national enterprise; the international character of science was evoked only together with its national definition. Not only were the processes of nationalization and internationalization interdependent in science, but the turn to nationalistic sciences from the late eighteenth century onwards codified scientific and scholarly activity in national terms, which subsequently led to the establishment of scientific internationalism.[5]

In the nineteenth century Central European intellectuals were constantly redefining and conceptualizing science, oscillating between describing it as a sphere of contact and as a sphere of division. Throughout the century scholars participated in the nationalization of science, which as an innate constituent of cultural well-being was to be distinguished from science in the imperial centres through language, networks of production and distribution, or content.[6] The process of nationalization did not, however, mean the rejection of intercultural contact; through references or quotations, participation in meetings and conferences, translations and foreign publications, Central European scholars followed the established scientific and scholarly *habitus*, but while doing so they particularized its direction and proclaimed goal. Science and scholarship was diverted from the goal of universalistic truth to culturally localized enlightenments; in other words, scientific knowledge was not to be produced for an abstract academic community, but used to enable the cultural and civilizational development of 'nations'[7].

The multiple meanings of the word 'science' had a massive influence on this development. The Polish, Czech and Ukrainian words for science – *nauka* and *наука* – mean both science and education, linking inherently the production of knowledge and its dissemination. (In Czech, however, the word *věda* is also used, analogically to German *Wissenschaft*, which encompasses both natural science and humanistic scholarship).

The transformation in the primary recipients of science, which took place after the French Revolution, resulted in a reconfiguration of scientific infrastructure, as scientific publications were supposed to become (even if only superficially) accessible to broader strata of society. Especially in Central Europe, this conception was intensified and linked with the so-called national revival in a positivistic conception of organic work; the betterment of society was to be achieved through the cultural and scientific improvement of the lower social spheres.[8] This nation-building ethos is visible, for example, in numerous scientific articles published in daily newspapers throughout the century,[9] as well as an abundance of popular scientific journals.

However, as soon as linguistic communities replaced the *République des lettres*, the main deficiency of the new system of circulation of knowledge was discovered – it actually prohibited horizontal, intercultural transfers. More importantly, it made them one-sided, since national languages were perceived to be of less importance than the 'world languages'; this was especially true in Central Europe, where the predominance of German as the acknowledged language of scholarship strengthened the impression of cultural dependence. At the very point that the nationalization of the sciences succeeded and scientific knowledge could be produced and dispersed in nationalized frameworks, the international presentation of national scientific achievements took on vital importance. This reopening towards internationalization, which took place at the end of nineteenth century, was codified in new terms. Science and scholarship became an issue of national pride, in which the scholar represented the community still more than himself.

In this chapter I follow the tension between national and international science as it manifested itself in the development of scientific languages in Central Europe. Being to a large extent a process of interrelation between social and cultural movements, nationalization did not happen synchronically throughout the region. Nevertheless the emerging national communities encountered similar problems and drawbacks on their way to their pronounced aim of 'pure nationhood'. While changes in scientific networks and the contents of science could also be drawn upon to demonstrate the validity of my thesis, I focus here on language because it worked at the meta-level, defining how and to whom knowledge was to be presented, and to a certain extent also influencing what was researched.

In the first part of the chapter I focus on the creation of vernacular scientific vocabularies beginning in the early nineteenth century. The codification of terminology was vital to the assertion of cultural development and maturity, language being the medium through which the autarchy of national cultures was to be codified and visualized. Although inscribed in the context of nationalization, this process was a continuous mediation between pronounced cultural puristic aims and the practice of international scientific communication. The second part is devoted to publication practices, which represented the tensions within Central European scholarship between the communication of results in different communities and between the personal and collective representation of achievements. Problems encountered during the establishment of linguistic and national publication networks, debates on the cultural (international) impact of their success and practices of cultural reconfiguration and reorientation in the late nineteenth and early twentieth centuries highlight here problematic perceptions of dependence and interdependence in 'international' science and consequences of 'internationalization' in Central European national contexts.

Between 'logic and rhetoric': language in the making

The process of German taking the place of Latin as the leading language of scholarship in Central Europe, which began at the end of the eighteenth century,[10] remained uncontested for only a short time. The new linguistic hierarchy was contested through countless 'apologies' for Slavic languages, which aimed at establishing or confirming the place of national languages under imperial circumstances.[11] Karel Thám (1763–1816) and Josef Dobrovský (1753–1829) for Czech and Onufry Kopczyński (1736–1817) for Polish established guidelines for the (re)invention of national languages; these included their historization and vernacularization, which were also intended to remove or lessen foreign influence on vocabulary and syntax. Their projects were both political and philosophical, closely linked to discussions among French and German linguists of the time. While questions of national unification and the establishment of a coherent language community were raised, science and education were also to profit from the changes. Influenced by Enlightenment theories on grammar and language, above all through the work of Étienne Condillac (1715–80) and Johann Gottfried Herder (1744–1803), these authors constructed a relation between language and cognition. Kopczyński, for example, asserted that '[t]he darkness of scholarship and mistakes of scholars nestle mostly in words.'[12] Similarly, Thám wrote that as 'reason rejoices in the fact that it understands and comprehends things as they are in themselves, so too language, when it expresses things as they naturally occur, such as the natural voice, ... seems almost an echo in our ears, by means of which its essence acquires its features and is formed in our minds.'[13] The ascription of naturalness and ease of communication to 'national' language, very much in the style of Herder or Johann Adelung (1732–1806), was linked here with idealized claims of philosophical perfection. Hence, an ancient idea of revolutionizing science through linguistic revolution[14] was remodelled in order to change communication structures from elitist forms to forms encompassing all classes of society. The Bohemian naturalist Jan Svatopulk Presl (1791–1849) wrote in this connection that science, once a privilege of the 'powerful and learned' (*mohovitých a učených*) should spill over to the common people (*národ*), since the power of a nation is not measured in the accomplishments of the elites, but by the breadth of the people's education.[15]

Jan Śniadecki (1795–1830), an influential universal scholar and empiricist (educated partially in Paris) and the most important propagator of Polish terminology in the sciences, defined various functions of language. He drew partly on Condillac's philosophy of language (its constitutive role in the formation of ideas from impressions, its autonomy from sensations), but refused to accept his analogy between language and mathematics, stating that from the mathematician's point of view Condillac was wrong to concentrate only on the analytical function of language and to reject the possibility and advantages of synthesis. Language cannot be seen as a purely logical system guided by laws, but neither can it be merely aesthetically driven rhetoric.[16] For Śniadecki language was in the first place a cultural resource, showing whether a nation was developed: 'when language is dark and coarse, modest in words ... a nation had not developed beyond the wilderness, had not entered the ranks of enlightened nations'[17]. But language also had a descriptive value for representing concepts: 'Language should serve thinking and not constrain and restrain it'[18]; first is the thought or object, not the (translated) word.

Śniadecki was, as far as language is concerned, a conventionalist; he rejected the tendency to translate all words into Polish, as 'a good macaronism [mixture of different languages] is better than a poor translation', and further 'language is not a work of metaphysical rummaging, nor a system of individual imagination; we should not search it in imitation of foreign folk, but in a national tongue, generally accepted and understood.'[19] Words for Śniadecki were representations of concepts (things and thoughts), and served a twofold function in language: 'one for the eye' (representation) and one 'for the ear, for presenting this knowledge in written and oral form, which means words and names in the national language.'[20] Thus the introduction of new nomenclature should be guided by principles that ensured a balance between epistemic precision and linguistic variability, drawing its models from 'old books written in candid Polish',[21] but at the same time should ensure that a relation to other words is kept – both to words in foreign languages (here, especially, Latin) and words in use (existing in popular or scientific language). In his examples from geography and astronomy, Śniadecki constructed scientific vocabulary on Polish principles, e.g. inertia = *bezwład*, velocity = *chyżość*, zenit = *punkt wierzchołkowy*; but he retained words that were already in use, like telescope = *teleskop* and astronomy = *astronomia* (for which other terms like *gwiazdoznawstwo*, literally: knowledge of stars, were in use at the time).[22]

The brother of Jan, Jędrzej Śniadecki (1768–1838), was also a philosopher and a chemist, and applied a new approach to chemical and medical vocabulary. He remained, however, sceptical about the use of the national language in every case. For example, during medical inspections with students, he argued that debate on the condition of the injured person should be conducted in Latin, so as not to bother the patient or have to leave his presence in order to carry out the debate (Latin was obviously the 'first of the learned languages').[23] Like his brother, Jędrzej Śniadecki sought words in the linguistic tradition and proposed an intuitive nomenclature of chemical substances based on existing vocabulary, though with analogies to the French classification of compounds, where he partially imitated the suffix and prefix system. In a discussion with Aleksander Franciszek Chodkiewicz (1776–1838), author of an introduction to chemistry (*Chemiia*, 7 vols. 1816–20), Śniadecki defended his use of neosemantisms, criticizing Chodkiewicz for simplifying chemistry through stress on the vernacular language, which resulted in a lack of correspondence with the findings of modern chemistry. He nevertheless stated that he was opposing mere imitation of French taxology, for example in the classification of acids. While Chodkiewicz proposed imitating French endings –*ique* (for acids) and –*eux* (for weak acids) with Polish endings, Śniadecki proposed a solution that would sustain vernacular terms without artificial endings. Hydrofluoric acid would thus be called *kwas solny* (because of the existing expression *żupa solna* – salt mine) and not *kwas solowy*; tartaric acid would be *kwas winny* (from *handel winny* – wine commerce) instead of *kwas winowy*; the distinction between acids and weak acids would be expressed in prefixes, for example *weak acid* = *podkwas* (literary – lower or under acid), e.g. *podkwas winny*.[24]

Slightly different was the philosophical basis of Czech purism, which came more from the German tradition than from the French. It was Josef Jungmann (1773–1847) who, through the translation of Milton's *Paradise Lost* (1811), fuelled the Czech linguistic revival by asserting the applicability of Czech to high literature. Jungmann, a student of the renowned linguist Dobrovský, nationalist and translator, was also the creator of the first comprehensive Czech–German dictionary (1834–39, 5 volumes). His linguistic conception was promoted as early as 1803 and included the merging of language and nation (*národ*). In *Talks on Czech Language* (*Rozmlouvání o jazyce českém*, 1806) Jungmann depicted a fictional dialogue in which he criticized Germanophone participants in the Czech nation. Through this position, obviously regarded as archaic, the author made connections with social status and proposed instead culturally – that is, linguistically – defined national boundaries (between Czechs, Germans and Jews in Bohemia). At the same time, he revalued languages with respect to their historicity, stating that determined effort for the betterment of Czech could outweigh the longer development of German.[25] He maintained that using national languages augmented productivity, and only thus could the Czechs acquire their place in a European concert of cultures and confront the Germans in Bohemia. Jungmann – strongly influenced by Herder[26] – also pleaded for a pan-Slavic conglomerate, in which national cultures had their political and geographical place as a communicational alternative to imperial, hegemonic culture.

Jungmann established the basis of Czech word formation as well, in which he rejected loan words and grammatical constructions, and argued that the accordance of words with historical Czech and the 'spirit' of the language should outweigh actual use. At the same time Jungmann did not contest borrowings from Polish and other Slav languages, and acknowledged that loan translations and loan renditions from German were indispensable for scholarship, although he assigned them the least important place in the hierarchy of vocabulary sources.[27]

The chemical vocabulary used by Jungmann's contemporary and follower Jan Svatopulk Presl reflects this practical purism. While imitating the suffix-prefix system used in standard German or French terminologies, Presl tended to create neologisms or borrow words from Slavic languages (mostly Polish and Russian), thus creating a non-vernacular purist Slavic terminology that had little in common with the actual spoken language.[28] He also used a number of direct translations from German (translating both parts of combined nouns) – for example *přírodoznanství* (botany, , from *Naturlehre*: *příroda* = Natur, *-znanství* = Lehre), síloskum (physics, from *Kraftkunde*: *síla* = Kraft, *-skum* = Kunde), dusík (nitrogen, from *Stickstoff*: dusit = Ersticken, *-ík* = Stoff), thus translating all terminological components into Slavic forms. Apart from fulfilling the purist credo, this also supported the analogy with foreign languages, though at the cost of numerous neologisms, many of which were rejected during the nineteenth century.[29]

This puristic trend is important for two reasons. At first, the purists argued that national culture should encompass all social strata – from common people to the nobility. Moreover, they maintained that it was ultimately the nobility and the new intellectual class that had to learn from the previously underprivileged strata, which they represented as the soil of the nation, having preserved the national culture from ancient times and protected it from the damaging influence of cosmopolitanism. Thus their word constructions were in theory reoriented towards vernaculars, as was the scholarship they represented. Their practice, though, led to the uncontrolled creation of a large number of short-lived neologisms and paleologisms, which in the case of scholarly language created an abundance of synonyms; this caused chaos in publications rather than cultural and scientific uniformity.[30]

Since the establishment of new institutions of scholarship, especially the Bohemian National Museum (Vaterländisches Museum in Böhmen/ Wlastenské museum w Čechách, 1818) in Prague and the Cracow Academic Society Linked with the Cracow University (Towarzystwo Naukowe Krakowskie z Uniwersytetem Krakowskim połączone, originally established as the Societatis Litterariae cum Universitate Studiorum Cracoviense

Conjunctae, 1815), one of the main aims of these intellectuals and scientists was to foster the development of national scientific languages; this led to the publication of several dictionaries and translations up to 1848. Similarly, a number of professional academic scholars – most notably anatomist Józef Majer (1808–99) in Cracow and brothers Karl and Jan Svatopulk Presl in Prague – devoted themselves to the creation of a standardized vocabulary for specialized disciplines.

The introduction of vocabulary was carried out mainly through textbooks, as Czech was not a language of education, and Polish had that status only for a short period. To a large extent, such books were only of secondary scholarly value, since their aim was to foster national languages through the reproduction of up-to-date knowledge, based most frequently on translations. Dictionaries, presenting the standardized version of vocabulary, began to be published in large numbers only in the second half of the nineteenth century; especially in the sciences they included comparisons of more than two languages, mostly French, German and Latin, and frequently another Slavic language as well. A dictionary of the first half of the nineteenth century was in fact not, or at least not only, intended to enable the understanding of a foreign language; its function was rather to catalogue, standardize, stabilize and archive vocabulary, also offering the possibility of learning a 'national' language by employing the 'foreign' tongues of the elites, who spoke mostly French or German.

For a long time, the purpose of vernacular textbooks was similarly unclear, as scholars were able to read foreign languages and scientific infrastructure remained elitist, failing to achieve the broader public dissemination of scientific knowledge. Even in the late 1880s, Władysław Natanson (1864–1937), who was then working on the first Polish compendium of theoretical physics, considered his work to be academically futile; he did not count on its having any influence at all, since learned men knew foreign languages and amateurs were not interested in the subject.[31]

Knowledge produced in this way was thus of very limited influence in terms of abstract international science. But work on national terminologies and vocabularies also led to an intensification of research in certain disciplines. Research carried out to establish vocabulary based on vernacular and historic examples included broad historical and ethnographical studies. Work on biological vocabulary was conjoined with wide research on regional fauna and flora. Pioneering works on terminology were thus not only lexical studies, but also descriptive-explorative enterprises involving the dense description of chosen regions, such as the (still bilingual) Czech-Latin *Czech Flora* (1819) written by the brothers Presl,[32] or Stanisław Bonifacy Jundziłł's Polish-language *Description of Lithuanian Plants According to the System of Linneus* (1811).[33] Also later in the century Józef Rostafiński (1850–1928) conducted wide research in Galicia and beyond, his aim being to establish a botanical nomenclature based on 'living folk names';

in the process his research interests deviated increasingly from mycology to the history of botany as well as ethnolinguistics and historical linguistics.[34] Biology was also one of the first disciplines in which nationalization brought an intensification of research, since amateurs could participate in classification on more or less equal terms through descriptions or short articles. Demands for broad participation in biological projects were also published in scientific journals, enlarging the range of people involved in scientific activity. Networks of collectors, whose knowledge was translated and codified through subsequent systemization in nomenclature-building atlases, considerably democratized the production of biological knowledge in Central Europe.

The creation of vocabulary involved not only democratization but also decentralization, since one of its main goals was not to imitate the politically problematic languages, German and Russian. Although emulation of those vocabularies was more frequent than desired by the purists, the main points of reference were publications in Latin and French, either for translations or as the bases of lexical constructions.[35] Work on vocabulary also intensified contacts and correspondence among Slavic scholars, who frequently compared and used one another's vocabulary for their publications. Jungmann, for example, used several Polish dictionaries for his work on biology; Jędrzej Śniadecki adopted some of Presl's vocabulary for his work on chemistry; and Vasyl' Volian (1832–99) compiled the first Ruthenian biological textbook as a commingling of Slavic termini. Thus, the scientific languages of Central Europe were clearly marked by inter-Slav reciprocity, even if purist tendencies throughout the century increasingly concealed this fact.[36]

While purism on the symbolical level was clearly perceived as a positive political process of national unification, it did not go uncontested from within. Such debates, beginning in the second half of the nineteenth century, nonetheless remained rare, as internationalization was not perceived to be influenced by language but by its use. For example, the use of Latin-based vocabulary in Polish medicine was rejected during debates at the Meeting of Polish Physicians in 1879, because the value of a largely imitative system with nationalized names was regarded as higher, notwithstanding the critique that such nomenclature restricted understanding of foreign medical texts.[37] Vernacularization went hand in hand with a relaxation of the 'internationality' of language – and also of terminological flexibility, since terminology was bound to the theory on which it was based. In the eyes of some scholars, therefore, working on language was superior to working with language; the confusion of termini, macaronisms, theories, etc. led to constant concentration on how something was written rather than what was written. The following are the words of Jan Baudouin de Courtenay (1845–1929), a well known systemic linguist, who took part in a discussion on Polish chemical language, proposing its internationalization in 1900 without success;[38] in his eyes, terminological theatre – meaning the invention of new terminology - was used to create 'national' sciences, but in vain:

> One 'Slav' philosopher translates the terminology of one of the German philosophical systems into Polish, creates the wildest words that nobody understands, and we have at once a system of 'Slav philosophy'. *Typhus* is translated as *dur*, catarrh as *sniffles*, syphilis as *chancre*, and so we have national medicine. ... Czechs have *hmota* and *latka* instead of 'matter', instead of 'physics' *silozpyt*, ... *počtařstvi* instead of 'arithmetic', *počtařstvi obecné* instead of 'algebra', *plyn* instead of 'gas', etc., and so they create national sciences.'[39]

Throughout the nineteenth century, however, the bulk of discussion concerning new scientific terminology remained one-sided. From various sides critiques were formulated based on the view that language or a given scientific terminology was not adequately national; these targeted both new terminology and compendia recreated from vernacular speech. Beginning with Chodkiewicz, mentioned above, pioneering works were mostly criticized for over-emulation of foreign terms, as well as for direct translations that sounded foreign or combinations that made no sense if spoken.[40] Within a nationalizing environment, even words used in practical contexts were to be changed so that they corresponded to the ideal of language purity.[41]

The main internal problem of work on scientific terminology remained the lack of centralized institutions that could carry through such endeavours. In a short period, through translations or textbooks, terminologies with widely varying vocabularies were proposed, each adopting a different basis for word-creation, which led not only to countless conflicts as to which choice was most apt for national and scholarly purposes, but also to great instability of terminology, which changed frequently over time and also in space. Especially within the ancient Commonwealth, distinct Polish terminological schools around Warsaw and Cracow developed, leading scholars in the second half of the nineteenth century to demand the strengthening of contacts and the establishment of obligatory rules for future development. For Józef Rostafiński in biology or Emilian Czyrniański in chemistry, for example, reading the publications of the other school seemed like learning the language anew.[42] In the latter discipline unified rules were adopted only in 1901, after more than two decades of emotive discussion.[43] In a short overview on Czech physical literature published in 1876, František Studnička had already mentioned similar problems, as various scientists used their own 'macaronisms', inhibiting clear understanding of the ideas presented.[44] The establishment of Ukrainian terminology took even longer. The existence of differing linguistic/national projects, the fact that the boundary between the Habsburg and Russian empires lay in the middle of the most successful 'Ukrainian' project, and strict opposition to the emergence of an independent Ukrainian language in the Russian Empire delayed the creation of unified terminology for the sciences. Only after the First World War were commissions for terminology established, which opted (although not unanimously) for the adoption of nomenclature based on international norms.[45]

Between 'culture and science'? language in practice

While the creation of local scientific vocabulary can be seen as a belated response to trends that were already adopted in German, French or Italian, the situation is slightly different with regard to the belated institutionalization of new 'national' languages and the political situation in Central Europe. While the other vernacular languages differentiated themselves from Latin, linking linguistic reform with a turn from scholastical science to Enlightenment philosophy,[46] in Central Europe Slavic language reformers encountered an already stabilized scientific communication network based on German. Paradoxically, most of the pioneers of language reforms in this region were at least bilingual; indeed, the 'national' language they advocated was not always identical with their mother tongue.[47] To a large extent, professional, academic scholarship remained faithful to the tradition of multilingualism, because it was accepted that smaller linguistic communities should remain in contact with the so-called world languages to inhibit cultural constriction.[48] Speaking and publishing in more than one tongue was, however, not only an advantage, but also a burden. Moreover, even the designation of whom the 'national' language should serve was problematic; as a result, several versions of Czech, Slovak and Czechoslovak languages, several competing versions of Ruthenian/Ukrainian, and much less successful Eastern Polish and Moravian languages were created.[49] Language issues were involved at almost every stage of the nationalization process. In several cases scholarship was precisely the means of solidifying one or other option.

Looking at the linguistic infrastructure of scholarship in Central Europe, the Polish language was in the most privileged position, because it could build on traditions of the Polish-Lithuanian Commonwealth and several existing literary-scientific periodicals in Polish.[50] Professional scientific journals in the vernacular included the proceedings of the Societies of Friends of Sciences in Warsaw (from 1802) and Cracow (1815). Even if political disturbances caused several breaks in their activity, they remained important journals for several decades.

Czech and Ruthenian took a different turn. The first journals in these languages were published without institutionalized 'national' infrastructures, remaining not only unsettled in the linguistic sense but also with respect to the particular national project they proposed. The first journal in Czech, *Krok. Public General Scientific Journal for the Educated People of the Czech-Slav Nation* (*Weřegný spis wšenaučný pro wzdělance národu Česko-Slowanského*), was published from 1821, and was directed , as the title indicates, towards an imagined Czech-Slav nation.[51] The first journal with institutional support, *Muzejník – Journal of the Bohemian/Czech Museum* (*Muzejník – Časopis Českého museum*), was published from 1827 with the support of the Bohemian Museum in Prague, at first jointly with a German-language periodical, later as the only publication of the Museum.[52] In Eastern Galicia several organizations published their popular scientific journals in different versions of Ruthenian/Ukrainian in the second half of the nineteenth century; the most important of these were the Russophile[53] Stavropihijs'kyj Instytut's *Annals* (*Vremennyk*), Moskwophile Tovarystvo im. M. Kačkovs'koho's *Science/Education* (*Nauka*), Halytsko-Ruska Matycja's *Scientific/Educational Collection* (*Naukovyi Sbornik*), and the Ukrainophile society Enlightenment (Prosvita), which published the journal *Truth* (*Pravda*). Still, only one society acquired political importance and also recognition as the official representative of Ukrainian scholars; this was the Shevchenko Scientific Society (Naukove tovarystvo imeni Ševčenka), which was closely linked to Prosvita. The strange choice of the poet Taras Shevchenko as patron of a scientific society was a political manifestation by Ukrainophiles, who thus merged national and scholarly projects into one. Through official recognition and financial support from the Galician Diet, the Society significantly contributed to the stabilization of a Ukrainian cultural nation. With several journals published, most importantly the *Literary-Scientific Herald* (*Literaturno-Naukovyj Vistnyk*) and *Notes of the Shevchenko Scientific Society* (*Zapysky Naukovoho Tovarystva imeni Ševčenka*), the Shevchenko Scientific Society acquired a strong position for serving unification needs, publishing not only in Galicia but also from the beginning of twentieth century in the Russian Empire. It was Myhailo Hrushevskyi who presided over the Society from 1897 to 1913, and also led the newly established Ukrainian Scientific Society in Kyiv in 1907; his editorial, scientific and political activity linked this 'transimperial' society, marking a milestone in his popularity, which finally led to his election as head of parliament in Ukraine in 1918.

One of the most important features of this publication system is that it predated not only the establishment of learned societies but also that of 'national' universities. Apart from a brief period between 1848 and 1853, Polish was not a language of academic instruction until the 1860s. Although Czech and Ruthenian were permitted languages of education after 1860, Czech remained clearly underrepresented prior to the establishment of a Czech university in 1882 (following the division of Charles University in Prague along nationality lines) and Ukrainian until the founding of the Free Ukrainian University in Vienna, which later moved to Prague, in 1921. Journals were henceforth published mostly for amateurs or scholars working in other languages, but the lack of qualified publics did not hinder them from producing high-level publications. Although history and languages occupied the foremost positions, literary-scientific journals[54] included from the beginning articles on work in abstract and natural sciences. Medicine was also one of the first disciplines with its own national specialized journals; most important were *Memoirs of the Warsaw Medical Society* (*Pamiętnik Towarzystwa Lekarskiego Warszawskiego*, 1837), *Medical Review* (*Przegląd Lekarski*), published by the Jagiellonian University from 1862, and *Journal of Czech Physicians* (*Časopis lékařů českých*, *ČLČ*), published in Prague from the same year.[55] With further specialized journals mushrooming in the second half of the nineteenth century, the transmission of knowledge in Central Europe became still more strongly confined to linguistic boundaries.

Still, the process of imperial nation-building followed its own rules, and the emergence of interested publics and scholars active in such publications came belatedly. While in Bohemia most active 'patriotic' scholars chose to publish in the new national media, many articles had to be translated, or at least corrected, so that their language would be the desired pure Czech. Editorial activity in the first years of *ČLČ* consisted to a large extent of translation from German, as the journal met with less interest from Czech-speaking scholars than desired – or perhaps their number had simply been overestimated.[56] After several years of publishing, the editors of the *Journal for Support for Mathematics and Physics* (*Časopis pro pěstování mathematiky a fysiky*) were still being asked to make more effort to homogenize the terminology and style of their journal.[57] Journals in Polish also strove for a long time to prove their right to exist; in the second half of the century periodicals in German or French were as popular as those in Polish, not only because they were more up to date, but also because they were cheaper. Like those in Bohemia, physicians in Polish lands preferred to publish in foreign languages, not only because of their prestige, but also for financial reasons.[58]

The unclear designation of the central aims of national scientific journals also caused problems. The largest sections of journals were devoted to extensive reviews of the foreign press, information on congresses and travel reports, while original articles took up only about half of the space.[59] What made sense for spreading knowledge within a monolingual reading public remained questionable for bi- or multilingual scholars, who thus paid for information that they had already acquired in more comprehensive form elsewhere. New journals also served in part as parallels to existing prestigious ones – for example at the Prague medical faculty, where *Practical Medicine Quarterly* (*Vierteljahrsschrift für die praktische Heilkunde*) was already being published in German with the participation of Czech scholars. Although the migration was gradual – one can find German scholars among subscribers and publishers in the first numbers of *ČLČ*, and few Czech scholars in *Vierteljahrsschrift* after 1862 – the increasing division along linguistic lines was clearly visible and grew stronger over time.

There was an unwritten rule – or a widespread practice – that scholars should write in the language of the institution with which they were affiliated. This had to do not only with questions of career or practicability, but also with the conviction that professional scholarship was a mission. This especially the case for scholars at universities. For example, the surgeon Jan/Johann Mikulicz-Radecki (1850–1905), appointed professor in Cracow in 1882, started publishing in Polish a year prior to his nomination, but stopped a year after he was nominated to Königsberg (now Kaliningrad).[60] Philologist Kyryl Studynsky (1868–1941) wrote his works in Polish while he was striving for habilitation in Lviv and Cracow; afterwards, as professor in Lviv with Ukrainian as the language of instruction, he published mostly in Ukrainian.[61] While he was striving for habilitation in Graz, physicist Władysław Natanson published in German. As soon as he was rejected and tried to habilitate in Cracow, he had to publish more in Polish. In many biographies of other young scholars, this break is also visible. The function of the language of publication as a sign of belonging to a national community remained important in the Habsburg as in other multicultural empires.

Pressure to publish in the national language was evident, especially because it marked and also helped to determine the political distinction between 'tribe' (*Stamm*) and 'nation'.[62] Use of the vernacular indicated a level of culture, which decided whether a nation was entitled to acquire an academy of sciences or a university. Looking at the discussions around the changes of the language of instruction at the University in Cracow from Polish to German in 1853 and vice versa in 1861, or the creation of national universities in Prague and Lviv, it can be seen that the main argument against the use of the national tongue was that the respective cultures lacked scientific publications in their languages, which was seen as an indication of the lack of potential for lectures in these languages. As soon as the existence of such literature could prove that a degree of cultural development had been achieved, universities were to be granted; otherwise, it was said, students would have to fight through insecure terminology (see above) and would lose contact with 'science'.[63] This argument – made mostly by people who did not know the other language at all – led to an increase in the pressure for national publications, even if the public to which they were addressed did not yet really exist. The main argument of the advocates of a given 'national' university was that once such an institution was established, educated scholars and also the interested public would make 'national' publications grow in number and achieve more terminological coherence.[64] This vicious circle of institutional and symbolical acceptance led to an intensification of patriotically motivated activity in the third quarter of the nineteenth century, as Central Europe's nationalities strove for official acknowledgement. Looking through the journals, one finds numerous scholars who published in 'national' periodicals even though they were active at distant institutions. To name only a few prominent examples: the Czech surgeon Eduard Albert (1841–1900) published in Czech while teaching in Innsbruck and Vienna; the Polish chemist Marceli/Marcel Nencki (1847–1901) sent his articles to Polish journals while he was teaching in Bern and St Petersburg and had rejected a chair in Cracow; after leaving Cracow for Graz and later Vienna, mathematician Franciszek/Franz Mertens (1840–1927) still published in Polish; physicist Ivan Pulyui (1845–1918) taught at the Technical School in Prague and published in both German and Ukrainian. In the front line in Bohemia and Galicia, serving the nation through publications was more than self-evident. A Ruthenian chemist at the Czech University in Prague, Ivan Horbačevs'kyi (1854–1943), represents perhaps the most extreme example of the varied demands a scholar had to fulfil, publishing in German, Ukrainian and Czech.[65]

Paradoxically, concentration on the internal development of 'own-culture' publications intensified the visibility of the dependence of Central European nations on 'world languages'. Concentration on national languages of publication and national publics disrupted the symmetrical circulation of (Latin-transmitted) knowledge within the *République des lettres*. Especially in the later nineteenth century, precisely when the scientific community was becoming more 'international', this issue began to be perceived as a serious problem. 'International' is here rather an elusive construction, since the dominant language of internationality in sciences at that time was German, which was inscribed as both a national and the imperial language in Central Europe. At international congresses, for example, Slavic scholars spoke German and were listed as citizens of their respective Empires; only later and after severe conflicts did they come to represent their respective nationalities.[66] While German journals were perceived as international, and therefore also as desired places of publication, Polish or Czech journals were national. This difference in the perception of what was and was not national or international strongly influenced publication patterns in the region.

As noted before, English, French, German and Russian publications were extensively reviewed in the Central European national scientific journals – especially since these languages were known locally. Many scholars had also gained linguistic competence from scholarship or from political or economic migration. However, reviews in the opposite direction did not take place, as knowledge of Slavic or Hungarian languages among German or French scholars was not normal. With multilingualism thinning out in Central Europe, it thus became the responsibility of a few to play the role of brokers between languages, publishing both reviews of foreign publications in national languages and of publications in national languages in international media. Alternatively, scholars themselves chose to publish for both national and international publics. Especially in the sciences, where priority and credit for discoveries were highly important issues – both individually and collectively –, this often put authors in a dilemma. Their choice was to serve the nation by publishing in the national language or to serve the nation through publishing in foreign languages. This was a dilemma which all new linguistic-national communities encountered, but which remained largely overlooked in those whose national tongue was regarded as a 'world language'.[67]

Still, there was occasional interest among the 'world cultures' in keeping intercultural circulation lively – visible, for example, in initiatives like the French *Slavic Archives of Biology* (*Archives slaves de biologie*, 1886–87). Such periodicals, edited by scholars in the respective 'centres', were short-lived, however, giving place to specialized, culturally undefined journals. The failure of such initiatives was compensated by national projects, although this did not occur not without controversies.

Shortly after the Revolution of 1848 voices were raised requesting the publication of 'national' scientific articles – already so called – in German or French, but only later in the century did the call for publications in 'world languages' gain major support, as their absence came to be seen as depreciating the visibility of scientific development in the learned world. August Seydler (1849–91) and Jan Evangelista Purkyně (1787–1869), for example, pleaded for a dual role for the Czech Academy of Sciences, which should intensify the development of national literature, but at the same time assure the spread of Czech scholarship and science in other languages as well – for Purkyně especially with other Slavs, for Seydler more internationally, in the first place in German and French.[68] Similarly, Tomáš Garrigue Masaryk (1850–1937) pleaded in a series of articles for an increase in non-Czech-language publications; in his view, however, this should serve the mutual understanding of nations within Bohemia. He therefore supported the bilingual Royal Bohemian Society of Sciences (Königliche böhmische Gesellschaft der Wissenschaften/Královská česká společnost nauk), which remained one of the few non-national institutions of that time but became less important during the second half of the nineteenth century on account of the creation of two competing national academies.[69] Similar discussion took place with respect to Polish publications at the end of the 1880s, when a young botanist, Maryjan (Marian) Raciborski (1863–1917), appealed to Polish scholars to publish more in foreign languages.[70] The natural right of every scholar to publish in his mother tongue notwithstanding, he argued, scientists should also prepare and publish longer abstracts in German, French or English, so that the scientific world could took notice of their achievements. This pursuit should also embrace Polish-speaking émigrés, who were supposed to write *comptes rendus* of Polish literature, as scholars of other small nationalities already did. Several responses to Raciborski's article strongly supported his idea but also underlined the factors that hindered such activity. As the disputants put it, Polish journals provided no free reprints or review copies to international journals, and German journals were sceptical about accepting articles from Poles, because they lacked qualified peer-reviewers.[71] Still, local projects were established in this regard; Raciborski extolled a commission of the Cracow Physicians Society (Towarzystwo lekarskie krakowskie) established to support reviews and publications of Polish articles in the 'journals of Virchow and Horsch'.[72] Mathematician Władysław Gosiewski (1844–1911) reported that Warsaw scholars Samuel Dickstein (1851–1939) and Marian Baraniecki (1848–95) even had contracts with the German-language *Yearbook on the Advances in Mathematics* (*Jahrbuch über die Fortschritte der Mathematik*) to regularly write contributions on Polish publications.[73] The last decades of the nineteenth century brought a series of endeavours for the international popularization of national science, so that shortly before the First World War, zoologist Ryszard Błędowski (1886–1932) could proudly announce that it 'is now customary' to display internationally the results of 'our scholars'.[74]

Most important were journals published by the respective academies of sciences. From 1894 the Czech Academy published an *International Bulletin. Abstracts of Presented Works* (*Bulletin international. Résumés des travaux présentés*; in fact multilingual, with a prevalence of German articles), and the Cracow Academy published from 1889 the *International Bulletin of the Academy of Sciences in Cracow* (*Bulletin international de l'Académie des sciences de Cracovie/Anzeiger der Akademie der Wissenschaften in Krakau*[75]), with both French and German articles. Both journals also published extended abstracts and excerpted articles, thus only contents that were also available in the proceedings of the respective academies of sciences in the respective national languages. Although they circulated widely, the fact that these publications covered a broad spectrum of disciplines made them certainly less read than specialized journals. With the exception of the *Review of Czech Medicine* (*Revue de médecine tchèque*, published in Prague from 1909 to 1912), specialized journals in foreign languages did not exist up to the First World War. This was because of the late institutionalization of specialized national journals in national languages, but also the lack of funds to assure wide circulation, correctness of language and payment for the authors; indeed, throughout the century journals in national languages did not pay their authors on account of their small distribution, while German-language journals did.[76] Only after the First World War did such journals become more securely established, once national scientific institutions had obtained larger funds from their governments. Many authors thus preferred publishing in foreign specialized journals from the beginning, if they wanted to reach an international audience.

Such dualistic structures in journals had a major influence on the publication patterns of scholars. Especially professionals with academic positions, who stood in the first rank of the nation-building process, were required to participate both in 'national' and 'international' scientific life. The bibliographies of scholars of the medical faculty at the Czech University in Prague show that, public pressure to work on and use Czech medical terminology notwithstanding, the number of publications in German rose during the second half of the nineteenth century from one-third to more than half of the respective totals.[77] The more differentiated statistics for the Jagiellonian University show similar patterns.[78] The most Western-oriented was the medical faculty, where some 30 per cent of publications were in German (10 per cent being dual publications in German and Polish). At the philosophical faculty, around 60 per cent of publications in the natural sciences were in Polish, with German and French making up about 20 per cent each, whereas in the humanities 80 per cent were published in Polish and 20 per cent shared by German, Latin, French and Russian.[79]

Increasingly in the nineteenth century, the trend of double publications caused professional articles to be published both in the respective 'national' and in an 'international' language. Even if most Central European scholars spent at least a few years on international travels, the additional workload was evident compared with that of scholars in Germany or France, for whom publication in their mother tongues sufficed. To a large extent professional academics were also active in popularizing science; a large proportion of articles consisted of popular science, published in a wide variety of journals.

Conclusion

This chapter has traced differing individual accentuations of nationality and internationality over the long nineteenth century. In the first half of the century, intensified work on the project of internal Enlightenment prevailed; by the century's end the internationalization of scholarship was more prominent. This trend coincides only overtly with political developments. While immediately after 1848 increasingly nationalized publics expected scholars to work on the national development of their cultures and countries, by the turn of the century the international arena had acquired at least the same importance. When the 'Polish' economic boycott of Prussia in 1908 was expanded to include culture, scholarship was explicitly excluded, because the impact of the scientific isolation that would result was considered to be detrimental.[80] Stanisław Madeyski-Poray (1841–1910), professor of civil law in Cracow and Habsburg Minister of Education and Religion from 1893 to 1895, put the nationalist view on this development in a nutshell:

> This is precisely the essence of the nation, that after reaching higher civilization through the use of the native tongue, it found means to develop to the fullest the spiritual forces that lie in the distinct national character. Given the continuous development of civilization, a nation can maintain its achieved position among the nations if it contributes together with them and with results in its own spirit to the general development of civilization.[81]

While the nineteenth century is usually described as a time when nationalism and internationalism drove apart, analysing the language issue in the scientific communities of the Habsburg Monarchy presents them as highly interdependent. In both the creation and the use of scientific languages, a turn from nationalism to internationalism can be observed, but this internationalism is highly different from the structure of the scientific community, or *République des lettres* in pre-national times. While scholarship became an international cultural process, scholars represented the national communities into which they inscribed themselves. At the time that communitarian objectivity was being accepted as a scientific standard, the alleged peculiarities of 'national styles' grew in importance. In this approach language was seen to have a pivotal role, linking nationalistic ideology and the alleged international character of the scientific community.

Language, in its intermediate position combining symbolic and communicative functions, clearly represented and indeed embodied this complex development in science and scholarship. After a period of nationalist inclinations and intensification of internal communication through the vernacularization of language and an abundance of publications confined to the respective national communities, belonging to the international community grew in importance by 1900. At this time bilingualism became less frequent, while the use of international terminology and publishing in foreign languages continued to intensify, reaching new levels after 1918. Still, this change followed an intended cultural path, in which the 'international' was only an extension of the 'national'.

Notes

Research for this chapter was carried out during my involvement in the PhD programme (*Initiativkolleg*) 'The Sciences in Historical Context' at the University of Vienna and also supported by a grant from the Austrian Research Council (ÖFG – MOEL-Plus-Förderungsprogramm). An earlier version of this chapter was presented at the ESHS conference in Vienna in 2008; parts were also discussed during talks in Boston, Budapest, Florence and Warsaw. I thank the audiences for their comments and questions.

1. For approaches to the relationship between the nationalization and internationalization of science, see Mitchell G. Ash (2000) 'Internationalisierung und Entinternationalisierung der Wissenschaften im 19. und 20. Jahrhundert – Thesen', in Manfred Lechner and Dietmar Seiler (eds.) *zeitgeschichte.at. Österreichischer Zeithistorikertag 1999* (Innsbruck: StudienVerlag), 4–12; Elisabeth Crawford (1992) *Nationalism and Internationalism in Science, 1880–1939: Four Studies of the Nobel Population* (Cambridge, New York: Cambridge University Press), esp. 79–106; Ralph Jessen and Jakob Vogel (2002) 'Die Naturwissenschaften und die Nation. Perspektiven einer Wechselbeziehung in der europäischen Geschichte', in idem (eds.) *Wissenschaft und Nation in der europäischen Geschichte* (Frankfurt am Main, New York: Campus Verlag), 7–37. An interesting approach to (inter)nationalism in physics can be found in Matthew Konieczny (2008) 'Science

and culture on the imperial periphery: the worldview of Władysław Natanson', in Arnold Suppan and Richard Lein (eds.) *From the Habsburgs to Central Europe: The Centers for Austrian and Central European Studies at the Universities of Stanford, Minneapolis, New Orleans, Edmonton, Jerusalem, Budapest and Vienna* (Vienna, Berlin: LIT Verlag), 113–29.

2. Andrzej Gawroński (1923) 'Nauka narodowa czy międzynarodowa' [National or international science?], *Nauka Polska. Jej potrzeby, organizacja i rozwój*, 4, 36–44; though different solutions were proposed during this debate, the internationality of science was a position on which most speakers agreed.

3. Gawroński supported his view with ideas of the Polish philosopher Wincenty Lutosławski; see especially Wincenty Lutosławski (1913) *Volonté et Liberté* (Paris: Félix Alcan), 266–68. Similar thoughts were expressed at the beginning of the twentieth century by the Czech philosopher Emanuel Rádl; see Tomáš Hermann (2003) 'Originalita vědy a problém plagiátu. Tři výstupy Emanuela Rádla k jazykové otázce ve vědě z let 1902–1911' [Authenticity of science and the issue of plagiarism. Contributions of E. Rádl to the language issues in science], in Harald Binder, Barbora Křivohlavá and Luboš Velek (eds.) *Místo národních jazyku ve výuce, vědě a vzdělání v Habsburské monarchii 1867–1918* [Position of National Languages in Education, Educational System and Science of the Habsburg Monarchy 1867–1918] (Praha: Výzkumné centrum pro dějiny vědy), 441–68.

4. Philipp Ther (2007) 'Das Europa der Nationalkulturen: Die Nationalisierung und Europäisierung der Oper im "langen" 19. Jahrhundert', *Journal of Modern European History*, 5, 39–66.

5. For methodological approach to nationalism and internationalism in Central Europe from the perspective of political history see also Kerstin Jobst (1996) *Zwischen Nationalismus und Internationalismus: Die polnische und ukrainische Sozialdemokratie in Galizien von 1890 bis 1914. Ein Beitrag zur Nationalitätenfrage des Habsburgerreichs* (Hamburg: Dölling und Galitz).

6. This process is analysed in Jan Surman (2009) 'Imperial Knowledge? Die Wissenschaften in der späten Habsburgermonarchie zwischen Kolonialismus, Nationalismus und Imperialismus', *Wiener Zeitschrift zur Geschichte der Neuzeit*, 9/2, 119–33.

7. The constructed character of the term 'nation' has recently been explored in a series of publications, none of which proposes an alternative, non-emotional designator. In this article all notions in quotes are thus actor-categories used by nationalists in nation-building discourse.

8. See the fundamental work of Barbara Skarga (1964) *Narodziny pozytywizmu polskiego, 1831–1864* [The Beginnings of Polish Positivism, 1831–1864] (Warszawa: Państwowe Wydawnictwo Naukowe); for Czech appropriation of the concept of organic work see Miloš Havelka (1998) '"Nepolitická politika": kontexty a tradice' ["Non-Political Politics": Contexts and Traditions], *Sociologický časopis*, 34/4, 455–66.

9. For example, in the 1870s the Cracow daily *Country (Kraj)* published the first translation of Charles Darwin into Polish and also one of the first social-Darwinist books in Polish, Ludwik Masłowski's *Prawo postępu: Studium przyrodniczo-społeczne* [The Law of Progress. A Social-Naturalist Study]; another Cracow daily, *Time (Czas)* published inter alia the inaugural lectures of professors at the Jagiellonian University and presentations in the Cracow Learned Society; alternating with literary texts, scientific literature was published, beginning on the front page, in parts over several weeks.

10. See Hans Lemberg (2003) 'Die Einführung der deutschen Unterrichtssprache in den deutschen Universitäten und ihre Auswirkungen auf Ostmitteleuropa', in Binder, Křivohlavá and Velek (eds.) *Místo národnich jazyku*, 169–82.

11. See, for example, Karel Ignac Thám (1783) *Obrana gazyka českého protí zlobiwým geho vtrhacům, též mnohým wlastencům, w cwičenj se w něm liknawým a nedbalým sepsaná* [Apology of the Czech Language Against Slenderers as well as many Countrymen Negligent and Indolent in the Practice of the Language] (Praha: J.F. ze Schönfeldu) and Josef Dobrovský (1791) *Über die Ergebenheit und Anhänglichkeit der Slawischen Völker an das Erzhaus Oesterreich* (Prague). For an English translation of Thám by Derek Paton with comments on the genre, see Balázs Trencsényi and Michal Kopeček (eds.) (2006) *Discourses of Collective Identity in Central and Southeast Europe 1770–1945, Vol. I. Late Enlightenment – Emergence of the Modern 'National Idea'* (Budapest: CEU Press), 205–9.

12. Onufry Andrzej Kopczyński (1804) *O duchu języka polskiego: wstęp. Na posiedzeniu publicznem dnia 16 listopada 1804* [On the Spirit of the Polish Language: Introduction. Read at the Public Meeting on 16. November 1804] (Warsaw), 10.

13. Karel Ignac Thám, *Obrana gazyka českého*, quoted in translation by Derek Paton (see note 11).

14. See Jürgen Schriewe (1998) *Die Macht der Sprache: Eine Geschichte der Sprachkritik von der Antike bis zur Gegenwart* (Munich: Beck).

15. Jan Svatopluk Presl (1837) *Nerostopis, čili, Mineralogia* [Decription of Minerals, or, Mineralogy] (Prague: Jan Spurný), iv.

16. Jan Śniadecki (1818) 'O logice i retoryce' [On Logics and Rhetoric], in *Pisma rozmaite Jana Śniadeckiego, Tom III. Zawieraiący listy i rozprawy o naukach* [Various Writings of Jan Śniadecki. Vol. 3 including letters and essays on the sciences] (Wilno: Józef Zawadzki), 185–203.

17. Jan Śniadecki, 'O ięzyku polskim' [On the Polish Language], ibid., 1–121, here 6.

18. Ibid., 66.

19. Ibid., 22.

20. Jan Śniadecki (1837 [1813]) 'O ięzyku narodowym w matematyce. Rzecz czytana na posiedzeniu Literackiem Uniwersytetu Wileńskiego dnia 15. Listopada roku 1813. v.s' [On National Language in Mathematics. Presented at a Literary Meeting at Wilno University on 15. November 1813], in Michał Baliński (ed.) *Dzieła Jana Śniadeckiego* [The Works of Jan Śniadecki], vol. 3 (Warszawa: Emanuel Glücksberg), 183–204, here 195.

21. Śniadecki, 'O logice i retoryce', 199.

22. Jan Śniadecki (1802) 'O obserwacyach astronomicznych' [On Astronomical Observations], *Rocznik Towarzystwa Warszawskiego Przyjaciół Nauk*, 1, 432–526; Jan Śniadecki (1818) *Jeografia, czyli opisanie matematyczne i fizyczne ziemi* [Geography, that is, Mathematical and Physical Description of the Earth] (Wilno: Józef Zawadzki). See also Jadwiga Waniakowa (2004) 'Historia polskiego podstawowego słownictwa astronomicznego na tle słowiańskim' [The history of Polish basic astronomical vocabulary on the Slavic background], *Studia z Filologii Polskiej i Słowiańskiej*, 39, 157–78.

23. Jędrzej Śniadecki (1840 [1830]) 'Przedmowa do Dziennika Medycyny, Chirurgii i Farmacji przez Cesarskie Towarzystwo Lekarskie w Wilnie roku 1830 wydawanego' [Introduction to the Journal of Medicine, Surgery and Pharmacy, published by the Imperial Medical Society in Wilno in 1830], in Michał Baliński (ed.) *Dzieła Jędrzeja Śniadeckiego* [Works of Jędrzej Śniadecki], vol. 3 (Warszawa: Emmanuel Glücksberg), 7–22, here 16.

24. Cf. Letter to the editor by Jędrzej Śniadecki in (1817) *Pamiętnik Warszawski, czyli Dziennik Nauk i Umieiętności*, 7, 385–401.

25. Josef Jungmann (1873) 'O jazyku českém' [On the Czech Language], in Národní Bibliotéka (ed.) *Josefa Jungmanna Sebrané drobné spisy: veršem i prosou* [Josef Jungmann's collected shorter works: in verse and prose], vol. I (Praha: I. L. Kober), 3–29.

26. For similarities and differences between Herder and Jungmann, see J.P. Stern (1989) 'Language consciousness and nationalism in the age of Bernard Bolzano', *Journal of European Studies*, 19, 169–89.

27. See, for example, Alois Jedlička (1948) *Josef Jungmann a obrozenská terminologie literárně vědná a linguistická* [Josef Jungmann and Literary-Scientific and Linguistic Terminology of the National Revival] (Praha: Česká akademie věd a umění); Teresa Wanda Orłoś (1980) *Polsko-czeskie związki językowe* [Polish-Czech Language Relations] (Wrocław: PAN, Ossolineum), esp. 39–44.

28. Similarly non-vernacular was the historically oriented biological terminology of Jan Svatopulk Presl. See Veli Kolari (1973) 'Notes on Jan Svatopulk Presl as terminologist', *Scando-Slavica*, 19/1, 187–95.

29. Cf. Jaroslav Batušek (1968) 'Zur Problematik der deutsch-tschechischen Beziehungen im Bereich der Geschichte der tschechischen physikalischen Terminologie', in Bohuslav Havránek and Rudolf Fischer (eds.) *Deutsch-tschechische Beziehungen im Bereich der Sprache und Kultur. Aufsätze und Studien II* (Berlin: Akademie-Verlag), 85–95; Jan Janko and Soňa Štrbáňová (2003) 'Uplatnění nového českého přírodovědného názvosloví na českých vysokých školách v průběhu 19. století' [Assertion of the new Czech scientific nomenclature at the Czech universities during the nineteenth century], in Binder, Křivohlavá and Velek (eds.) *Místo národnich jazyku*, 297–312.

30. Vladímir Macura (1998) 'Problems and Paradoxes of National Revival', in Mikuláš Teich (ed.) *Bohemia in History* (Cambridge, NY: Cambridge University Press), 182–96, goes so far as to state that the Czech literature of the time created its own semiosphere (Juri Lotman). For the broader context of this thesis see Vladímir Macura (1995) *Znamení zrodu: česke národní obrození jako kulturní typ* [Signs of the Birth. The Czech National Revival as a Cultural Type] (Praha: H&H), especially the criticism of the abundance of Czech word formations from the viewpoint of structuralism, 51–2.

31. Władysław Natanson to Ludwik Gumplowicz, 13 December 1888, Jagiellonian Library, division of manuscripts, signature 9007 III, vol. 6.

32. (Jan Svatopulk Presl and Karl Bořiwog Presl) Joanne Swatopluko Presl and Carolo Bořiwogo Presl (1819) *Flora čechica: indicatis medicinalibus, oeconomicis technologicisque plantis/Květena česká: s poznamenánjm lékařských, hospodářských a řemeslnických rostlin* (Prague: J. G. Calve).

33. Stanisław Bonifacy Jundziłł (1811) *Opisanie roślin litewskich według układu Linneusza* [Description of Lithuanian Plants according to the Linnaean System] (Wilno: Józef Zawadzki).

34. Alicja Zemanek, 'Koleje życia Józefa Rostafińskiego' [Life of Józef Rostafiński] in idem (ed.) *Józef Rostafiński. Botanik i humanista* (Kraków: PAU), 19–99.

35. See, for example, Emilian Czyrniański (1853) *Polskie słownictwo chemiczne* [Polish chemical vocabulary] (Kraków: Czas).

36. Orłoś, *Polsko-czeskie związki językowe*; on Volian see Michael Moser (2005) 'Some Viennese contributions to the development of Ukrainian terminologies', in Giovanna Brogi-Bercoff and Giulia Lami (eds.) *Ukraine's Re-integration into*

Europe: A Historical, Historiographical and Politically Urgent Issue (Alessandria: Edizioni dell'Orso), 139–80, here 154–75.

37. Teresa Ostrowska (1973) *Polskie czasopiśmiennictwo lekarskie w XIX wieku (1800–1900): Zarys historyczno-bibliograficzny* [Polish medical journals market in the nineteenth century (1800–1900). Historical-bibliographical outline] (Wrocław: Zakład narodowy im. Ossolińskich), 177.

38. Jan Baudouin de Courtenay (1900) *Głos członka Akademii J. Baudouina de Courtenay w sprawie słownictwa chemii* [The opinion of Academy member J. Baudouin de Courtenay on chemical vocabulary] (Kraków: Akademia Umiejętności).

39. Ibid., 12.

40. See Andrzej Grabowski (1900) *Polskie słownictwo chemiczne* [Polish chemical vocabulary] (Warsaw: Warszawskie Towarzystwo Akcyjne Artystyczno-Wydawnicze); Emilian Czyrniański (1881) *Słownictwo chemiczne* (Kraków: Akademia Umiejętności).

41. See (1865) 'Kritika: "Slovnik lékařské terminologie"' [Review: Dictionary of medical terminology], *Časopis lékařův českých*, 6, 46–57; 7/4, 55, here 55. The anonymous author meant that the dictionary was still full of German words, and should be corrected even if such words were widely used in practice (46).

42. Emilian Czyrniański (1872) *O słownictwie chemiczném polskiém: Odbitka z Bibljoteki umiejętności przyrodniczych* [On Polish chemical vocabulary. Reprint from Biblioteka umiejętności przyrodniczych] (Kraków: A. Dygasiński, W. Tomaszewicz); Józef Rostafiński (1887) 'Kilka słów o naszéj nomenklaturze i terminologii botanicznej na tle historii botaniki w Polsce' [A few remarks on our botanic nomenclature and terminology on the background of history of botany in Poland], *Wszechświat*, 9/6, 138–40.

43. Mieczysław Bąk (1984) *Powstanie i rozwój polskiej terminologii nauk ścisłych* [The creation and development of Polish terminology of exact sciences] (Wrocław: Zakład narodowy im. Ossolińskich, PAN), 154.

44. F[rantišek] J[osef] Studnička (1876) 'O rozvoji naši literatury fysikální za posledních padesáte let' [On the development of our physical literature in the last fifty years], *Časopis musea království českého*, 50, 35–46.

45. For the development of a Ruthenian/Ukrainian scientific language see, for example, Marina Höfinghoff (2008) 'Entwicklung der chemischen Terminologie in Galizien (Mitte des XIX.–Anfang des XX. Jh.)', *Zeitschrift für Slawistik*, 4/53, 403–37; Iryna Romanivna Procyk (1999) *Ukrajinska fizyčna terminolohija druhoji polovyny XIX – peršoji tretyny XX stolittja. dyssertacya kand. filol. nauk* [Ukrainian physical terminology of the second half of the nineteenth and the first third of the twentieth century. PhD Dissertation] (L'viv: Nacional'nyj Universytet imeni Ivana Franka).

46. Jürgen Schriewe (1998) *Die Macht der Sprache* (cit. note 14), esp. 86–97.

47. Derek Sayer (1996) 'The language of nationality and the nationality of language: Prague, 1780–1920', *Past and Present*, 153, 164–210, here 193.

48. The issue of cultural constriction was differently coded in national and imperial discourse. See Jan Surman (forthcoming) 'Symbolizm, komunikacja i hierarchia kultur: Galicyjski dyskurs hegemonii językowej początku drugiej polowy XIX wieku' [Symbolism, communication and cultural hierarchy: Galician discourse of language hegemony in the early second half of the nineteenth century], *Historyka. Studia Metodologiczne*.

49. For Czech and Slovak languages see [Jan Kollár] (1846) *Hlasowé o potřebě jednoty spisovného jazyka pro Čechy, Moravany a Slowáky* [Voices on the necessity of a

unified literary language for the Czech, Moravians and Slovaks] (W Praze: Nákladem Českého museum: V kommissí u Kronbergra i Řiwnáče); for Moravian language separatism see Ondřej Bláha (2005) 'Moravský jazykový separatismus: zdroje, cíle, slovanský kontext' [Moravian linguistic separatism: Origins, aims, Slavic context], *Studia Moravica. Acta Universitatis Palackianae Olomucensis Facultas Philosophica Moravica*, 293–99; for Ruthenian/Ukrainian in general see Alexei Miller and Oksana Ostapchuk (2009) 'The Latin and Cyrillic alphabets in Ukrainian national discourse and in the language policy of empires', in Georgiy Kasianov and Philip Ther (eds.) *A Laboratory of Transnational History: Ukraine and Recent Ukrainian Historiography* (Budapest, New York: Central European University Press), 167–210; see also below for a description of the main currents. For Polish and the competition between two ideas of the Polish language, one based on central dialects and one on eastern dialects, unsuccessfully proposed by Piotr Semeńko (Semenko) to unite Poles and Ruthenians, see Włodzimierz Borodziej, Błażej Brzostek and Maciej Górny (2005) 'Polnische Europa-Pläne des 19. und 20. Jahrhunderts', in Włodzimierz Borodziej and Heinz Duchhardt (eds.) *Option Europa: deutsche, polnische und ungarische Europapläne des 19. und 20. Jahrhunderts, Band 3* (Göttingen: Vandenhoeck & Ruprecht), 43–166, here 63; Brian Porter (2000) *When Nationalism Began to Hate: Imagining Modern Politics in Nineteenth-century Poland* (New York: Oxford University Press), 19–21; and in general Daniel Beauvois (1996) *Histoire de la Pologne* (Paris: Hatier), 169–91.

50. Kazimierz Opałek (1977) 'Oświecenie' [Enlightenment], in Bogdan Suchodolski (ed.) *Historia nauki polskiej. T. 3, 1795–1862* [History of Polish Science. Vol. 3, 1795–1862] (Wrocław: Zakład Narodowy im. Ossolińskich, Polska Akademia Nauk, Zakład Historii Nauki, Oświaty i Techniki), 233–465, here 322–329.

51. Jan Evangelista Purkyně (1787–1869) had been the lead editor of the journal between 1821 and 1840; *Krok* means 'step' but is also a reference to the mythological figure of the ancient Czech ruler (sometimes magician or arbitrator) from Kosmas' *Chronica Boemorum* (1119–1125).

52. The journal appeared at first as a dual publication: *Časopis Společnosti wlastenského museum w Čechách* and *Monatsschrift der Gesellschaft des Vaterländischen Museums in Böhmen* [Journal of the Society of Bohemian National Museum], the two parts aiming at different publics and including different articles. On account of low readership, the German-language journal was abandoned 1832; the Czech one was renamed *Časopis Českeho Museum*, and because of financial problems, was put under the patronage of The Czech Foundation (Matice česká), an autonomous branch of the Museum concerned with support for Czech literature and at the same time a printing house for a number of texts that played a significant role in the nation-building process.

53. Roughly, Russophiles were proposing a dialect based on Old Church Slavonic and close cooperation with Russia; Moskwophiles were opting for a nation dependent on the Russian Empire with a language close to Russian; Ukrainophiles were striving to establish an independent nation with a language based on the Poltava dialect of Taras Shevchenko, uniting Ukraine (that is, a part of the Russian Empire) and Eastern Galicia. This terminology also makes a distinction between Russophile and Moskwophile, which is not widely used in historical literature, where these are treated as synonyms; although these two groups are close to each other, their alliance dates only from the late nineteenth century. See John-Paul Himka (1999) 'The construction of nationality in Galician Rus': Icarian flights in almost all directions', in Ronald Grigor Suny and Michael D. Kennedy (eds.)

Intellectuals and the Articulation of the Nation (Ann Arbor: University of Michigan Press), 109–64.

54. And *Zapysky* of the Shevchenko Scientific Society, which was based on the structure of writing of 'western' academies.

55. Medical journals in Ruthenian/Ukrainian appeared, however, only at the end of the nineteenth century, after journals for other disciplines.

56. Karel Chodounský (1911) 'K padesátiletí "Časopisu lékařův českých"' [Fiftieth anniversary of the Journal of Czech Physicians], *Časopis lékařův českých*, 50/53, 1602–04.

57. K. (1880) 'Časopis pro pěstování mathematiky a fysiky, kterýž se zvláštním zřetelem k studujícím reediguje Dr. F. Studnička a vydává Jednota českých mathematiků' [Review: Journal for the Support of Mathematics and Physics, edited with special consideration for students by F. Studnička and published by the Union of Czech Mathematicians], *Časopis musea království českého*, 54, 367–68.

58. Ostrowska, *Polskie czasopiśmiennictwo lekarskie*, 21, 120.

59. Depictions of congresses were often rich in detail and included personal opinions on their organization, and thus can be seen as extended versions of contemporary conference reports. Reports from travels included in the first place descriptions of laboratories, manual procedures and styles of research at particular institutions, and were often supported by scholarship and grant organizations.

60. Waldemar Kozuschek (2003) *Jan Mikulicz-Radecki 1850–1905. Współtwórca nowoczesnej chirurgii/Johann von Mikulicz-Radecki. Mitbegründer der modernen Chirurgie* (Wrocław: Wydawnictwo Uniwersytetu Wrocławskiego).

61. Uljana Jedlins'ka (2006) *Kyrylo Studyns'kyj, 1868–1941: žyttjepysno-bibliohrafičnyj narys* [Kyrylo Studyns'kyj, 1868–1941: Biographical-bibliographical outline] (L'viv: Naukove tovarystvo im. Ševčenka).

62. This distinction was frequent not only in German (*Stamm* vs. *Nationalität*) but also in Polish (*szczep* vs. *naród/narodowość*).

63. Such arguments were raised in Cracow in 1853, when the replacement of Polish by German as the language of instruction was supported by the Academic Senate of Jagiellonian University. From the 1860s onwards this position was advanced against the idea of a Ruthenian university (by Poles) and against the idea of a Czech university. See for Galicia Jan Surman (2009) 'Die Figurationen der Akademia. Galizische Universitäten zwischen Imperialismus und multiplen Nationalismus', in DK Galizien (ed.) *Galizien – Fragmente eines diskursiven Raums* (Innsbruck, Wien, Bozen: StudienVerlag), 17–40; for Bohemia, for example, (1882) 'Das Ende der deutschen Universität in Prag', *Wiener Medizinische Wochenschrift*, 7, 197–98, and documents reprinted in Jaroslav Goll (1908) *Rozdělení Pražské university Karlo-Ferdinandovy roku 1882 a počátek samostatné University české* [The division of the Prague Charles-Ferdinand University and the beginning of a separate Czech university] (Prague: Nakl. Klubu historického).

64. See the literature quoted in the previous note, and for the Ruthenian university especially the petition of intellectuals for the establishment of a Ruthenian university in L'viv: Ivan Bartoševskyj, Myhajlo Hruševskyj, Ivan Dobrjanskyj, Stanyslav Dnistrjanskyj, Oleksander Kolessa, Josyf Komarnyckyj, Tyt. Myškovs'kyj, Petro Stebel'kyj, Kyrylo Studyns'kyj (1907) 'Zajava ruskyh profesoriv universytetu y L'vovi' [Petition of Ruthenian professors of the University in Lviv], *Ruslan*, 6/19. The most direct formulation is in Hrushevskyi's long article on the relations between scholarship and nation: Myhajlo Hruševs'kyj (2002) 'Sprava ukrajins'kyh

katedr i naši naukovi potreby' [The question of Ukrainian chairs and our scientific needs], in Jaroslav Daškevyč, Ihor Hyryč, Hennadij Borjak and Pavlo Sohan' (eds.) *Myhajlo Hruševs'kyj. Tvory v 50 tomah, serija 'Suspil'no-polityčni tvory'*, vol. 1. *1894–1907* (L'viv: Svit), 458–84, here 473–74; the article was originally published in 1907 in *Literaturno-Naukovyj Vistnyk*, 37, 52–57, 213–20 and 408–18.

65. Ivan Holovac'kyj (ed.) (2005) *Naukovi praci, dokumenty i materialy profesora Ivana Horbačevs'koho* [Scientific works, documents and materials of professor Ivan Horbačevs'kyj] (L'viv: Naukove tovarystvo im. Ševčenka, Biohemična komisija).

66. Permission to present under a 'national' banner differed from discipline to discipline and country to country. Congresses held in France were the first to divide speakers from the Habsburg Monarchy into nationalities, not without protests from imperial scholars. In the Russian Empire this was an additional platform for Russian–Ukrainian national conflict; see Serhii Plokhy (2005) *Unmaking Imperial Russia: Mykhailo Hrushevsky and the Writing of Ukrainian History* (Toronto: University of Toronto Press), 49–53.

67. By the end of the nineteenth century, Czech historians, for example, were asked to publish more in German, because readership of Czech historiography was becoming confined to Czech publics. Of course, history was an international battlefield, in which international reading publics could be drawn towards the patronization of national claims. Especially Jaroslav Goll strove to compete with German historians in their language, in order to lend support to the Czech viewpoint on history and thus strengthen the popularity of the Czech national narrative. This fact was mentioned by Antonín Kostlán in his unpublished presentation 'To be a good son of one's nation…Czech historians between national program and scientific style' at the XXIII International Congress of History of Science and Technology, Ideas and Instruments in Social Context, 28 July–2 August 2009, Budapest, Hungary.

68. August Seydler (1890) 'Akademie česká a Společnost nauk' [Czech Academy and the Scientific Society], *Athenaeum. Listy pro literaturu a kritiku vědeckou*, 3/8, 65–69; excerpts from and an analysis of Purkyně's most interesting publication in this regard, *Akademia* (1861), can be found in Milan Kratochvíl (ed.) *Jan Evangelista Purkyně a jeho snahy o reformu české školy* [Jan Evangelista Purkyně and his efforts for a reform of the Czech/Bohemian schools] (Prague: SBN).

69. The Royal Bohemian Society of Sciences continued to exist until 1952, when it was merged with the Czech Academy of Sciences and the Arts (Česká akademie věd a umění císaře Františka Josefa I.) to form the Czechoslovak Academy of Sciences (Československá akademie věd). In addition to the Czech Academy (founded as a private institution in 1890), its counterpart, the Society for the Advancement of German Science, Art and Literature in Bohemia, was established in 1891 as the Gesellschaft zur Förderung deutscher Wissenschaft, Kunst und Literatur in Böhmen.

70. Maryjan Raciboski (1888) 'Do naszych przyrodników' [To our naturalists], *Wszechświat*, 42/7, 668–70.

71. The most condensed argumentation is in August Wrześniowski (1888) *Wszechświat*, 47/7, 748–49.

72. Raciboski, 'Do naszych przyrodników', 670.

73. Władysław Gosiewski (1888) *Wszechświat*, 47/7, 749.

74. Ryszard Błędowski (1912) *Szkic dziejów zoologii w Polsce od początku wieku XIX (Odbitka z Wszechświata)* [Outline of the history of zoology in Poland since the

beginning of the nineteenth century (Reprint from Wszechświat)] (Warsaw: L. Bogusławski).

75. The question why the name of the Kraków Academy is abbreviated here is also a language issue, since the translation of the Polish world *umiejętność* could be either 'sciences and arts' or 'sciences'; in the first case it refers to the Enlightenment concept of science being both a theoretical and a practical activity, in the second to modern science.

76. Ostrowska, *Polskie czasopiśmiennictwo lekarskie*, 120–21.

77. Ludmila Hlaváčková (2003) 'Čeština v medicíně a na pražské lékařské fakultě (1784–1918)' [The Czech language in medicine and at the Prague Medical Faculty (1784–1918)], in Binder, Křivohlavá, Velek (eds.) *Místo národních jazyku*, 327–44.

78. Author's calculation on the basis of annually published lists of professors of the University 1888–1913 in *Chronicles of Jagiellonian University* (Kroniki Uniwersytetu Jagiellońskiego).

79. Because of the inclusion of popular articles in this calculation, the actual proportion of foreign scientific publications should be regarded as higher.

80. See the discussions in *Przegląd Powszechny* 1908 (vol. 100, 1*–24*), with several scholars commenting on the issue; most speakers agreed that the boycott should not include scientific publications, instruments, participation in congresses and organizations, or studies at German universities.

81. Ibid., 24*.

3

'Staatsnation', 'Kulturnation', 'Nationalstaat': The Role of National Politics in the Advancement of Science and Scholarship in Austria from 1848 to 1938

Johannes Feichtinger

As Gary Cohen has written, in the Habsburg Empire national politics served 'in fundamental ways' as both an 'emancipatory, centripetal process' that 'fuelled a popular revolt against the traditions of state absolutism' and helped to transform the state which developed after the *'Ausgleich'* or 'Compromise' of 1867 between Austria and Hungary, and also as a 'destructive, centrifugal process'[1] that eventually destroyed the Habsburg Empire, which had already been divided into halves by the recognition of Hungary as a seemingly autonomous nation-state. In Austria national politics most definitely served as a source of conflict between progressive- and conservative-minded political publics. In the mid-nineteenth century, the liberal Hungarian statesman and writer Joseph von Eötvös (1813–71) gave an idea of how national politics worked: 'The basic principle of all national efforts is the sense of higher capability, but the intended purpose is dominance.'[2] When viewed against the background of national power politics in the Habsburg Empire, science and scholarship became decisive factors in politics, and politics conquered the scientific and scholarly field.

In the course of the nineteenth century, the concepts of *'Staatsnation'* (nation-state) and *'Kulturnation'* (cultural nation) constituted diametrically opposed major political 'camps', one of which scholars and scientists were expected to commit to if they wished to use political resources to establish personal academic careers in a scientific institution and to validate their scientific work. It should be noted that these terms were only rarely employed in contemporary discourse.[3] However, the progressive Habsburg political economist Gregor von Berzevicy (1763–1822) testified to the very early use of the term *Staatsnation* when he noted in 1817 from his allegedly 'disinterested position' that there were no original nations in Europe any more; that language no longer served as proof of nationhood; that the term nationality could be found neither in roots, nor in origins, nor even in language, but in public law, that is, in the *state*. In fact this led him to the term 'state-nation' by his recognizing that 'unity in government and administration gives rise to the equality of customs, laws and ways of thinking; and this unites different peoples ("nationalities") – despite speaking various languages – into *a single* state-nation,' with which one should be satisfied, since the original nation has been lost.[4] Be that as it may, historians of the Habsburg Empire are more familiar with the notion of the *'Gesamtstaat'* pioneered by the jurist and historian of law Hermann Ignaz Bidermann (1831–82), which could be described as the Austrian counterpart of the *Staatsnation* concept.[5] Those subscribing to the idea of *Gesamtstaat* usually attacked the adherents of *Kulturnation* as nationalists, or language nationalists. In fact both positions strove for nationhood, though conceived in opposite terms. Proponents of *Kulturnation* constructed their specific national understanding by using culture – i.e. cultural difference – as a means of demarcation. As to *Staatsnation*, identity and commitment were primarily based on the principle of dynastic rule over the Habsburg territory. Following the collapse of the Habsburg Empire in 1918, any interpretation of scientific activities in context also has to consider the *Nationalstaat* (national state), a new factor confronting Austrian scholars and scholarship.

In this essay political, institutional, personal and epistemological aspects of the conduct of scientific and scholarly research will be brought together from an actor-oriented perspective. In doing so, I will highlight the impact of individual scholars upon the above-mentioned varieties of discourse surrounding the nation. Examples from the humanities will serve to illustrate the present thesis. The problem at hand can be exemplified in the following way: in the later Habsburg Monarchy scholars and scientists operated in a *relatively* autonomous academic field in which they struggled for professional opportunities. However, historical interpretation should not be permitted to ignore that there were indeed specific levels of political involvement. Denying self-implication is a distortion particularly obvious, for instance, in the work produced on the history of the Academy of Sciences after 1945 from an apologetic perspective. Such a line of reasoning ignores the fact that the advancement of science and scholarship in Austria and Hungary during the late Habsburg Monarchy was at least as strongly determined by political factors as in Germany. Strictly speaking, there had been 'no science without premises' (*'Voraussetzungslosigkeit der Wissenschaft'*).[6]

Nevertheless, a deceptive, defensively apolitical image of the emergence of the Austrian academic landscape was to become relatively widespread in a post-Second World War scientific scene that had to confront the serious question of the involvement of its own scholarship and science in the rise and success of National Socialism. Richard Meister (1881–1964), at that time vice-president of the Austrian Academy of Sciences, is a typical representative of this attitude, which he put as follows in his official address on the occasion of the centenary celebrated in May 1947. Meister emphasized 'that the Academy took pride in adhering to that supreme law which guides all scientific enquiry, namely objectivity, through all vicissitudes. Everybody, without exception, who served the Academy acceded to this very law', which is the fundamental principle – he announced – we will serve in the future.[7]

Autonomous and heteronomous scholarship

Given the fact that from that period that marks the beginning of modern industrial society in Austria, scholars and scientists became engaged in political activities, it was still possible to distinguish between those who explicitly tried to defend and extend scientific autonomy and those who did not hesitate to use it in order to promote political aims. According to Pierre Bourdieu two typological forms of involvement may be identified.[8] First, there were the so-called *heteronomous* scholars, i.e. those who advanced their careers by building strategic alliances with non-academic powers, e.g. with the Roman Catholic Church and/or entrepreneurship or, as in the present specific case, those who wished to achieve this end by engaging in national politics in order to achieve and improve their academic status. Second, there were *autonomous* scholars, who sought to defend and extend the independence of scientific research to the greatest extent possible. Whereas the former strove for the advancement of institutional infrastructure and their personal careers in scholarship and science, usually at the expense of scientific autonomy, the latter often fought for absolute individual and scientific independence. However, both took the risk of falling victim to the whims of political interests – heteronomously acting scholars through what Herbert Mehrtens and Mitchell G. Ash call 'self-mobilisation',[9] autonomously acting scholars by producing scientific results from which political decision-makers could benefit for certain political purposes.

In what follows, the terms 'autonomy' and 'heteronomy' will be used consistently as the typological terminology to highlight the strategies pursued by individual scholars. In addition, the author wishes to suggest the analytical term 'relative autonomy'. Introducing this term, Bourdieu does not suggest complete independence of higher education from the fields of economic and political power that determine societal power relationships; in fact, relative autonomy describes how the members of the scientific field cope with and respond to the demands of external forces. In our specific case, relativity does not signify the degree of autonomy from politics but rather how members of the scientific field deal and dealt with it. According to Bourdieu, autonomy becomes manifest in the ability the scientific field acquires to refract every external determination 'like a prism'.[10] This process of refraction is predicated on how the scholarly field is structured by autonomous and heteronomous positions. The term 'scientific field' is understood as an 'exceptional social universe'[11] endowed with specific practices, logics and laws of functioning within the field of power. Scholars who strive for a relatively autonomous position aimed at widening the scope of their independence. Thus, 'autonomy is not a given, but a historical conquest, endlessly having to be undertaken anew.' In addition, 'to say that the field is relatively autonomous with respect to the encompassing social universe is to say that the system of forces that are constitutive of the structure of the field ... is relatively independent of the forces exerted on the field'.[12] According to this view, there are positions that we have to distinguish in the way Bourdieu has proposed, namely the autonomous, the heteronomous and relatively autonomous approaches: 'The decisive indicator hinting at the degree of autonomy of a given field is its power of refraction' – an insight particularly pertinent to the scientific field – i.e. 'its power of translation. Conversely, the heteronomy of a field is marked by the fact that external issues, namely political, are expressed without distortion.'[13]

However, Mitchell G. Ash refers to another important fact, namely that 'a science which has been assured autonomy by an authoritarian hand can serve a political regime more efficiently under the guise of granted autonomy'. On the one hand, the state 'mobilize[s] scientists as resources in the interest of achieving certain political aims; on the other hand, scientists can convert themselves into such resources (or claim that they are doing so)'; finally 'both things can happen at once.'[14] In fact, these significant practices, as shown by Ash in many examples, revealed that some scientists' purported autonomy, i.e. their rejection of external demands and their striving for the strict division of science and politics, nonetheless served political purposes.

Scholars and scientists who defended the relative autonomy of the scientific field did not surrender the goal of achieving political objectives for society. However, they did not act to legitimize ongoing political issues conformistically, but intervened critically on the basis of scientifically generated concepts. By calling this specific relationship of science to politics 'scholarship with commitment',[15] Bourdieu emphasizes that it was not in the interest of scientists who subscribed to this position to describe the field of science as absolutely autonomous. If science remains absolutely autonomous – however illusory this might be –, it is excluded from having impact on other fields, such as politics. On the other hand, if political claims are transmuted into scientific agendas, science loses its autonomy and, as Bourdieu makes abundantly clear, thereby its authority. Science retains its relative autonomy by accumulating scholarly authority, and ensures its independence in terms of 'competence and authority' and its impact on the 'world of temporal/profane purposes'.[16]

In what follows, the term relative autonomy will be applied to that group of scholars who strove for the strict division of the two fields – science/ scholarship and politics – but who nevertheless remained committed to achieving political objectives for society. In the late Habsburg Monarchy some sort of relatively autonomous attitude becomes manifest, for instance, in those scholars who developed theories of democracy and of how collective identities should be constructed in culturally heterogeneous state entities. They should be regarded neither as 'Mandarins'[17] (Fritz K. Ringer) nor (in the typology proposed by Jonathan Harwood for the German case) as 'specialists' or 'experts'.[18] The scholars of whom we are speaking unmasked the idea of pure science as fiction, demonstrated political attitudes while remaining unaffiliated to party politics, and strictly distinguished between politics (as a function of will) and science (as a function of cognition). However, through developing theories which they classified applicable to settle conflicts and to identify means of social interaction and integration, they also invested the results of their scholarly and scientific work to achieve political goals. Outside the scientific community, they tended to public affairs of the utmost importance.

A practice- and process-oriented perspective will be used to shed light on the problem at hand. With the advancement of empirically based research, the structure of modern science tended towards independence from other limiting state and political factors – at least as perceived by the scientists themselves. Of course, scholarship in fact remained subject to political developments and influence, as is definitely the case for the nations of Central Europe, and the ways by which scholarly work was carried out equally remained heavily fraught with political overtones. In this sense scientific practice constituted a political challenge to both the scientists and the politicians. It goes without saying that scholars' ability to stay in tune with political demands was decisive for institutional and academic advancement, as well as for personal careers. Put briefly, the potentially autonomous scientific community and the politicians viewed each other reciprocally – in Mitchell G. Ash's words – as 'resources for one another'.[19]

Nationalisms – Habsburg's challenge

If we examine the emergence of a modern academic landscape in nineteenth-century Austria-Hungary, Ash's conception may bring us a step closer to answering the question of the relationship between the specific forms of national politics and scholarship/science. Wherever scholars entered into strategic alliances with politics to establish themselves institutionally and to validate their scientific work, they ran considerable risks. In the Habsburg Empire, however, they found themselves involved in the polemics of the two diametrically opposed political projects referred to above: the dynastic, transnational *Staatsnation* (nation-state) and the liberal-bourgeois *Kulturnation* (cultural nation). The term nation-state stood for the *all-inclusive* narrative of the transnational empire, whereas cultural-nation referred to a specific strategy shaped by two interconnected developments – the nationalist activists of the Czech, Polish, Slovenian, Ruthenian and Italian minority populations of the Habsburg Empire, who were already developing an awareness of their own independent political identity on the one hand, and the German and Hungarian activists who represented the politically dominant majority in their respective halves of the Habsburg Monarchy on the other – 'to create new...boundaries' within 'multi-lingual communities' in regions 'where few such boundaries had traditionally existed'. The building of national identities needed the construction of cultural differences, which then had to be imported into neighbourhoods where trans-cultural practices characterized daily life. Thus, activists 'worked hard' to delineate the cultural boundaries that they claimed separated national communities in pluricultural areas.[20] Their hard work permeated all levels of society and was supported by the media, the arts and members of the academic community (e.g. national historiographies, economics and philologies), all playing their parts in reducing cultural heterogeneity to a singular predominant national narrative. The 'Guardians of the Nation' defined and legitimized these *Kulturnationen* through the creation of a common cultural, linguistic and/or ethnic heritage. Ethnic differences were therefore largely an effect, as Peter Judson put it, and not a cause of nationalist activism.[21] The concept of the cultural nation thus fulfilled an *exclusive* function.

No matter which decision university-based scientists took in order to advance their own scholarly status and that of their disciplines,[22] there were distinct advantages and disadvantages. If they sided with the *Staatsnation*, they served an integrative purpose for the Empire. At the same time they aroused suspicion that they were helping to stabilize a system already infamous for obstructing democratic developments. Scholarship that propagated the concept of the *Kulturnation* served an integrative, potentially democratic purpose, but only for a chosen nationality; for the *Gesamtstaat* – the entire Monarchy – *kulturnational*-oriented scholars were viewed as acting disruptively when they supported political activists to make territory/space, language and people (*Volk*) as congruent as possible within that social form called *Kulturnation*. Its rise caused what the Austrian statesman and historian Joseph Redlich (1869–1936) dubbed the *'Staats- und Reichsproblem'* because it weakened two guiding principles upon which the Monarchical order rested, namely the legitimacy of the Habsburg dynasty and the cohabitation of the Monarchies' inhabitants in administrative and political units defined legally and historically, but not in terms of ethnicity.[23] From the vantage point of cultural-nationalists, the envisaged 'nations' did not live in entities corresponding to their national identity. No 'nationality' inhabited a compact territory, and many members of the same *Volksstamm* (nationality) – the official designation – lived in dispersed groups throughout various provinces of the Habsburg Monarchy. When activists began to aspire to a fusion of territoriality and nationality in this 'most colourful mix of people Europe ever saw', fragmentation seemed inevitable: the crownlands (Dominions of the Crown) of which the Monarchy consisted 'tear apart the nations', Karl Renner (1870–1950) wrote in 1918; thus 'it is scarcely surprising that the nations wish to tear apart the Länder.'[24]

Scholars' chances for advancement

It goes without saying that the alternative options of political involvement had a profound impact both on the structure of the given scientific field and on the professionalization of academic life in the Habsburg Monarchy. There seems to be a connection between increasingly professionalized scholarship and the risk or chance of supporting national activism. To put it briefly: in Austria the process of professionalization of science and scholarship does not necessarily stand for the absolute rejection of political activism, and the rise of autonomy, i.e. the process of science becoming independent from external factors and thus more self-governing, did not preclude politically informed and responsible agency within the scholarly field. The question remaining relates to the justification for this political commitment: Was it a commitment made in order to advance academic research? To what purpose? And of course: What function did it fulfil? Whereas some scholars harnessed national politics affirmatively as a vehicle for their self-promotion, others – among them jurists, psychologists, philosophers and students of culture – reflexively reconfigured the relationship between science, politics and culture.[25] With Jürgen Habermas we may distinguish two varieties of manoeuvring in the Central European field of scholarship, namely the *reflexive-inspiring* and the *politically-legitimizing* variety.[26] The latter – either directly or indirectly – forged alliances with politics in order to enhance its own position, which was attuned to the respective national idea, while the former cautiously maintained a degree of detachment. 'Relative autonomy' was necessary, enabling scholars to formulate proposals for coping with national heterogeneity and with the challenge of contriving an overarching identity. This kind of activity was what Pierre Bourdieu would later call 'scholarship with commitment'.[27]

In Austrian history there are many examples of both relative autonomous and heteronomous scholarly action. The latter tendency is symptomatic for the era of the new absolutist regime, established in December 1851, as well as for some areas of advanced scholarship in the late nineteenth and the first half of the twentieth century. However, around 1900 a few scholars adopted a more critical attitude. They strove to defend the autonomy of science in order to accumulate the necessary authority to intervene into disruptive social processes, not in a legitimatory, but in a delegitimatory, highly reflexive and inspiring way.

Thun's window of opportunity

In the 1850s, the new style of authoritarian regime combined a 'strongly centralized ministerial government with a clear mandate to proceed with a vigorous program of social management and economic development'.[28] In this period, dubbed neo-absolutist, state-nationalism became the driving force of modernization. To this end, state bureaucracy was suddenly increased to enable the government to assume a leading role in determining the policies, structures and personnel of university life. Against this political background Count Leo Thun-Hohenstein (1811–88) – a Bohemian aristocrat, Catholic, and enemy of liberal thinking who was appointed imperial minister of education in July 1849 – sponsored 'a very progressive educational reform package', initially culminating in the Higher Education Organization Act of 1849.[29]

One century earlier, school and university affairs had been almost entirely a central issue of the Roman Catholic Church, and – on account of their efforts in bringing about the Counter-Reformation – of the Jesuits in particular. In the course of the state reform movement of the eighteenth century, the influence of the Church in educational politics had been continuously curtailed and, ultimately, Empress Maria Theresia (who ruled from 1740 to 1780) and her son, the Emperor Joseph II (1780–1790), had placed the entire educational system under the control of the increasingly efficient centralist state administration. In the so-called *Vormärz*, the Studienhofkommission – a court commission – kept the political opinions of both professors and students under surveillance. Lectures had to follow a prescribed script, and the educational system did not necessarily aim at a broadening of knowledge but rather at a preservation of its established state.[30] Students' forays to foreign universities, and particularly to German higher institutions, were largely prohibited in order to 'protect' the future élite of the state administration from insidious national and 'liberal' ideas. University reform – particularly the issue 'freedom of teaching and learning' (meaning: freedom from state control and supervision) – was one of the central demands for which Viennese students fought in the revolution of 1848.[31] The students' demand was promptly met: academic freedom was proclaimed at the University of Vienna by the vice-director of medical-surgical studies, Ernst von Feuchtersleben (1806–49), on 19 March.[32] A week and a half later, on 30 March 1848, the first imperial minister of education, Franz Freiherr von Sommaruga (1780–1860), publicly announced the reformation of university life according to the principle of 'freedom of teaching and learning',[33] which had already been formally recognized in Prussia. The principle was included in Franz Serafin Exner's (1802–53) publication of the 'basic principles of public education in Austria' on 21 July 1848, and finally made law in the Higher Education Organization Act of 30 September 1849. When the newly appointed minister Thun-Hohenstein also accepted 'freedom of teaching and learning' as a basic principle, the separation of the two fields – science and state politics – was accepted, at least rhetorically. The reform was inspired by the so-called 'German' university model,[34] but what the higher education reformers adopted from Germany depended on uniquely Austrian circumstances: the reform-oriented party had to appease both the Catholic Church and the reactionary political elite as well as patriotic late Josephinian opposition.[35] After minister Thun-Hohenstein had overcome their resistance, the university reform of 1848/49–1853/55 brought a considerable modernizing boost.

However, one must not ignore the fact that Thun's reform aimed at the 'purposeful' (zweckmäßige)[36] organization of university life. What precise purpose – academic or political – was to be served remained to be seen. Historians have provided convincing evidence that regardless of the censorship policy in the Vormärz (a period extending from 1815 to 1848), a hitherto unknown politicization of academic matters took place in the decade during which Count Leo Thun-Hohenstein served as minister of education.[37] Whereas professors of liberal conviction had gained a foothold at the faculty of law of the University of Vienna prior to 1848, in the first decade after the revolution candidates of dubious political convictions were kept strictly off the academic staff. In practice, university autonomy remained ambivalent – with 'varying interpretations of freedom'.[38] On the one hand, minister Thun set limits to academic freedom with regard to maintaining the loyalty of the universities to both church and state as well as to the political reliability of the professors, since it was their task to educate obedient civil servants for senior positions in the state administration. On the other hand, the universities were granted a relatively high degree of independence (relative 'self-administration'), including the right to pre-select candidates for vacant professorships and propose them to the minister. However, university life remained controlled by and accountable to the state. In the end, it was Count Leo Thun and the officials of his ministry who appointed the professors and decided on their remuneration, and thus they were representatives of the system to which they owed their loyalty – if only for their salaries. It would appear that the minister of education was fully aware that the universities, science and scholarship had to be allowed to reach a specific degree of autonomy in order to serve the advancement of science and learning, and thus serve as a prolific resource for state politics.

When the renowned Austrian anatomist Joseph Hyrtl (1810–94) addressed the 32nd convention of German Naturalists and Doctors in 1856, he emphasized how closely interwoven 'the newly formed shape of the state' and the 'life of science' were, as reciprocally connected 'forms of life necessary' under the auspices of the state. Vesting his address in a distinctly post revolutionary garb, Hyrtl paid tribute to the 'union of all scientific interests under an autonomous supervising power' and the readiness of all organs of government to promote the flourishing of science.[39] Hyrtl, we may surmise, sincerely believed that he was witnessing the progress of the sciences in his lifetime, a progress due to Thun's reshaping of the educational system: 'The safest means to promote progress of the sciences' was – according to this regime-loyal anatomist – 'to remove the obstacles which impeded its free unfolding.' The reform allegedly abolished the 'law of compulsion' that had hindered the free and autonomous development of science.[40] Hyrtl was one of those who portrayed the pre-1848 system as reactionary in order to praise the newly enshrined system of free teaching and learning as progressive – a reform not as liberal as frequently supposed. However, he failed to mention that the autonomy granted by the prevailing neo-absolutist regime was an integral part of the politically purposeful organization of university life, since state politics had a vital interest in the advancement both of science, which served to improve public health, economy and technology, and of scholarship, which provided knowledge for the stabilization of the authority of the nation-state. Both could be achieved only by the recognition of relative academic freedom.

One of the key motivations of minister Thun-Hohenstein, again, may have had its origins in his commitment to the Monarchy's national heterogeneity. The author suggests that Austrian scholarship was permitted to develop relatively freely, i.e. autonomously, against the background of the rising danger of the *Kulturnationen*, irrespective of cultural and linguistic nationalism, which was increasingly considered to be destructive and a menace to the integrity of the *Staatsnation*. However, there were some exceptions: scholarly disciplines such as jurisprudence, which were committed to producing loyal servants for higher office, remained as fully controlled as, for example, philosophy and the national philologies, which were primarily suspected (besides other fields of scholarly research) of operating to advance cultural nationalism.

Subjectivism versus objectivism: styles of thinking – modes of agency

Both principles – the conservative *staatsnational* and the progressive *kulturnational* – were in one way or another correlated with a specific epistemological orientation. We can preliminarily describe the former as objectivist and the latter as subjectivist in nature.[41] The subjectivist style of enquiry associated with the latter epistemic trend prominently situates the individual subject as the focal point of cognitive and intellectual agency. The cognizing subject conceived as autonomous by Kant was given a historical bent, highlighting the historicity of the 'subject', and diluted to a collectively hypostasized agent by different followers of Kant. In nineteenth-century thought this gave rise to the notion of a collective subject. Objectivism rejects these propositions of historicity and bases cognition on a meta-subjective level. Thus it entails and emphasizes formal and classificatory analyses that largely preclude 'contextualist' explanations.

Both – subjectivism and objectivism – can be regarded as epistemological orientations that gained importance within national political contexts. In Prussia and beyond, 'subjective idealists' hypostasized the subject conceived as autonomous by Kant and elevated it to the level of the collective 'we'. This was a crucial epistemological shift in the construction of the so-called German 'national character'.[42] The 'most natural state', as Johann Gottfried Herder (1744–1803) had pointed out in his *Ideas on the Philosophy of the History of Humanity* (1784), is based on 'One People, with One National Character',[43] and – according to Johann Gottlieb Fichte (1762–1814) – the perpetual, uninterrupted use of one language, the national language.[44] A 'spiritualized' collective subject laid the foundation for later collectivist concepts like the '*Volksgeist*'.

The idealist hypostasis, transforming the subject into a collective 'we', gave a significant boost to the German unification process and to statehood. However, in the multilingual Habsburg Monarchy this *kulturnationalistisch* subjectivism was bound to have disruptive consequences. The Austrian imperial administration regarded this blossoming of national identities with dismay, seeing it as a considerable threat to the integrity of the state. Conversely, the activists of nascent nations perceived it as an excellent opportunity to emancipate their respective nations, conceptualized on the basis of linguistic and cultural difference. Scholarship could be used to serve both ends. But this was no one-way process, as scholarship could also reasonably expect to gain recognition and support from lending ammunition to politics. In the case of imperial-centralist politics, this functioned by supporting epistemological objectivism; in the case of *kulturnationalistisch* attitudes, it worked by backing subjectivism.

The objectivist approach was actively supported during Count Leo Thun's tenure as minister of education, whereas epistemological tendencies linked to the culturally conceived nation were attacked as subjectivist. Objectivism seemed a perfect device to divest scholarship of its potentially nationalist implications. Thus the ministry permitted scholarship to proceed in an objectivist, formalistic manner. At the same time, Thun and his advisers were at pains to nominate professors who would safeguard the realization of this integral, imperial aim.

Epistemological objectivism in scholarship as a desire of politics

The dominance of objectivism as a politically desired orientation of scholarship can be demonstrated, for instance, in the case of the advancement of Slavic studies and philosophy. Slavic studies could potentially be used for both purposes: to enhance the national status of the different Slavic populations within the Habsburg Empire, or to advance the idea of the *Staatsnation*. It is worth noting that in Austria, Slavic studies tended to associate themselves with the mission of the nation-state. The chairs established at the University of Vienna in 1849 were professorships of philology, and not primarily dedicated to research on national myths and narratives. The linguist Franc Miklošič (1813–91) and the literary scholar Ján Kollár (1793–1852), who was described as 'loyal to the state' (*bewährt staatstreu*),[45] were appointed professors. Through their linguistic orientation, in particular the task of classifying and standardizing the legal-political terminology, Slavic studies were established as a means of increasing the control of a centralist state government oriented towards tightening the hegemony of all that was culturally and/or linguistically German. In July 1849 the minister of the interior, Alexander von Bach (1813–93), called together scholars in the field and established a state commission dedicated to the standardization of the legal-political terminology in the Slavic languages.[46] The writer and Slavic scholar Pavel Josef Šafařík (1795–1861) was appointed head of the commission. Starting with research on Old Church Slavic, the newly appointed scholars invested a great deal of work in comparative linguistics. The comparative approach allowed the retardation of the vernacularization process in favour of a standardized Slavic language.[47]

Despite this objectivist thrust, the establishment of Slavic studies in Austria was precariously poised between centralist rule and the movement that provided arguments in favour of linguistic nationalism. We must not neglect the initial efforts undertaken to establish Slavic studies as a field of scholarly research in Austria, but the modernizing of Slavic languages nevertheless stood in close relationship to cultural nationalist aims. The achievement of terminological standardization and grammatical uniformity in these languages prepared the 'scholarly consolidation'[48] of Slavic studies.[49] Other disciplines, such as German philology and history, would follow suit in their efforts, spurred by the 'characterization of the national spirit'.[50] However, the German-nationalist and Czech- or Polish-nationalist varieties of historiography remained at loggerheads with a dynastic-integralist style of historiography represented, for instance, by Václav Vladivoj Tomek (1818–1905) (appointed professor at the University of Prague in 1850),[51] Antoni Walewski (1805–76) and Heinrich von Zeißberg (1839–99) (appointed professors at the universities of Cracow and Lviv, respectively).[52] The same politically induced objectivist orientation of scholarship holds true for other disciplines, especially for philosophy as a newly established scholarly discipline.

In Austria objectivism had two fathers, namely the logician Bernhard Bolzano (1781–1848) of the University of Prague and the German philosopher Johann Friedrich Herbart (1776–1841). Both rejected the subjectivist transfiguration of the individual. Whereas Bolzano appealed to reason and *vérités du raison*, which he considered to be objective facts not subject to concrete verification,[53] Herbart tried to mediate between the cognisant subject and the object to be cognitively grasped.[54] Both systems lent themselves well to the aims of Austrian politics, as they rejected what was perceived as Fichtean hypersubjectivism, his 'metaphysics of a world-creating I'.[55]

Bolzano's favourite protégé, the would-be Herbartian Robert Zimmermann (1824–98), fused these tenets in order to establish an Austrian 'philosophy of the state', which gained pre-eminence at the Monarchy's universities and schools.[56] The inherently objectivist conceptual system was designed to give stability to the overarching nation-state and to deflect centrifugal tendencies. Zimmermann's system responded to national-particularist, contextualist and relativist challenges with universalism, formalism and value-absolutism. When he was appointed professor of philosophy at Prague university in 1852, Zimmermann unambiguously announced in his inaugural lecture: 'The sickness of our time is subjectivity'.[57]

It is difficult to overlook the political mission with which this new academic discipline was imbued. A glance at its long-term development makes it obvious that the purely objectivist orientation gradually failed to persuade in an increasingly liberal climate, although it remained a venerated theoretical tradition. The Austrian philosopher and mathematician Carl Siegel (1872–1943) called 'objectivism' the 'main characteristic of Austrian scholarship in its overall advancement [Gesamtentwicklung]'.[58] However, if objectivism fulfilled this function, historical analysis must not lose sight of subjectivist tendencies within the objectivist framework. Both became manifest in the studies of public law, one of the most political of the academic professions.

Constitutional law: balancing objectivism and subjectivism

In the second half of the nineteenth century, jurisprudence began to focus attention on developing a state theory that transcended politics. But even the positivistic studies of public law were not as unpolitical as they were supposed to be: 'Among the legal disciplines', Michael Stolleis, the renowned German legal historian, notes, 'constitutional law remained the most political profession'.[59] – not only in Germany, but also in Austria.

In the 'Austrian' half of the Habsburg Dual Monarchy the newly established chairs of constitutional law pretended to proceed 'purely legally' – objectively. However, it goes without saying that their occupants acted in favour of the *Staatsnation*. This would not be surprising if one did not know of the subjectivist bias of state-nationalism towards all that was German – *deutsch-national*. Austrians of German origin saw themselves as the leading cultural force, committed to maintaining the structure and the power of the central state.

Two examples. In 1883 the German-Bohemian lawyer Joseph Ulbrich (1843–1910) published the first system of Austrian constitutional law.[60] As a representative of the supposedly apolitical positivistic *Begriffsjurisprudenz* of Labandian origin, he kept up the appearance of acting as an autonomous scholar who stood above state politics while using the supposedly neutral legal method to protect traditional power relationships. Since the national interests of the Czechs ran counter to Ulbrich's political attitude, he coupled his scientific work with a defence of the supremacy of the German nationality in Austria. Ulbrich is remembered as a constant fighter against the idea of 'Bohemian state law'.[61]

Another Austrian legal scholar, Friedrich Tezner (1856–1925), used historical arguments to defend the idea of the overarching imperial Austro-Hungarian sovereignty. Due to this, he fought for a realist terminology for what he called the 'Austrian Monarchy',[62] which he named a 'state sui generis'.[63] He refused to categorize the Austro-Hungarian Empire as a 'confederation' or a 'federal state', which were typological concepts developed by theoreticians of the state, e.g. by Georg Jellinek (1851–1911), the Austrian-born liberal law scholar in Heidelberg. Since they had classified indeterminate or hybrid state forms as 'organic abnormities' (*organische Mißbildungen, Abnormitäten*), Tezner dubbed their scholarly work *'Toilettenkunst'* – 'cosmetic art'.[64] For him sovereignty was not to be considered as a matter of two states, Austria and Hungary, but of one empire.

Tezner's story illustrates his rejection of state theory as defined by theoreticians in the German nation-state for political reasons. He intervened in a controversial debate about the extent of Hungary's national independence. As Hungarian lawyers had legitimized Hungary's striving for greater independence within the Habsburg Empire by applying German concepts of the state, Tezner accused them of the crime of weakening the idea of the *Staatsnation*. Both apparently acted heteronomously. In Austria Tezner was characterized in an obituary written by the Viennese jurist Hans Kelsen as a scholar with a 'strong political mind'.[65]

Both Ulbrich and Tezner defended the Austrian *Staatsnation* at the expense of the autonomy of the juridical field, but with different motives and different goals. Ulbrich, a German liberal in Bohemia, defended existing power relationships, since political autonomy for the Czechs could be achieved only at the expense of the centralist Habsburgian nation-state and of German supremacy in that country (clearly Ulbrich's priority). In his protracted battle against the idea of Bohemian state law, 'which completely ran counter to his theoretical orientation', Ulbrich, as Ernst Mischler (1857–1912) put it, defended the 'legal standing of his German-Bohemian homeland'. Mischler described Ulbrich as one of the 'most influential Bohemian politicians', whose importance manifested itself 'in his scholarly work' as an 'intellectual [*geistig*] politician'.[66] Tezner, a Viennese monarchist, brought in his scientific authority to reject demands to increase the political autonomy of Hungary.

Both examples show that in constitutional law, political involvement paved the way for the accumulation of scientific and professional capital: in 1884 Ulbrich was appointed full professor at the German-language Charles

University of Prague, where he became rector in 1897/98. Tezner's professional career took off much more slowly, probably since he was of Jewish origin; he worked as a bank official and as an advocate. Having received the venia legendi for administrative law and state law, Tezner was appointed adjunct professor at the University of Vienna. The academic chairs in Austrian state law had already been occupied. In 1921 he was appointed *Senatspräsident* in the Austrian higher administrative court (*Verwaltungsgerichtshof*).[67]

Critical-reflexive concepts versus epistemological hypersubjectivism

After the collapse of the Austro-Hungarian Monarchy in 1918, scholarship was confronted with a new challenge: inventing Austria, one of the successor states of the now defunct Empire, as a purely German state.[68] This was achieved at least in part by downgrading minorities to 'ethnic' groups: Jews, Czechs, Slovaks, Hungarians, Poles and other peoples who spoke a language other than German. Since politics no longer saw any need to promote the objectivist orientation of science in the newly created and self-defined, but in no way truly monocultural, republic, the scientific field also gradually drifted into national subjectivism. Thus, a new cohort of Austrian scholars gained significant opportunities, of which they took advantage. Some were quick to adopt the new orientation. Again, scholarship served as a political resource.

In Austria, the years between the two World Wars witnessed an increase in nationalist activity among that part of the Austrian population which was ethnically German or felt drawn to the German language. To build a specific Austro-German identity, political activists increased an awareness of difference (which they called cultural, ethnic or racial) among those of non-German and/or non-Christian origin. At the same time and on the academic level, scholars such as the prehistorian Oswald Menghin (1888–1973), the legal historian Karl Gottfried Hugelmann (1879–1959) and the ethnologist Wilhelm Schmidt (1868–1954) invented theories explicitly for the purpose of legitimizing the exclusion of the Jewish population from the so-called *deutschen Volksgemeinschaft* (community of German people) in Austria. With the acceptance of these theories by the politicized nationalist public, these scholars' Jewish competitors were gradually eliminated from the fierce competition for positions in academe and the way was now free for them to work their way up to the highest academic ranks. Convinced as these and other anti-Semites may have been of their scientific autonomy, this claim did not go undisputed, for there was also an academic counter-elite which could legitimately claim to extend the autonomy of science and scholarship. This elite remained unaffiliated to nationalist politics, while strongly rejecting scholarship that purported to be autonomous. Rather, they tried to keep science as far removed as possible from politics. If their aim was to extend the autonomy of the scientific field, they pursued a specific goal: the more independent science was, the more authority it had to intervene scientifically in disruptive, anti-democratic and inhuman social processes. Their objectivist approach to scholarship made it possible to scrutinize the above-mentioned disruptive processes of social in- and exclusion.

Activities in this direction can be demonstrated, for instance, in the work of Ernst Mach, Sigmund Freud and Hans Kelsen, who had developed their critically-reflexive and inspiring scholarly positions in the late Habsburg Monarchy. It is argued here that these scholars, path-breaking in their fields, were the first to dissolve essentialist foundations of individual and collective (e.g. national) identity. They argued on a transdisciplinary plane, weaving together, for example, physiology, psychology and philosophy, logic and linguistics, aesthetics and history, and so forth; and they smashed the conceptual apparatuses of their own disciplines.[69] As both a physicist and a philosopher of science, Ernst Mach (1838–1916) disapproved of so-called substance concepts. He strongly criticized the achievements of science in the seventeenth century as 'metaphysical obscurities'[70] unusable to build basic scientific assumptions. As a physicist, Mach, the radical adherent of scientific positivism, rejected Newton's postulates of absolute time, space and motion. Mach refused to acknowledge the physical reality of entities such as atoms, quanta or electrons, while understanding such basic notions of physics as abstractions that could perform only auxiliary (practical) functions for analysis. As a physiologist of the senses he consistently described the physical and psychical world as complexes of sensations, interdependent of each other. According to Mach's new line of reasoning, even the 'self' (ego) was only 'a practical unity, put together for purposes of provisional survey'.[71] He not only denied the substance concepts (body and ego, matter and soul/mind) still prevalent around 1900, but also the dualism between the ego and the world of matter, phenomenon and object, the physical and the psychical: 'The ego must be given up'; as a substance it was 'unsalvagable [*unrettbar*]'. 'In the long run,' he wrote, 'we shall not be able to close our eyes to this simple truth', which would help to 'arrive at a freer and more enlightened view of life, which will preclude the disregard of other egos and the overestimation of our own.' According to Mach, this understanding of the ego would reject 'the ideal of an overweening Nietzschean "superman", who cannot' and, Mach hoped, 'will not be tolerated by his fellow men.' 'If we regard the ego as a real unity,' he concluded,

we become involved in the following dilemma: either we must set over against the ego a world of unknowable entities ... or we must regard the whole world, the egos of other people included, as comprised in our own ego .... But if we take the ego simply as a practical unity, put together for purposes of provisional survey, or as a more strongly cohering group of elements, less strongly connected with other groups of this kind, questions like those above discussed will not arise, and research will have an unobstructed future.[72]

Sigmund Freud (1856–1939) no longer acknowledged a basic conception that had become hardened over time, namely the notion of the soul as a substance beyond the body. He denied not only the soul as a real entity, but also the notions of the ego, the group and mass as permanent substances, reifications he classified as 'achievements' of a metaphysical tradition. Whereas for Mach the ego, bodies and the world did not consist of mysterious entities ('*Wesen*'), but of sensations (colours, sounds, spaces), which he defined as the ultimate elements and whose given connection he investigated, Freud proposed that the individual mind had no substantial character; rather it was structured and made up of three components – the 'id', the 'ego' and the 'superego' – which needed to be well balanced. After focusing on the analysis of the conflict between the id and the superego that has to be negotiated by the ego, Freud increasingly investigated the individual in its relations to the world outside. He also declared this relationship constitutive for the so-called 'unconscious' and for the individual mind. Thus, the unconscious was formed by both instinctual impulses and social relationships: 'In the individual's mental life someone else is invariably involved, as a model, as an object, as helper, as an opponent.' In his essay *Group Psychology and the Analysis of the Ego* (1921, English 1922), Freud concluded that 'from the very first, individual psychology...is at the same time social psychology as well',[73] and vice-versa, that social and narcissistic processes would fall wholly within the domain of individual psychology. For Freud, the contrast between these mental acts was not well calculated to differentiate individual from a social or group psychology. Differentiation would rather be a cheap disguise for the dangerous attempt to identify the group as an organism, a substance, equipped with a 'collective psyche' whose functions corresponded to that of the individual. In fact, Gustav Le Bon and other important representatives of the rising discipline of sociology had defined the social group as a provisional being that displays specific characteristics. The sceptical individualist Sigmund Freud distanced himself sharply from the assumption of the group or mass as a 'collective body' endowed with a 'collective psyche' or a 'collective unconscious' (a term popularized by C.G. Jung), since such reifications showed a tendency to become politically functionalized. With regard to Le Bon's notion of the 'unconscious', Freud noted clearly and briefly that it 'more especially contains the most deeply buried features of the racial mind [*Rassenseele*], which as a matter of fact lies outside the scope of psychoanalysis.'[74]

It is one of Sigmund Freud's merits to have redefined the concept of the social group dominant at the beginning of the twentieth century. To Freud the 'group' was not to be considered as such a substance insusceptible of dissection; such a term was merely to be considered and used as an abstract notion to investigate the way individuals were united. It was this bond – the libidinous attraction – which Freud recognized as the characteristic of groups or masses.[75]

The Viennese legal scholar Hans Kelsen (1881–1973) strongly appreciated Freud's progressive approach, while criticizing him for diagnosing the constitutive criterion of the state as lying in social cohesion. For Kelsen the notion of the state as 'person', a concept of substance providing for '"the unity of a diversity of human behaviour", an example of a "reification", was to be replaced by the recognition of the state as a legal function'.[76] In his perspective the state was a mere abstraction – the formalization of law, but by no means bound up with societal substance.[77] Kelsen chose the term 'hypostasis' to analyse scholarly the political interventions from which he maintained a critical distance, namely those which essentialized and substantialized abstract notions like 'state', 'people' and 'nation' – in fact nothing but 'things in the mind [*Gedankendinge*]',[78] as Kelsen put it. The underlying substance showed the tendency to lend credibility to notions of difference in essential being (*Wesensverschiedenheit*) and at the same time, according to Kelsen, stabilize national identity: nation, people and state are naturalized, invested with a body and a will.[79] In Austria's complex pluricultural environment,[80] Herder's slogan of 'one state, one people, one national character' collided with the reality of intermingling, and thus this kind of conceptual hypostasis furnished politics with a tool for social and cultural exclusionism and even racism.

Kelsen, however, was but one representative of the group of scholars who may take credit for several advancements. First, they put substance concepts, frequently used in science and scholarship, into (historical) perspective, revealed their legitimizing function, and delegitimized scientific endeavours that purported to define the elements which should constitute collective identity in Austria instead of analysing the processes leading to the identification of groups with a specific '*Volksgemeinschaft*' (people's community), '*Ethnizität*'(ethnic group) or '*Rasse*' (race) (or leading to their exclusion). Second, they brought constructive suggestions into academic discussion, going beyond the critique of language. Basic assumptions in science were replaced by new, critically-reflexive positivistic approaches, which Kelsen once described as characteristic for the above-mentioned counter-elite: Mach invented a 'physics without force (interaction) [*Physik ohne Kraft*]', Freud inaugurated a 'psychology without a soul [*Seelenlehre ohne Seele*]' and Kelsen a 'science of the state without a state [*Staatslehre ohne Staat*]'.[81]

Third, they reassessed the interrelatedness of science, politics and culture and made science/scholarship a public 'player' attuned to democracy. By developing basic notions free of substance, they countered the attacks of the adherents of hypersubjectivism on objectivism, but strove for the recognition of the self-conscious subject, since they qualified the latter as a precondition for democracy.

If one follows the reverberations of Freud's, Mach's and Kelsen's diagnoses of modernity, coupled as they were with their formulation of the autonomy of science as well as with their analysis of the essentialist notion of the world, one can note that these assessments indubitably affected the take on these questions, e.g. by the art historian Alois Riegl (1858–1905), the philosopher Ludwig Wittgenstein (1889–1951) and the social scientist Otto Neurath (1882–1945), in their striving for desubstantialization and for liberation from the dichotomy between the 'autonomous' and 'engagé' intellectual. Although a precise mapping of these contours is well beyond the constraints of this chapter, these thinkers' endeavours demonstrated that scholars could act with engagement without violating the scientific rules of their scholarly field or relinquishing their autonomy. Since they aligned their research with the purpose of settling conflicts, integration and conciliation, they too made political commitments. However, they observed the rules of the scientific field and made these commitments on the basis of critical academic reasoning and in a delegitimatory attitude.

To put it briefly, scholars who involved themselves in Austrian politics had two options. If they did not take the division of the political and scientific field for granted, politics and scholarship could take advantage of each other. However, in this case scholars always ran the risk of serving as a resource for power politics. Only if scholars rejected alliances with politics could they help to create the necessary academic freedom to intervene successfully in the politically desired and scientifically justified unreflexive processes of national identity-building. Thus, striving for 'relative' autonomy (in Bourdieu's sense) offered the chance to develop new, sophisticated scientific concepts to solve the destructive effects of nation-building on the basis of so-called scientific findings.

In 1934 Kelsen, the Austrian jurist and 'architect' of the constitution of the First Republic, reflected on the historical process of science becoming independent. He noted critically in his pure theory of law: 'If science was able to enforce its claim of autonomy from politics, [this is] only because politics had a vital interest in the advancement of technology. That could be guaranteed only by relative academic freedom.'[82] The more the distinguished Viennese scholar Kelsen became aware of the political-academic nexus, the more he recognized that science had not only to defend autonomy, but also to proceed in a politically responsible manner. For him, autonomy was not an end in itself. As a famous representative of the academic counter-elite in Austria in the interwar period, Kelsen adopted an attitude of conducting scientific research in a politically responsible way without giving up autonomy, an attitude which can be formulated in the terms of Max Weber, Berthold Brecht and Pierre Bourdieu, who have reminded us that autonomy does not mean autarchy.[83]

Notes

1. Gary B. Cohen (2007) 'Nationalist politics and the dynamics of state and civil society in the Habsburg Monarchy, 1867–1914', *Central European History*, 50, 241–78, here 274–78.

2. Joseph von Eötvös (1851), *Ueber die Gleichberechtigung der Nationalitäten in Oesterreich* (Wien: Jasper, Hügel & Manz), 17: '*Die Grundlage aller nationellen Bestrebungen ist das Gefühl höherer Begabung, ihr Zweck ist Herrschaft.*'

3. The terms '*Staatsnation*' and '*Kulturnation*' were later used by Friedrich Meinecke for the analysis of the emergence of the German national state. Idem, *Weltbürgertum und Nationalstaat. Studien zur Genesis des deutschen Nationalstaates* (Munich-Berlin, 1911 [1908]), here 1–20.

4. Gregor von Berzeviczy (1817) 'Etwas über Nationen und Sprachen', in Joseph von Hormayr (ed.) *Archiv für Geschichte, Statistik, Literatur und Kunst*, 8, 287–89, here 288: '*Wohin leiten diese Beobachtungen den unparteyischen Forscher? Dahin, daß es in Europa keine ursprünglichen Nationen mehr gibt; daß die Sprache kein Beweis der Nationalität ist; daß der Begriff der Nationalität jetzt nicht im Stamm, Ursprung, auch nicht in der Sprache, sondern eigentlich staatsrechtlich im Staate zu suchen sey. — Einheit der Regierung, der Verwaltung führt zur Gleichheit der Gebräuche, Gesetze und Denkart; und dieß macht verschiedene Völker, wenn sie auch verschiedene Sprachen reden, zu einer Staatsnation; und mit dieser Nationalität mag man sich begnügen, da man die ursprüngliche verloren hat.*' Many thanks to Franz L. Fillafer for this quotation.

5. Hermann Ignaz Bidermann (1889 [1867]) *Geschichte der österreichischen Gesammt-Staats-Idee 1526–1804*, 2 vols. (Innsbruck: Wagner). In 1853 Joseph Alexander Helfert (1820–1910), a Viennese historian and high-ranking civil servant loyal to the state, used the term '*Osterreichische Nationalgeschichte*' to specify the history of the Austrian '*Gesamtstaat*' (including all provinces) and '*Gesamtvolk*' (encompassing all the different nationalities of the Habsburg Empire). See Joseph Alexander Helfert (1853) *Über Nationalgeschichte und den gegenwärtigen Stand ihrer Pflege in Oesterreich* (Prague: Calve), 1–4.

6. See Hans Lentze (1959) 'Graf Thun und die voraussetzungslose Wissenschaft', in Helmut J. Metzler-Andelberg (ed.) *Festschrift. Karl Eder zum siebzigsten Geburtstag* (Innsbruck: Wagner), 197–209.

7. Richard Meister (1948) 'Festvortrag des Vizepräsidenten Richard Meister *Die Geschichte der Akademie der Wissenschaften in Wien 1847–1947*', *Almanach [of the Austrian Academy of Sciences] für das Jahr* 1947 (Vienna: Rudolf M. Rohrer), 196–216, here 216: '*So steht die Akademie vor Ihnen heute mit dem Bekenntnis, von sich sagen zu dürfen, daß sie durch alle wechselvollen Zeiten hindurch dem Gesetz, unter dem alle Forschung steht, dem der strengsten Sachlichkeit und Objektivität, nie untreu geworden ist. Diesem Gesetz hat jeder gedient, ausnahmslos [sic!], der in diesem Haus gewirkt hat, und diesem Gesetz wollen auch künftig alle Mitglieder der Akademie und die Akademie als Ganzes dienen.*'

8. Pierre Bourdieu (1998) *Vom Gebrauch der Wissenschaft. Für eine klinische Soziologie des wissenschaftlichen Feldes* (Konstanz: UVK Univ.-Verl. Konstanz), 16–31, here 26–31.

9. Herbert Mehrtens (1994) 'Kollaborationsverhältnisse. Natur- und Technikwissenschaften im NS-Staat und ihre Historie', in Christoph Meinel and Peter Voswinckel (eds.) *Medizin, Naturwissenschaft, Technik und Nationalsozialismus – Kontinuitäten und Diskontinuitäten* (Stuttgart: Verlag für Geschichte der Naturwissenschaften

und der Technik), 13–32; Mitchell G. Ash (2002) 'Wissenschaft und Politik als Ressourcen für einander', in Rüdiger vom Bruch and Brigitte Kaderas (eds.) *Wissenschaften und Wissenschaftspolitik. Bestandsaufnahmen zu Formationen, Brüchen und Kontinuitäten im Deutschland des 20. Jahrhunderts* (Stuttgart: Steiner), 32–51, here 40.

10. Pierre Bourdieu (1993) *The Field of Cultural Production. Essays on Art and Literature*, ed. and introduced by Randal Johnson (London: Polity Press), 164–65.

11. Pierre Bourdieu (1999) 'The specifity of the scientific field and the social conditions of the progress of reason', in Mario Biagioli (ed.) *The Science Studies Reader* (New York, London: Routledge), 31–50, here 39.

12. Pierre Bourdieu (2004) *Science of Science and Reflexivity*, transl. by Richard Nice (London: Polity Press), 47.

13. Bourdieu, *Vom Gebrauch der Wissenschaft*, 19: '*Halten wir also fest, dass mit zunehmender Autonomie eines Feldes seine Brechungsstärke umso größer ausfällt; äußere Zwänge umso stärker, oft bis zur Unkenntlichkeit, umgestaltet werden. Der entscheidende Hinweis auf den Grad der Autonomie eines Feldes ist also seine Brechungsstärke, seine Übersetzungsmacht. Umgekehrt zeigt sich die Heteronomie eines Feldes wesentlich durch die Tatsache, daß dort äußere Fragestellungen, namentlich politische, halbwegs ungebrochen zum Ausdruck kommen. Das bedeutet, daß die "Politisierung" eines wissenschaftlichen Faches eben nicht auf eine große Autonomie des Feldes schließen läßt.*'

14. Ash, 'Wissenschaft und Politik als Ressourcen für einander', 50–51.

15. Pierre Bourdieu (2004) 'Forschen und Handeln', in Joseph Jurt (ed.) *Forschen und Handeln. Recherche et Action. Vorträge am Frankreich-Zentrum der Albert-Ludwigs-Universität Freiburg (1989–2000)* (Freiburg: Rombach), 93–101, here 100.

16. Pierre Bourdieu (1991) *Die Intellektuellen und die Macht*, ed. by Irene Dölling (Hamburg: VSA), 43.

17. See Fritz K. Ringer (1969) *Decline of the German Mandarins. The German Academic Community, 1890–1933* (Cambridge, Mass.: Harvard University Press).

18. See Jonathan Harwood (2002) 'Forschertypen im Wandel 1880–1980', in vom Bruch and Kaderas (eds.), *Wissenschaften und Wissenschaftspolitik*, 162–68, here 162.

19. Ash, 'Wissenschaft und Politik als Ressourcen für einander', 32–36 and 50–51.

20. See Pieter M. Judson (2006) 'Constructing nationalities in East Central Europe', in Pieter M. Judson and Marsha L. Rozenblit (eds.), *Constructing Nationalities in East Central Europe* (New York, Oxford: Berghahn Books) (Austrian studies 6), 1–18, here 4; Pieter M. Judson (forthcoming) 'The limits of nationalist activism in Imperial Austria: creating frontiers in daily life', in Johannes Feichtinger and Gary B. Cohen (eds.) *Understanding Multiculturalism and the Central European Experience* (New York, Oxford: Berghahn Books) (Austrian and Habsburg Studies 15).

21. Ibid.; Pieter M. Judson (2006) *Guardians of the Nation. Activists on the Language Frontiers of Imperial Austria* (Cambridge, Mass., London: Harvard University Press).

22. See Jan Surman (2008) 'Supranational? Die cisleithanischen Universitäten im Spannungsfeld zwischen *république des lettres* und *république des nations*', *Moderne. Kulturwissenschaftliches Jahrbuch*, 4, 213–24.

23. Fillafer, Franz L. (2012) 'Enlightenment's legacy and early nineteenth-century liberal reformism in the Habsburg Monarchy', in Jürgen Elvert (ed.), *The proceedings of the European Academy of Yuste* (Frankfurt am Main: Peter Lang), 12–38.

24. Karl Renner (1918) *Das Selbstbestimmungsrecht der Nationen in besonderer Anwendung auf Oesterreich, Erster Teil: Nation und Staat* (Leipzig, Wien: F. Deuticke), 72: '*Die Länder zerreissen die Nationen, kein Wunder, dass die Nationen die Länder zerreissen wollen.*'

25. See Johannes Feichtinger (2010) *Wissenschaft als reflexives Projekt: Von Bolzano über Freud zu Kelsen. Österreichische Wissenschaftsgeschichte 1848–1938* (Bielefeld: Transcript).

26. See Jürgen Habermas (1976) 'Können komplexe Gesellschaften eine vernünftige Identität ausbilden?', in Jürgen Habermas, *Zur Rekonstruktion des Historischen Materialismus* (Frankfurt am Main: Suhrkamp), 92–126, here 107–08.

27. Bourdieu, 'Forschen und Handeln', 100.

28. John Boyer (1981) *Political Radicalism in late Imperial Vienna. Origins of the Christian Social Movement 1848–1897* (Chicago: University of Chicago Press), 18.

29. Peter Wozniak (1995) 'Count Leo Thun: a conservative savior of educational reform in the decade of neoabsolutism', *Austrian History Yearbook*, 26, 61–82, here 63.

30. See Peter Stachel (2000) 'Das österreichische Bildungssystem zwischen 1749 und 1918', in Karl Acham (ed.) *Geschichte der österreichischen Humanwissenschaften.* Vol. 1: *Historischer Kontext, wissenssoziologische Befunde und methodologische Voraussetzungen* (Vienna: Passagen), 115–46.

31. See (1848) 'Petition der am 12. März 1848 in der Aula der Wiener Universität versammelten Studierenden', in Carl Heintl (ed.) *Mittheilungen aus den Universitäts-Acten (vom 12. März bis 22. Juli 1848)* (Vienna: Leop. Sommer), 1.

32. In July 1848 Ernst von Feuchtersleben was appointed undersecretary (*Unterstaatssekretär*) of the newly established ministry of education. See Herbert H. Egglmaier (1998) 'Reformansätze vor der Thunschen Reform. Feuchterslebens und das Konzept einer genuin österreichischen Unterrichtsreform', *Mitteilungen der österreichischen Gesellschaft für Wissenschaftsgeschichte*, 18, 59–85, here 60–61.

33. See 'Petition der am 12. März 1848', 10.

34. It is still not definitely clear how the so-called 'German' model (in particular the faculty system, the so- called 'Ordinarienuniversität' and the ideal of 'freedom of teaching and learning') was adopted and transferred to Austria, given the fact that the higher education reformers had no access to Humboldt's unpublished memoranda. See Mitchell G. Ash (2006) 'Bachelor of what, master of whom? The Humboldt myth and historical transformations of higher education in German-Speaking Europe and the US', *European Journal of Education* 41, 2, 245–67, here 246–49. However, one may assume that the architect of the university reform, Franz Serafin Exner, was as well informed about higher education in Prussia as the Berlin school reformer Hermann Bonitz (1814–88). Having been appointed professor of ancient philology at the University of Vienna, Bonitz participated in Austria's reform process as the closest collaborator of Exner at the ministry of education. Both were strong supporters of the German pedagogue, philosopher und psychologist Johann Friedrich Herbart, who had also been involved in the Prussian school reform.

35. See Walter Höflechner (2010) 'Nachholende Eigenentwicklung? Der Umbau des habsburgischen Universitätssystems nach der Mitte des 19. Jahrhunderts', in Rüdiger vom Bruch (ed.) *Die Berliner Universität im Kontext der deutschen Universitätslandschaft nach 1800, um 1860 und um 1910* (Munich: Oldenbourg), 93–103, here 96–100.

36. A[lfred] Fischel (1907) 'Nationalitäten', in Ernst Mischler and Josef Ulbrich (eds.) *Österreichisches Staatswörterbuch. Handbuch des gesamten österreichischen öffentlichen Rechtes*, Vol. 3 (Vienna: A. Hölder), 676–702, here 686.

37. See Lentze, 'Graf Thun und die voraussetzungslose Wissenschaft'; Karl Eder (1955) *Der Liberalismus in Altösterreich. Geisteshaltung, Politik und Kultur* (Vienna, Munich: Herold) (Wiener Historische Studien 3); Werner Ogris (1999) *Die Universitätsreform des Ministers Leo Graf Thun-Hohenstein. Festvortrag anläßlich des Rektortages im Großen Festsaal der Universität Wien am 12. März 1999* (Vienna: WUV) (Wiener Universitätsreden N.F. 8).

38. See Karl Heinz Gruber (1982) 'Higher education and the state in Austria. An historical and institutional approach', *European Journal of Education* 17, 3, 259–70, here 261–62.

39. See Josef Hyrtl (1856) *Einst und Jetzt der Naturwissenschaft in Oesterreich. Eröffnungsrede der 32. Versammlung deutscher Naturforscher und Aerzte in Wien, am 16. September 1856* (Vienna: Auer), 5–10.

40. Ibid., 5–7.

41. See Jürgen Mittelstraß (ed.) (2004) *Enzyklopädie. Philosophie und Wissenschaftstheorie*, vol. 2 (Stuttgart, Weimar: Metzler), 1054–55 and vol. 4, 128–31.

42. See Franz L. Fillafer, Jürgen Osterhammel (2011) 'Cosmopolitanism and the German Enlightenment', in Helmut W. Smith (ed.) *Oxford Handbook of Modern German History* (Oxford: Oxford University Press), 119–43.

43. Johann Gottfried Herder (1989) *Ideen zur Philosophie der Geschichte der Menschheit*, ed. Martin Bollacher (Frankfurt am Main: Deutscher Klassiker Verlag) (Werke 6) (Original 1784–91), 369–70.

44. See Johann Gottlieb Fichte (2008) *Reden an die deutsche Nation*, ed. Alexander Aichele (Hamburg: Felix Meiner) (Original 1808), 62–76, 211–12.

45. Günther Wytrzens (1969) 'Ján Kollár', in *Österreichisches Biographisches Lexikon 1850–1950*, vol. 4 (Vienna: Verlag der Österreichischen Akademie der Wissenschaften), 85; Anton Slodnjak (1975) 'Miklosich Franz von', in *Österreichisches Biographisches Lexikon 1850–1950*, vol. 6 (Vienna: Verlag der Österreichischen Akademie der Wissenschaften), 281–82.

46. See Fischel, 'Nationalitäten', 686.

47. Comparative linguistics played a preponderant role in Thun's nominations, as shown, for instance, by August Schleicher (1821–68) and Georg Curtius (1820–85). Both were appointed professors at the University of Prague: Schleicher, who attempted to reconstruct a proto-Indo-European language, as professor of comparative linguistics, Curtius as professor of classical philology. Many thanks to Jan Surman for this information.

48. Stanislaus Hafner (1997) 'Der Beitrag der Österreichischen Slawistik für das Erkennen und für den Aufbau der slawischen Nationalkulturen', *Die Slawischen Sprachen*, 55, 7–18, here 10. See also Otto Kronsteiner (1998) 'Sprachgeschichte, politische Geschichte, und ihre Ideologien', *Die Slawischen Sprachen*, 56, 5–15.

49. See Heinz Miklas (2003) 'Zur Rolle der Wiener akademischen Institutionen in der Geschichte der Slawistik des 19. Jahrhunderts', in Antonia Bernard (ed.), *Histoire de la Slavistique. Le rôle des Institutions* (Paris: Institut d'études slaves), 37.

50. August Sauer (1907) *Literaturgeschichte und Volkskunde* (Prague: J.G. Calve), 15–17. See also Jürgen Fohrmann and Wilhelm Vosskamp (eds.) (1991) *Wissenschaft und Nation. Studien zur Entstehungsgeschichte der deutschen Literaturwissenschaft* (Munich: Wilhelm Fink).

51. See Jiří Štaif (2003) 'The Image of the Other in the Nineteenth Century. Historical Scholarship in the Bohemian Lands', in Nancy M. Wingfield (ed.) *Creating the 'Other'. Ethnic Conflict and Nationalism in Habsburg Central Europe* (New York, Oxford: Berghahn Books) (Austrian and Habsburg Studies 5), 81–102.
52. See Jan Surman (2009) 'Imperial Knowledge? Die Wissenschaften in der späten Habsburger-Monarchie zwischen Kolonialismus, Nationalismus und Imperialismus', *Wiener Zeitschrift zur Geschichte der Neuzeit*, 9/2 [Special issue: Wissenschaft und Kolonialismus], 119–33, here 125–30.
53. See Bernard Bolzano (1985) *Wissenschaftslehre*, ed. by Jan Berg, vol. 11.1 (Stuttgart, Bad Cannstatt: Frommann) (Bernard Bolzano-Gesamtausgabe. Reihe 1. Schriften).
54. See Andreas Hoeschen, Lothar Schneider (2006) 'Herbartianismus im 19. Jahrhundert. Umriss einer intellektuellen Konfiguration', in Lutz Raphael and Heinz-Elmar Tenorth (eds.), *Ideen als gesellschaftliche Gestaltungskraft im Europa der Neuzeit. Beiträge für eine erneute Geistesgeschichte* (Munich: Oldenbourg) (Ordnungssysteme. Studien zur Ideengeschichte der Neuzeit 20), 447–77.
55. Heinrich Scholz (1961) 'Die Wissenschaftslehre Bolzanos. Eine Jahrhundert-Betrachtung (1937)', in Heinrich Scholz, *Mathesis universalis. Abhandlungen zur Philosophie als strenger Wissenschaft*, ed. by Hans Hermes, Friedrich Kambartel, Joachim Ritter (Basel, Stuttgart: Schwabe), 219–67, here 224: '*Metaphysik des welterschaffenden Ich*'.
56. See Peter Stachel (2000) 'Leibniz, Bolzano und die Folgen. Zum Denkstil der österreichischen Philosophie, Geistes- und Sozialwissenschaften', in Acham (ed.), *Geschichte der österreichischen Humanwissenschaften*, vol. 1, 253–96, here 275–96.
57. Robert Zimmermann (1852) *Was erwarten wir von der Philosophie? Ein Vortrag beim Antritt des ordentlichen Lehramts der Philosophie an der Prager Hochschule gehalten am 26. April 1852* (Prague: Credner & Kleinbub), 12.
58. Karl Siegel (1935) 'Philosophie', in Eduard Castle (ed.) *Deutsch-Österreichische Literaturgeschichte. Ein Handbuch zur Geschichte der deutschen Dichtung in Österreich-Ungarn. Unter Mitwirkung hervorragender Fachgenossen. Dritter Band: 1848–1890* (Wien: C. Fromme) (Deutsch-österreichische Literaturgeschichte 3), 17–48, here 48.
59. Michael Stolleis (1996) *Staatsrechtslehre und Politik* (Heidelberg: C.F. Müler) (Heidelberger Universitätsreden 12), 6.
60. Joseph Ulbrich (1883) *Lehrbuch des Oesterreichischen Staatsrechts. Für den akademischen Gebrauch und die Bedürfnisse der Praxis* (Berlin: T. Hofmann). The academic textbook was reedited several times under the title 'Österreichisches Staatsrecht'.
61. See Ernst Mischler (1912) 'Josef Ulbrich. Ein Lebensbild', in *Sammlung Gemeinnütziger Vorträge*, ed. by the Deutsche Vereine zur Verbreitung gemeinnütziger Kenntnisse in Prag, (Prague: Calve), 1–16.
62. Tezner attracted resentments of Hungarian law scholars, when he dubbed the officially named 'Austro-Hungarian Monarchy' the 'Austrian Monarchy'.
63. Friedrich Tezner (1916) 'Das ständisch-monarchische Staatsrecht und die österreichische Gesamt- oder Länderstaatsidee', *Zeitschrift für das Privat- und Öffentliche Recht der Gegenwart*, 42, 1–136, here 135: '*So ist denn die österreichische Monarchie ein Staatswesen sui generis und deckt sich weder mit dem doktrinären Typus der Real- oder Personalunion, noch auch mit dem eines wahrhaft konstitutionell monarchischen Staates.*'

64. Friedrich Tezner (1905) *Die Wandlungen der österreichisch-ungarischen Reichsidee. Ihr Inhalt und ihre politische Notwendigkeit* (Vienna: Manz), 49.
65. Hans Kelsen, 'Friedrich Tezner', *Neues Wiener Tagblatt*, 14 June 1925, 5–6: '*Tezner war – seinem ganzen Charakter nach – ein vornehmlich aufs Historisch-Politische gerichteter Geist.*' Hans Kelsen was on friendly terms with Tezner despite strong scholarly differences.
66. Mischler, 'Josef Ulbrich', 11–12.
67. See Friedrich Wilhelm Kremzow (1971) 'Friedrich Tezner. Ein Beitrag zur Geschichte der österreichischen Verwaltungsrechtswissenschaft', *Acta Universitatum. Zeitschrift für Hochschulforschung, Kultur- und Geistesgeschichte* 1, 2–3, 23–41.
68. See Dieter A. Binder and Ernst Bruckmüller, *Essay über Österreich. Grundfragen von Identität und Geschichte 1918–2000* (Vienna, Munich: Verlag für Geschichte und Politik Oldenbourg), 101–03.
69. See Johannes Feichtinger (2006) 'Das Neue bei Mach, Freud und Kelsen. Zur Aufkündigung der Legitimationsfunktion in den Wissenschaften in Wien und Zentraleuropa um 1900', in Johannes Feichtinger, Elisabeth Großegger, Gertraud Marinelli-König, Peter Stachel and Heidemarie Uhl (eds.), *Schauplatz Kultur – Zentraleuropa. Transdisziplinäre Annäherungen* (Innsbruck, Vienna, Bozen: Studienverlag) (Gedächtnis – Erinnerung – Identität 7), 297–306.
70. Ernst Mach (1919) 'Preface to the first Edition (1893)', in Ernst Mach, *The Science of Mechanics. A Critical and Historical Account of Its Development* (Chicago: Open Court Publishing), ix.
71. Ernst Mach (1914) *The Analysis of Sensations, and the Relation of the Physical to the Psychical*, transl. from the first German edition by C.M. Williams. Revised and supplemented from the fifth German edition by Sydney Waterlow (Chicago, London: Open Court Publishing), 28.
72. Ibid., 24–28.
73. Sigmund Freud (1960) *Group Psychology and the Analysis of the Ego* (New York: Bantam Books), 3.
74. Ibid., 9.
75. Ibid., 26–32.
76. Clemens Jabloner (1998) 'Kelsen and his Circle. The Viennese Years', *European Journal of International Law*, 9, 368–85, here 383.
77. See Hans Kelsen (1922) *Der soziologische und der juristische Staatsbegriff. Kritische Untersuchung des Verhältnisses von Staat und Recht* (Tübingen: J.C.B. Mohr), 1–3 and 208–11; Hans Kelsen (1934) *Reine Rechtslehre. Einleitung in die Rechtswissenschaftliche Problematik* (Leipzig, Vienna: Franz Deuticke) [first English translation of the second revised and enlarged German edition: (1967) *Pure Theory of Law* (Berkeley, Los Angeles: University of California Press); a translation of the first edition of the Reine Rechtslehre or Pure theory of law of Hans Kelsen by Bonnie Litschewski Paulson and Stanley L. Paulson (1992) *Reine Rechtslehre. Introduction to the Problems of Legal Theory* (Oxford, New York: Clarendon Press, Oxford University Press)].
78. Kelsen, *Der soziologische und der juristische Staatsbegriff*, 208.
79. See Johannes Feichtinger and Sabine Müller (2009) 'Kelsen im wissenschaftshistorischen Kontext. Das reine Recht und die 'Freunde der Demokratie', in Tamara Ehs (ed.) *Hans Kelsen. Eine politikwissenschaftliche Einführung* (Stuttgart, Vienna: Nomos), 209–35.

80. On pluriculturalism see Anil Bhatti (forthcoming) 'On culture and diversity', in Feichtinger and Cohen (eds.) *Understanding Multiculturalism.*
81. Kelsen, *Der soziologische und der juristische Staatsbegriff*, 208.
82. Kelsen, *Reine Rechtslehre*, vii: '*Wenn die Naturwissenschaft ihre Unabhängigkeit von der Politik so gut wie durchzusetzen vermochte, so darum, weil an diesem Sieg ein noch gewaltigeres soziales Interesse bestand: das Interesse an dem Fortschritt der Technik, den nur eine freie Forschung garantieren kann.*'
83. Bertolt Brecht (1993) *Arbeitsjournal. Erster Band. 1938 bis 1942*, ed. by Werner Hecht (Frankfurt am Main: Suhrkamp) (st 2215), 125 [entry of '24.8.40']. In 1940 Brecht noted in respect to art that it must be regarded as '*ein autonomer bezirk ... wenn auch unter keinen umständen ein autarker*'. This might also be true for scholarship and science.

4

National 'Consensus' As Culture and Practice: The Geological Survey in Vienna and the Habsburg Empire (1849–1867)

Marianne Klemun

The Jewish cosmopolitan Franz Werfel understood the Habsburg Empire as a supra-national state, a kind of harmonious, bourgeois 'Great Switzerland', a picturesque and well organized mosaic presenting a variety of natural spaces, which connected to each other 'the Tyrolian Alps, the lakes of the Salzkammergut, the gentle horizons of Bohemia, the wild highlands of the karst, the sumptuous landscapes on the Adriatic Sea, the palaces of Vienna, the churches of Salzburg, the towers of Prague...the wide steppes of the Puszta...the mountain pastures of the Carpathian Mountains and the plains along the Danube with all the wonders of its valley, with its wild meadows full of birds, and the large islands of its tributary, the Theiß.'[1] The geological survey of the country, which was successfully completed for all territories of the Habsburg Empire in a very short period, between 1849 and 1867, also guarantees a view which connects all natural spaces.

However, what Claudio Magris stated many decades ago – a view that is still often quoted today and that is based on a range of examples[2] – that the *belles lettres* contributed essentially to the idea of a supra-nationally defined ideal as the pillar of the Habsburg Empire, has so far been completely neglected by research on the history of science, as far as the involvement of the natural sciences is concerned. The reasons for this are clear: whereas literature and the study of history contributed genuinely to the process of nation-building or owe their vitality to it, the natural sciences seemed to consciously avoid this area of reference, simply because of their postulate of being international. Accordingly, William Johnston[3] included philosophy, psychology, law and economics in his overall view of the formation of the 'Austrian' identity, but he did not mention the natural sciences. And even if one could accuse the other European powers of a similar a gap in their research, it is particularly conspicious in the case of the 'multinational Empire' that such studies are still missing.[4]

A history-of-science analysis might pursue questions in two opposing directions: in what way did the natural sciences also contribute to establishing the myth, and in what way did they themselves benefit from this?[5] A third approach is based on understanding cultural and discursive common grounds as the foundations of scientific action and allows also cultural paradigms to become manifest in the epistemic space. This is the approach that will be taken in this study.

Referred to by contemporaries as a 'conglomeration of nations' (Friedrich Hegel) and a team of 'absurdly tied horses' (Franz Grillparzer),[6] the Monarchy faced an uncertain future after the revolution of 1848. The fact that in subsequent years the overall state (*Gesamtstaat*) also worked towards a contemporary consciousness of unity through the medium of the geological survey, apart from other meaningful constructs, proves to be rooted in the paradigm of treating all lands of the Monarchy as scientifically equal. This procedure, based on the equal principles of geological mapping, was supposed to contribute to the reconciliation of different powers and nations. My analysis will deal with the 'how' of the interconnections of politics, practice and episteme at different levels of meaning – the political rhetoric of scientific texts, the self-presentation of the geological survey and the cooperation of different protagonists (documented by letters and reports), as well as the relation between administration and logistical application. A further step in my argument will refer to the activities of the geologists themselves, that is, to the practice of the geological survey. In this context a subject of discussion will be the way in which these activities were organized within a specific institutional situation.

There are many aspects which lead me to assume that in the period of neo-absolutism geology and political culture were particularly close allies. If we want to put it somewhat more cautiously, at least they were not out of touch with each other. This does not mean that one of them happened against the background of the other, but rather that science and the state were related to each other and interwoven. Both in the state and within the sciences, a reorientation towards 'cooperation' took place after the failure of the revolution of 1848. To give a rough sketch of the further analysis, it may be stated that during this phase it was not the retreat from but the active engagement with existing political conditions that can be grasped both from the scientists' self-attributions and their ways of proceeding. To borrow a term from the handwritten records of active geologists, I call this jumble of attributions of meanings within the institutional frame of the Geologische Reichsanstalt (Imperial Geological Survey) a culture of 'consensus'.[7] This forms the context for the steps in my treatment, and it is the heart of my analysis.

In almost all European countries there appeared, in the course of the nineteenth century, state geological survey centres, the task of which was to map geologically the territories of the national governments. This development started in Great Britain, where the first institution of this kind was established in 1835. Its imitators showed different organizational characteristics.[8] Despite all the institutional differences, stratigraphy experienced its first peak and became an essential subject of geology.

In this study, knowledge, which is a term of many meanings, will not be dealt with in the sense of secured factual contents but will be characterized as an element that develops only in the course of the discourse, to follow the approaches of Michel Foucault. In this context, this concept of knowledge is not about describing 'what was the general assumption'[9] but about a situation at a time when the subject, a group or a society was considered to be a sufficient informer about a particular question, or complex problem. What is rejected is the idea of the direct representation of objective knowledge. Instead this concept of knowledge focuses on elements that are introduced by way of a discursive practice, are organized and 'formed' by it. In the context of the discourse, this kind of knowledge does not generate what is actually concrete but what is supposed to be or might be. The fact of being in advance rather than backward in comparison with the other European states, including additional requirements and progressive thinking, is the fertile ground from which discourses in Vienna originated.

The quintessential task of the Geologische Reichsanstalt in Vienna, founded in 1849,[10] was clearly defined as the geological investigation of all territories, which meant the comprehensive incorporation of *all* of the territories of the Habsburg Crown through stratigraphy, the language of geology. Covering this enormous space, with its great geological diversity, within fourteen years yielded the desired consistency of the mapping, a unified overall image incorporating the territories from Lombardy to Bukovina and from Dalmatia to the Elbe gorge.

Let us examine the self-presentation of the newly founded Imperial Geological Survey, which on the one hand developed a scientific method, an objective, systematic and coherent approach, and on the other hand made rhetorical reference to the political slogan 'viribus unitis'.[11] My principal consideration is that such a striking self-image could not be without influence on the inner life of such a project and on the stabilization of knowledge in its work. The political pressure to assure the supra-national unity of the multi-ethnic state and the desired goal of the survey map, both of which were related to the practice of joint action, must ultimately have had a disciplining effect on the cooperation and coordination of the geologists' work: this is the main proposition of this paper. The result, in the form of a map (*Geological Survey Map of the Austro-Hungarian Monarchy*, 1867),[12] was of course preceded by an elaborate set of negotiated relationships for the fieldwork in which many geologists took part, and this involved a practical culture of 'mixing' or 'agreeing' or 'consensus'. This geological work, more-over, was produced within a political framework of agreement concerning one national identity. To what extent was the political requirement of unity rooted in practice as a style of thinking? And may we speak of a unique and typical phenomenon in the case of the survey in Vienna? These are the core questions of my paper. I shall present my argument under three headings:

1. 'Viribus unitis' as a piece of politico-scientific and action-creating rhet-oric: its function for the Imperial Geological Survey;
2. Scientific method: institutional procedure and the geologists' practice of reciprocal scientific reference as the search for 'agreement';
3. Conclusion: state policy, institution and map-production as pillars of a local and communal style of thinking that was typical of the Imperial Geological Survey.

'Viribus unitis' as a Piece of Politico-scientific and Action-creating Rhetoric: its Function for the Imperial Geological Survey

After the failure of the revolution of 1848, neo-absolutism brought about an opposition between an imperialist communal consciousness and the national founding myths of different nationalities. Relations of unity gained weight in the context of progressive thinking. Not coincidentally, the core period of neo-absolutism is the same as the phase of the first geological survey ordered by the state (1852–1863/67): neo-absolutism includes both reaction and reform, which means a constitutional-political setback and at the same time a process of innovation in administration and science. If we leave out the fact that all decision-making power lay with the Emperor, this programme of rule was defined above all by the establishment of a modern bureaucracy. That the latter must be viewed 'in the context of the overall instruments of political and social structural organization'[13] is something that was demanded for the science of history by Fritz Fellner in 1984:

> Until the most recent past, the defamation of the neo-absolutist decade from 1852 to 1860, which is due to the view of victorious parliamen-tarism, has prevented us from recognizing the innovative push this decade meant for the Habsburg Empire, not least because the histories of individual fields of life were treated separately. But the restructuring of sciences in the decades after the middle of the century must be paralleled to the reform of administration.[14]

But the establishment of geology as an element of bureaucratization of the public/state order makes the same ambivalence obvious as the new freedom of the sciences, as part of a process of liberalization in the context of neo-absolutism, no matter of what kind. The parallelism of heterogeneous phenomena gave neo-absolutism a markedly exciting potential, which has hitherto been little discussed in the history of the sciences. Nowadays, this contradiction cannot be corrected, even more so as we must abandon the 'fiction of a unity of history'.[15] Although it is a consensus of history that the 'non-unity of history must be accepted and be productively employed for science', much work has still to be done to 'realize' the many histories 'precisely for the sake of their contradictions, their non-unity, their differences'.[16]

The crisis of 1848 led to a successful reorganization of Austrian central government, which can be roughly characterized as the transition from cooperative decision-making 'in cabinet' without a clear division of responsibilities to the creation of seven different ministries, each headed by ministers responsible for and personally committed to their portfolios. The establishment of the Geologische Reichsanstalt must also be seen against the background of this development. It originates from the Montanistisches Museum,[17] established by the Hofkammer für Münz- und Bergwesen (Imperial Chamber of Coining and Mining), where from 1846 on Wilhelm Haidinger (1795–1871)[18] held lectures for the training of the higher-ranking mining staff. Owing to the general reorganization of the administration, the organization of mining and of the iron and steel industry was at first under the supervision of the ministry of public works, but after May 1849 of the Ministerium für Landescultur und Bergwesen (ministry of agriculture and mining).[19] The latter's minister, Ferdinand von Thinnfeld (1793–1868), had Wilhelm Haidinger present him with a plan for the reorganization of the Montanistisches Museum. Thinnfeld, being the owner of estates and mines in Styria, was well able to decide positively on the relevance of the new discipline of geology to coal-mining as well as to the iron and steel industry, since he had heard the first lectures on mineralogy outside the mining academies by Friedrich Mohs at the Joanneum in Graz. The fact that on account of his marriage with Haidinger's sister, Maria Klara Sidonie, he was connected with the matter not only topically but also through his family meant further support for these developments. Being a committed mentor of this field of knowledge (mining and earth sciences), Thinnfeld made sure at the beginning of 1849 that the institution was successfully changed into a state institute, which was supposed to take on completely new tasks: the communication of knowledge was replaced by its direct production through research, and arbitrary documentation by the surveying of all countries of the monarchy, directed from Vienna.

In a letter to Johann Kundernatsch (1819–56),[20] mining administration adjunct in Steierdorf near Orawitza in the Banat (today a part of Romania), Franz Hauer (1822–99),[21] the first mining assessor at the newly founded state institution, took stock of the early foundational period. The extent to which he considered the status of geology to be part of the political development is shown in this account:

> To me, the various questions sound almost comical, which proves that you have almost no idea of the gigantic changes that the status of the sciences in Austria, apart from the radical political change, have undergone. Thus, to stay with our subject, geology, I will give a short account of what has been happening here in the past few years. [...] By decree of the Ministry of Agriculture and Mining, the establishment of an Imperial Geological Institution has been decided, and the latter has been provided with a budget of 31,000 fl a year. [Wilhelm] Haid[inger][22] has been appointed its Director, bearing the title of 'Sektionsrath', I myself have been appointed First Mining Assessor, and Czjzek[23] has been appointed Second Mining Assessor. Next summer, the Monarchy's best experts are supposed to be occupied with the works, accordingly Reuss,[24] Lipold,[25] Simony,[26] Ehrlich[27] from Linz and Zigno[28] in Padua will contribute, and probably you will receive your call some days after having received this letter. One will start with investigating the Alps north of Vienna. But as early as 1851 we hope to be able to start with the important region of Banat, and this will result in a further sphere of activity for you, in the matter of supervising and starting the works. [Concerning the university, I must tell you that] most of the old professors have been fired; in brief there has been a solid start [and], God willing, the Austrian sciences will no longer be Europe's Cinderella.[29]

The price that critical scientists nevertheless had to pay for the state's concessions in favour of reviving the sciences was that as a consequence, because of political pressure, they not only had to submit to neo-absolutism but also to justify it in their work. Many representatives of the educated and business middle class changed their critical attitude towards state authority, having already distanced themselves from their democratic ideas in the course of the revolution.[30] The multinational state proved to be a common house, based on three pillars – the dynasty, the bureaucracy and the armed forces.[31] Most of all for the sciences this extended structure seemed to provide opportunities for a variety of activities. In any case the business middle class wished for an extended, unitary economic area where they could rely on standardized law and administration as well as protection against foreign competitors. The Germans' attitude of supremacy looked convincing at the time, and this is supported by the fact that Vienna, being the centre, benefited by receiving information from all parts of the Empire and that this concentration of information also became the basis of their continued self-confidence. Neo-absolutism confronted the national founding myths of the Hungarian, Czech and Italian nations with the consciousness of an Empire community. At the same time the cultural construct of a plausible state unity combined with progressive thinking gained weight.

It was not only politics but also science that generated this supra-national consciousness and became an embodiment of the construct of a supra-national state identity. The fact that after the 1848 revolution science contributed to maintaining the multinational Habsburg Monarchy and immediately provided support for the project of supranational harmonization became particularly obvious in the work undertaken at the statistical department by Karl Freiherr von Czoernig. In his *Ethnographic Map of the Monarchy* (1857), he made it clear that the individual Austrian Crown Lands did not constitute homogeneous ethnic and national territories. The fact that the population of Austria was ethnically and nationally mixed and that no nation had an exclusive claim to a particular territory corresponded to a higher state interest. The deciding concepts with which Czoernig argued were those of 'mixture'[32] and harmony. This reference to unity was also regularly invoked in the Imperial Geological Survey by the public reports, diaries and letters of the protagonists. The President of the Zoologisch-Botanischen Gesellschaft, Eduard Fenzl, put it as follows:

> Gentlemen, hold steadfastly to one principle: the interest in the smallest thing that each of us gives... Our joint meetings must remain the living cement of this reciprocal bonding of interests.... Show the world that Austria's men of science, in spite of all differences of nationality and language, rank and status, are swifter to grasp higher goals, more decided in their choice of means, more united in their pursuit and more resolute in adhering to their aims, than others who could rightly boast of being of one tribe, who became and have remained united, but who have never been able to boast.[33]

To the extent to which the geological recording of the territories was seen as a civilizing mission, this project seemed to put the Habsburg Monarchy in the forefront of civilization.

In the period of neo-absolutism, national academics often spoke with one voice, as if they had agreed to one attitude. For example, in 1853 Joseph Alexander Freiherr von Helfert (1820–1910), Permanent Under-Secretary at the ministry of education and a representative of Habsburg-loyal Conservatism, expressively demanded 'national history' as an overall history of the Austrian Imperial state. He understood the latter to be 'the history of a population which belongs territorially and politically together, is tied together by the same authority, under the protection of the same law. For us, Austrian national history is the history of the entire Austrian state and all of its people, all the tribes which are different by origin, education and customs moving on the wide territory of the Empire appearing as its organically intertwined parts.'[34]

In addition, the way in which contemporaries and scientists established and carried out (natural) scientific activities in the context of this policy of homogenization and, beyond this, gave that policy a higher meaning, had an effect not only on the state's domestic situation but also on its position in the increased competition among the European powers.

To the same degree to which geological surveying was considered a civilizing project,[35] as a product of collective knowledge providing most difficult, unique and useful insight into the depth 'of fertile ground with all its varieties, with all its parts,' as Minister Thinnfeld put it in his report to the Emperor[36], the Habsburg monarchy was able to reach a higher level of civilization.

'Viribus unitis', this bodily metaphor that made the state a hero and science its servant, and as a 'promise that Austria wears on its brow',[37] were described by the Liberal Wilhelm Haidinger, the founder of the Imperial Geological Survey – freed, of course, from this heroic metaphor, but nonetheless committed to the idea of unity – in his introduction to the first publication of the K.k. Geologische Reichsanstalt's Royal and Imperial Geological Survey: 'May the Emperor's promise, the word of *magna Austria*, the true basic condition for the existence of human society, also unify immutably, by the individual works in the year-book presented here, the friends of science and its application for the advance of knowledge in our beautiful Fatherland.'[38] 'Harmony', 'agreement', 'consensus', 'unified forces' and 'cooperation': these are the terms that represent a cultural paradigm that we encounter repeatedly, first in a political but then also in a scientific context. But how did this paradigm become manifest in geological practice?

Scientific Method: Institutional Procedure and Geologists' Practice of Reciprocal Scientific Reference as the Search for 'Consensus'

In concrete terms, how did the Geological Survey in Vienna work?[39] The participants, who were brought together in sections and deployed in parallel in different areas, always worked in a number of teams, consisting of three field geologists. Within this framework, one leading person always collated the results for a larger area. During the collection, diaries of fieldwork cycles and profiles were kept and stratigraphic discoveries were immediately entered into the copied maps of the general staff using an agreed colour-coding. After completion of this stage of the work, these materials were sent from the fieldwork site to Vienna. The senior geologist at the Geologische Reichsanstalt in Vienna checked the diaries and maps and made enquiries by letter in cases of unusual discoveries or gaps in the information. Dubious results were checked in a second fieldwork cycle the following summer.

Controversies on particular issues were first discussed by the two cooperating field geologists, and differences in results levelled out, and then the link to neighbouring territories was made by the section leader. Finally, every-thing was discussed and coordinated during winter meetings in Vienna. The individual geologists also published their travel reports independently. The modulation of observations – one might also speak of collating results – took place, therefore, at different levels and different stages of the work: within the teams, among the assistant geologists and the section leader, and at headquarters. This procedure guaranteed the quality control for which the institution was responsible. In my opinion it was here that a culture of 'consensus' came into practical existence.

What I am interested in here is not the system of recording[40] nor the mapmaking procedure used by the individual cartographers, nor even the question of how the data were obtained by the field geologists. I am concerned, rather, with the institutional-collective process by means of which the various observations – at first subjectively obtained and uncer-tain – were compared with each other and brought together to produce a result. I will show the process by which observations, in the form of collective knowledge, gained acceptance and validity.

Constellations of observations were brought into line with each other, and this led to modifications of interpretations. The system theorist Niklas Luhmann[41] has provided us with a theoretical concept for this kind of collective procedure. He discusses sequencing procedures as creating hier-archies, defining and channelling the scope of activities. Such modes of work are typical of institutions where there are interrelations between either action-restricting or action-linking commandments or prohibitions. Sequencing rules and selection criteria are generated by way of institutional procedures. These prescribe forms of decision making, in which sequences begin as being open-ended and still lead to binding results. Such controlled sequencing leads to a reduction in complexity,[42] and it is only by virtue of the goal of achieving agreement that actions that are dependent on each other become possible at all.

This kind of collating procedure guarantees institutionally based quality assurance, and in sum it produces what the protagonists themselves repeat-edly style a culture of 'consensus' or culture of arrangement. This reference to 'arrangement' or 'consensus' is indeed not merely rhetorical; it characterizes the substance of the work processes, the typical feature of this disciplining structure. For any standardization must be binding on all participants. If one of the surveying geologists does not keep to the agreement – whether because of lack of knowledge, ignorance or stubbornness – the result will be useless. To take an example: Franz Foetterle, travelling along the Piave, wrote that after renewed checking of the preceding survey by Dionys Stur (1827–93) he had come to the conclusion that it was not the best, if compared to the guide-lines: 'and on the 27th I had already completed work up to the Piave water-shed; then I made some excursions to the Carnia, Stur's area of the preceding year. Just imagine my dismay: Raibler Layers, Hallstatt Chalk and Guttenstein

C., everything, Guttenstein Chalk, also Werfen L.[ayers] and sandstone from Raibler, all mixed up, so that the complete record is almost useless'.[43]

However, because new findings may result in renewed differentiation that affects or negates definitions that have already been laboriously worked out, the revision process happens only gradually. It will occur only if it is sanctioned by the leadership, after the arguments have been carefully evaluated according to the steps described above. That is why the leadership keeps a particularly watchful eye on differing approaches. Contradictory results within a survey are discussed and compared to 'make all mutual observations compatible with each other'.[44] This orientation is clearly most important; again and again we find it in the sources.[45]

During the winter of 1854, for example, there was a revision as described in a letter from Johann Czjžek to Franz Hauer: 'Farther north, between Sieggraben and Forchtenau, I have made the mutual observations compatible with each other. The other corrections of Bernstein's observations have been discussed earlier and are now to be found in the ordnance survey map included here.'[46]

At the same time, the intended ideal logistics of institutional consistency were interpreted in very different ways by different people, who independently changed the sequence according to their status and self-confidence. During fieldwork, for example, an assistant geologist found himself confronted with insoluble tasks, which he wanted to discuss immediately not only with the section leader but also with the institution's senior geologist, Franz Hauer. By proceeding in this way, mistakes were excluded right from the beginning, and the length of the discussion time was reduced. At this time the sequence of stratigraphic layers was being continuously differentiated. As a result of intensive geological surveying, the classification of Alpine Chalks was constantly differentiated, which raised questions of attribution that had to be decided by the leadership. For example, on 20 June 1854, Dionys Stur wrote directly and in a subordinate manner to Franz Hauer, because he felt insecure about his current work:

Dear Mr. Mining Assessor, you see the many doubts plaguing me here. It is in your hands to throw light upon them and to drive them away. Apart from fossils, the first chest you will receive from me does not contain much. Thus, I feel free to ask you, Dear Mining Assessor, to look through it and then, but as soon as possible, kindly inform me about the result. This will help to improve my investigations.[47]

By improvement, Stur meant institutional assurance under the responsibility of the head of the institution, the culture of consensus. We encounter this attitude in countless letters and reports by the protagonists. Self-confident geologists such as Vinzenz Lipold or Carl Peters, however, chose a more individual path. Being independent researchers, they insisted first on their own opinions, which they wanted to defend against those of others, in order also to distinguish themselves within this community. Vinzenz Lipold, for example, at once came to the notice of the leadership as an active geologist. Upon being informed about a change in the generally binding guidelines, Franz Foetterle, who was temporarily acting as coordinator in Vienna instead of Hauer because all the other geologists were in the field, at once informed Hauer, who was also making use of the summer days to work in the field:

> Lipold, who has been staying here [in Vienna] for more than a week because of his haemorrhoidal suffering, has discovered not only the intermediate position of the Gervilles and the Hierlach Layers but also the equalisation of the Adneth and Hallstatt Layers. As the best locality in this respect he cites Aussee and the Schwarzenbach. Already in his information, the reasons given for the latter are not convincing. But to sort out the matter fundamentally, Haidinger wants you to make the return journey via Aussee, to have a look at the matter and discuss it in situ.[48]

Now I would also like to exemplify this complex procedure in concrete terms by looking at a particular terrain. Using a comparison of letters, diaries, published reports and the final product – the map – I would like to trace the steps in the process of collation. By a comparison of statements at each stage, my thesis of 'matched activity' may be substantiated and its scientific consequences determined.

From his fieldwork during the geological mapmaking process, the assistant geologist Carl Peters wrote to his superiors in Vienna, giving a report on the progress of his data collection. He was very concerned with obtaining detailed results, but was worried by the fact that he was confronted with a variety of findings from his colleagues. Peters had already taken account of the reduction in complexity required by the institution, since he had the final product, the map, in sight. Different views of his team leader and his colleague on the classification of gravel led him to an ambiguous comment: 'I am far removed from all these beauties and I am swimming, if not in an ocean then at least in enormous rivers of gravel.'[49]

This was a hidden reference to Leopold von Buch's theory of mud flow (and possibly also to the theory of rock flow), which says that powerful floods brought mud and rocks from their original locations to some other terrain, where both were deposited, having been randomly mixed.[50] This was a modification of Charles Lyell's 'Drift Theory' of 1834. Lyell postulated a 'diluvial sea', in which icebergs floated that had become detached from the formerly more extensive glaciers in Scandinavia or Britain and deposited northern rock material in the north of Germany. This theory was widely acclaimed. Geologists in the German-speaking territories – Heinrich Georg Bronn, Carl Friedrich Naumann and Bernhard Cotta – were among its supporters, and they were decidedly against the glacier theory. In the Habsburg territories, too, there was no support for the Ice Age theory that went back to the Swiss geologists and was supported in Austria only by Adolphe Morlot (1820–67).[51]

For this reason, Peters weighed carefully whether he could use the Ice Age theory, which implies an explanation of gravel through early glaciation, and expected assistance from headquarters. But clearly his appeal in this connection fell upon deaf ears. Avoidance of the Ice Age theory had become an institutional commandment.

Shortly before this, in his first published report on the subject of the Salzburg Alps,[52] Peters had expressed his scepticism concerning this theory, which defined gravel deposits as being of marine origin. He diminished his rejection, however, in the published version of his work in Carinthia: he stressed that the scattered gravel in Carinthia was difficult to explain in terms of assumptions about rivers of the 'youngest tertiary period.'[53] Here he came down decidedly against the opinion of his team colleague Dionys Stur,[54] but without undermining his interpretation at a personal level. He was therefore obliged to postpone a definitive assessment:

> whatever enormous rises and falls have had an effect on the mountain mass or on individual mountains, they could not remove alien rolling fragments without trace, and it seems to me relatively unimportant whether we claim that they have been rolling around in the mountains since the Miocene epoch, or whether we assume that this has happened since the Cretaceous period. But the necessity for a Miocene ocean that flooded the Alps is – hard to reconcile with the organic life of the real Miocene Ocean that left such excellent deposits in the hollows – and which would explain these phenomena: this necessity I now feel less strongly on account of the latter assumption. ... Let us have our terrain flooded, in certain north-west to south-east strips, in some ancient epoch or other. Then the massive displacements in the mountain ranges, which consist of these alien and indigenous gravels, of sand, sometimes even of clay, will cause us no further trouble.[55]

In this way Peters did not have to renounce his original idea, but to make discursive space in his publication for the unsolved question, the lack of a satisfying explanation for the existence of gravel and erratic blocks of stone. It was only the final product – the map – that required unambiguous colouring and the clear allocation of layers. This reduced the range of possible interpretations and suggested consensus. And Peters wrote to Hauer:

Honoured Councillor!
    Having completed two cycles of excursions, I take the liberty of submitting an unofficial report on my activity, in the hope that you will not find

it disagreeable to learn something of our work in Carinthia. By a north-bound and then a westbound tour from Klagenfurt I have completed almost one third of my terrain, which is regrettably much less rich in interesting material than I had anticipated. In folds of up to five undula-tions, the grey and green semi-crystalline slates – beneath so-called mica slates [*Tonglimmer*] – spread from north to south over the whole country. Extremely subordinate are those slates that could be classified as grey-rock slates, on account of their connection to the red sandstone [*Werfen Sandstone*] and its dolomite, and which clearly constitute the upper layer of the entire range of mountains. The ore conduits, which to a certain extent were predominant in the Salzburg region, are not so here. The iron ore layer is sometimes found within the limestone, the non-prominent crystalline axes – mica slates – and sometimes in the grey slate, whose limestone layer is partly crystalline, partly thick and mostly dolomite. This last difference may, with some justification, be considered definitive concerning the question of the nature of these slates; to put it honestly: it is only a matter of the colour that should be used for these surfaces.[56]

To emphasize the point once more: we see here the impact of unspoken but binding rules of a decision-making process that begins as an open-ended sequence and yet leads to binding results.

Let us again turn to the scientific conflict that concerned Peters and that he articulated more dramatically in his letter than in the published report. He could not agree either with his fieldwork colleagues in the team or with his team leader. Peters wrote:

It is indeed remarkable how this country is covered with gravel ....I have sent an *exposé* on this to Lipold, but this has caused little pleasure, since he considers all these deposits diluvial: that is to say, colours them all the same and requires the same of me. Since I am now convinced that Stur will show 4/5 of them as tertiary, as the middleman in every sense I am completely desperate. Apart from the fact that this wretched gravel complicates the treatment and makes it difficult in detail, without the hideous tedium of the basic mountain range causing any cheerfulness – apart from that fact I fear differences of opinion on this matter a hundred times more than in previous years, because these large deposits are spread over the whole country and play an important role on the map. It will be difficult to reach a *consensus*, and much as I am trying to activate my sense of subordination and organic rules – this year more than ever – I shall continue to make a distinction between high gravel, which occurs scattered in as yet unmeasured altitudes (above 3200'), more rarely in large deposits, and the diluvium, which has its materials mostly from that and in countless stages reaches up to considerably above 2000'. ....I shall soon be in a position to formulate my maxim appropriately, yet

I find the whole business very depressing and must confess that my work has never given me less pleasure than this year.[57]

In fact Peters was ultimately unable to make his ideas prevail against those of his colleague Dionys Stur and his section leader Markus (Marco) Vinzenz (Vincenc) Lipold, and he had even less support from Franz Hauer, the principal geologist at the Imperial Survey. Fifty years later it was he who was found to be right, in the outstanding work of Albrecht Penck,[58] the primary authority on the Quaternary Period.

If Stur attributed the high mountain gravel to the late tertiary age,[59] this happened under the influence of the fact that in the large valleys of the Eastern Alps, particularly those of the Drau, the Save and the Mur, genuine late tertiary deposits do occur; these can be distinguished from quaternary[60] gravel masses only by the irregular surface formation which Stur, unlike Peters, did not consider in his calculations. The two, therefore, used different criteria for decision-making. And on the basis of different theories they observed differently. To explain this supposed appearance, Stur developed complicated theories about the oscillations and movements of the tertiary oceans, which Peters generally rejected. An alternative solution could have been provided only by the Ice Age theory, but that was not invoked because of the rejection decreed by the upper levels of the institution's hierarchy. The fact that Stur was familiar with the Ice Age theory is confirmed by his diary; otherwise, he would never have written there of his assumption of a 'vanished glacier'[61] on the subject of his Almsee inspection in the valley of the Enns. In the printed report, furthermore, there is only an observation about moraines, but no reference to a possible vanished glacier.[62] Those interpretations, on which there was no consensus, fell victim to the process of collation.

The collective rejection of the Ice Age theory on the part of the leadership of the Imperial Survey presumably left no scope in discussions for innovative reinterpretations of traditional stratigraphy, and this is documented by the final product, the map, which depended on and expressed the greatest possible agreement. The leading authority on the Ice Age in the eastern Alps, Albrecht Penck, commented on the subject as follows:

> Through Franz von Hauer's survey map of the Austrian Monarchy sheet [BL.] VI, these erroneous interpretations of the glacial deposits in the Eastern Alps, particularly the Drau districts, as tertiary formations have won widespread support. If the theoretical standpoint of those older geologists was far away from today's, they at least observed with sharp eyes. But from their reports the detailed representation, from Peters onwards, of the spread of erratic events is given particular emphasis.[63]

The attribution of the quaternary gravels to the tertiary age remained in place for many years, even in new editions of the map.[64]

Discussion and Conclusion

Via the roundabout route of the geological record of the Empire, undertaken according to standardized principles, the unified nation was supposed to be represented or indeed united. But does it not sound like a crude or naively simplified relationship, if we understand this political rhetoric as an intrinsic element of geology? Another mode of thinking, which is diametrically opposed to this – namely that of a pure science that is being politically misused – is no less simplifying or naive, because science was explicitly involved in the process of giving the Habsburg Monarchy the image of a unified state after 1848. The scientists themselves were aware of the interweaving of power, politics and science, and oriented their activities to the requirements of the state. A uniform transfer of the crownlands to a systematic and geologically coherent national territory, modelled as a unified entity and also scientifically defined, was the goal. From the outset this project was seen by its protagonists themselves as a great opportunity to undermine the political borders of the crownlands within the Monarchy by relating them all to a primeval age, and thus to bring about an internal and naturalistic unification of the territories of the state: stratigraphy provided a map with an abstract temporal dimension that was both a record of nature and, at the same time, profoundly political. As Haidinger proudly described this obvious matter: 'One could take one Crown Land after the other. But the nature of the mountains is such that a division according to artificial borders cannot be carried out.'[65] Now, by means of their mapping activities, the geologists were referring to the total state (*Gesamtstaat*), which they regarded as a *tabula rasa*, the internal integrity of which was in their hands, and the significance of which, moreover, depended on consensus and thus on the practice of agreement, which allowed no scope for creative excursions.

A structural consensus between political administration and scientific activity – in this we find a coherent relationship between state power, political culture and geology (stratigraphy), which manifested itself in its procedures and ultimately its products. The compact cartographic picture (*Geological Survey Map of the Austro-Hungarian Monarchy*, 1867) suggests a unity of results that had some scientific basis. But the final product was preceded by a complicated network of negotiation processes that was anchored in a practical culture of 'consensus'. I propose that we regard this as the culture of a local community at the centre of calculation of the Habsburg Empire in Vienna.

Notes

1. Franz Werfel (1937) *Twilight of a World* (New York: The Viking Press); see also Claudio Magris (1966) *Der habsburgische Mythos in der österreichischen Literatur* (Salzburg: Otto Müller Verlag).

2. Magris, *Der habsburgische Mythos*.
3. William M. Johnston (2006) *Österreichische Kultur- und Geistesgeschichte. Gesellschaft und Ideen im Donauraum 1848 bis 1938*, 4th ed. (Vienna, Cologne, Weimar: Böhlau Verlag); William M. Johnston (1972) *The Austrian Mind. An Intellectual and Social History 1848–1938* (Berkeley, Los Angeles, London: University of California Press); as well as recently William M. Johnston (2010) *Der österreichische Mensch. Kulturgeschichte der Eigenart Österreichs* (Vienna, Cologne, Graz: Böhlau Verlag).
4. One exception is Tatjana Buklijas's study on medicine and particularly the example of surgeon Theodor Billroth. See Tatjana Buklijas (2007) 'Surgery and national identity in late nineteenth-century Vienna', *Studies in History and Philosophy of Biological and Biomedical Sciences*, 38, 756–74.
5. Ralph Jessen and Jacob Vogel (2002) 'Die Naturwissenschaften und die Nation. Perspektiven einer Wechselbeziehung in der europäischen Geschichte', in Ralph Jessen and Jacob Vogel (eds.) *Wissenschaft und Nation in der europäischen Geschichte* (Frankfurt am Main, New York: Campus Verlag), 7–37.
6. See Hubert Lengauer and Primus-Heinz Kucher (2001) 'Vorwort', in Hubert Lengauer and Primus-Heinz Kucher (eds.) *Bewegung im Reich der Immobilität. Revolutionen in der Habsburgermonarchie 1848–1849. Literarisch-publizistische Auseinandersetzungen* (Vienna: Böhlau), IX–XVIII.
7. This term is found in Peters, see Library of the Geological Survey [Bibliothek der Geologischen Bundesanstalt, Wissenschaftliches Archiv], Inv.-Nr. A00209-B.112, letter from Peters to Hauer [Brief von Carl Peters an Franz Hauer], Klagenfurt, 10.7.1854. For further discussion, see below.
8. Simon J. Knell (2007) 'The sustainability of geological mapmaking: the case of the Geological Survey of Great Britain', *Earth Sciences History, Journal of the History of the Earth Sciences Society*, 26, 13–29.
9. Michel Foucault (1972) *Archaeology of Knowledge* (London: Tavistock Publications), 258.
10. On this subject cf. Die Geologische Bundesanstalt in Wien (1999) *150 Jahre Geologie im Dienste Österreichs (1849–1999)*. published by the Geologische Bundesanstalt (Vienna: Böhlau Verlag), esp. 93.
11. A number of visual and lyrical works of the period employ this motto. The basic idea is already to be found in Franz Grillparzer's well known poem in homage to Radetzky. *'Treue und Eintracht der österreichischen Völker, Viribus Unitis'* is the title of a lithograph by Franz Kollarz from 1849 (after a drawing by Josef A. Hellich). This is an allegorical depiction of the eighteen peoples living within the Monarchy. Men in national costumes are assembled around Emperor Franz Joseph; above them, in the ornamental band, the seven ministries are symbolized, and below, the coats of arms of all crownlands. See Siegfried Nasko (ed.) (1979) *Österreich unter Kaiser Franz Joseph I., Historische Sonderausstellung im Schloß Pottenbrunn* (St. Pölten: Verlag des Magistrats der Stadt St. Pölten), esp. 102.
12. Franz Hauer (1867–74). *Geologische Übersichtskarte der Österreichisch-Ungarischen Monarchie nach den Aufnahmen der k.k. geologischen Reichsanstalt 1: 576.000* (Vienna).
13. Fritz Fellner (1984) 'Geschichtswissenschaft', in Harry Kühnel (ed.) *Das Zeitalter Kaiser Franz Josephs. 1 Teil: Von der Revolution zur Gründerzeit 1848–1880. Beiträge* (Katalog des Niederösterreichischen Landesmuseums, N.F. 147) (Vienna: Amt der NÖ Landesregierung), 374–79, esp. 374.
14. Ibid.

15. Karin Hausen (1998) 'Die Nicht-Einheit der Geschichte als historiographische Herausforderung. Zur historischen Relevanz und Anstößigkeit der Geschlechtergeschichte', in Hans Medick and Anne – Charlott Trepp (eds.) *Geschlechtergeschichte und Allgemeine Geschichte. Herausforderungen und Perspektiven* (Göttinger Gespräche zur Geschichtswissenschaft, Bd.5.) (Göttingen: Wallstein Verlag), 35.

16. Ibid.

17. On this see esp. Wilhelm Haidinger (1869) *Das kaiserlich-königliche Montanistische Museum und die Freunde der Naturwissenschaften in Wien in den Jahren 1840–1850. Erinnerungen an die Vorarbeiten zur Gründung der kaiserlich-königlichen geologischen Reichs-Anstalt* (Vienna: Braumüller).

18. For biographical information see esp. Karl Kadletz (2000) 'Wilhelm Haidinger (1795–1871)', in Gerhard Heindl (ed.) *Wissenschaft und Forschung in Österreich. Exemplarische Leistungen österreichischer Naturforscher, Techniker und Mediziner* (Frankfurt am Main: Lang), 9–30; Tillfried Cernajsek (1996) 'Wilhelm Ritter von Haidinger – der erste geowissenschaftliche Manager Österreichs', *Abhandlungen der geologischen Bundesanstalt*, 53, 5–13.

19. See Otto Guglia (1869) 'Das Ministerium für Landeskultur und Bergwesen 1848–1853', *Burgenländische Forschungen, Festschrift für Heinrich Kunnert, Sonderheft*, 2, 54–65.

20. See Wilhelm Gümbel (1883) 'Kudernatsch, Johann', in *Allgemeine Deutsche Biographie*, vol. 17 (Leipzig: Duncker & Humblot), 292.

21. Franz Hauer was Haidinger's successor as Director from 1866 to 1885. For his biography see (1959) 'Hauer, Franz von', in *Österreichisches Biographisches Lexikon 1815–1950*, vol. 2 (Vienna: Verlag der Österreichischen Akademie der Wissenschaften), 211.

22. On Wilhelm Haidinger see note 18.

23. Johann Čžjžek (1806–55) was Second Geologist in Chief and Paleontologist from 1850 to 1855.

24. August E. Reuss (1811–73) studied medicine in Prague and became a full professor of mineralogy there. From 1863 to 1873 he was professor of mineralogy in Vienna.

25. Marcus Vinzenz Lipold (1816–83) was Second Geologist in Chief from 1849 to 1867. After this he was the Director of Mining in Idria (Idrija, today a part of Slovenia).

26. Friedrich Simony (1813–96) was a surveying geologist from 1850 to 1852, and at the same time the Curator of the Museum of Klagenfurt. From 1850 he was professor of geography in Vienna.

27. Franz Karl Ehrlich (1808–86) worked as Curator of the Museum of Linz from 1841 to 1879 and contributed to the survey.

28. Achille de Zigno (1801–92) was an Italian palaeobotanist who worked mainly in Padua.

29. Library of the Geological Survey Inv.-Nr. A00209-B.77, letter from Franz Hauer to Johann Kudernatsch, Vienna, 1.1.1850.

30. Library of the Geological Survey, Inv.-Nr. A00209-B.59, letter from Franz Hauer to Hingenau, Vienna, 22.1.1849.

31. Ernst Bruckmüller (1984) *Nation Österreich. Sozialhistorische Aspekte ihrer Entwicklung* (Studien zur Politik und Verwaltung 4) (Vienna, Cologne, Graz: Böhlau), esp. 145.

32. On this point cf. Brigitte Fuchs (2003) *'Rasse', 'Volk', 'Geschlecht'. Anthropologische Diskurse in Österreich 1850–1960* (Frankfurt am Main, New York: Campus Verlag), 154.

33. *Verhandlungen der Zoologisch-botanischen Gesellschaft* 1853, 2, 4f.

34. Joseph Alexander Freiherr von Helfert (1853) *Ueber Nationalgeschichte und den gegenwärtigen Stand ihrer Pflege in Oesterreich* (Prague: Calve), 1–2. See also Ernst Bruckmüller (2001) 'Die österreichische Revolution von 1848 und der Habsburgermythos des 19. Jahrhunderts. Nebst einigen Rand- und Fußnoten von und Hinweisen auf Franz Grillparzer', in Hubert Lengauer and Primus Kucher (eds.) *Bewegung im Reich der Immobilität. Revolutionen in der Habsburgermonarchie 1848–1849. Literarisch-publizistische Auseinandersetzungen* (= Literaturgeschichte in Studien und Quellen 5) (Vienna, Cologne, Weimar: Böhlau), 1–33, esp. 18.

35. In his report to the Emperor, Thinnfeld emphasizes that 'any kind of industrial production is a treasure trove meeting the countless needs of higher civilization'. See (1850) *Jahrbuch der kaiserlich-königlichen Geologischen Reichsanstalt*, 1, 1.

36. Ferdinand Edlen von Thinnfeld (1850) 'Allerunterthänigster Vortrag des treugehorsamsten Ministers für Landescultur und Bergwesen', in *Jahrbuch der kaiserlich-königlichen Geologischen Reichsanstalt*, 1, 2–5, here 2.

37. (1862) *Transactions of the Zoological Botanical Association*, 12, VIII.

38. Wilhelm Haidinger (1850) 'Vorwort', *Jahrbuch der kaiserlich-königlichen Geologischen Reichs-Anstalt* 1, p. 2.

39. This analysis is based on insights from many letters and reports from geologists, which for reasons of space cannot be developed here.

40. Cf. Friedrich A. Kittler (1990) *Discourse Networks 1800/1900* (Stanford: Stanford University Press).

41. Niklas Luhmann (1980) 'Temporalisierung von Komplexität. Zur Semantik neuzeitlicher Zeitbegriffe', in Niklas Luhmann (ed.) *Gesellschaftsstruktur und Semantik 1* (Frankfurt am Main: Suhrkamp), 235–300.

42. On this phenomenon see David Gugerli and Daniela Speich (2002) *Topografien der Nation. Politik, kartografische Ordnung und Landschaft im 19. Jahrhundert* (Zürich: Chronos), esp. 114, 132.

43. Library of the Geological Survey, Inv.-Nr. A00209-B.37, letter from Foetterle to Hauer, 7.7.1856.

44. Ibid., letter from Čžjžek to Hauer, 10.2.1854.

45. It is not possible to discuss all references here.

46. Library of the Geological Survey, Inv.-Nr. A00209-B.37, letter from Čžjžek to Hauer, 10.2.1854.

47. Library of the Geological Survey, Inv.-Nr. A00209-B.141, letter from Stur to Hauer, 24.6.1854.

48. Library of the Geological Survey, Inv.-Nr. A00209-B.37, letter from Foetterle to Hauer, 8.8.1852.

49. Library of the Geological Survey, Inv.-Nr. A00209-B.1, letter from Peters to Hauer, Klagenfurt, 10.7.1854.

50. Cf. Otfried Wagenbreth (1999) *Geschichte der Geologie in Deutschland* (Stuttgart: Georg Thieme Verlag), 119–20.

51. Marianne Klemun (2008) 'Questions of periodization and Adolphe von Morlot's contribution to the term and the concept "Quaternär" (1854)', in *History of Geomorphology and Quaternary Geology*, Geological Society, London, Special Publications, 301, 19–31.

52. Karl Peters (1854) 'Die salzburgischen Kalkalpen im Gebiet der Saale', *Jahrbuch der Geologischen Reichsanstalt*, V, 116–42.

53. Karl Peters (1855) 'Bericht über die geologische Aufnahme in Kärnten 1854', in *Jahrbuch der kaiserlich-königlichen Geologischen Reichsanstalt*, VI, 508–80, here 550.

54. Dionys Stur was a Slovak who was Director of the Imperial Geological Institute from 1885 to 1892.

55. Karl Peters (1855) 'Bericht über die geologische Aufnahme in Kärnten 1854', 551.

56. Library of the Geological Survey, Inv.-Nr. A00209-B.1, letter from Peters to Hauer, Klagenfurt 10.7.1854.

57. Ibid.

58. Cf. Albrecht Penck and Eduard Brückner (1909) *Die Alpen im Eiszeitalter*, 3 vols (Leipzig: Hirzler).

59. Cf. Dionys Stur (1854) 'Die geologische Beschaffenheit der Centralalpen zwischen dem Hoch Golling und Venediger', *Jahrbuch der Geologischen Reichsanstalt*, V, 818–51.

60. Cf. William A. Berggren (1998) 'The Cenozoic Era: Lyellian (chrono)stratigraphy and nomenclatural reform at the millenium', in Derek J. Blundell and Andrew C. Scott (eds.) *The Past is the Key to the Present*, Geological Society, London, Special Publications, 143, 111–32, here 111.

61. Diary of Dionys Stur, Library of the Geological Survey, Inv.-Nr. A00057-B, entry for Thursday 27 May 1852, fol. 6.

62. Dionys Stur (1854) 'Die geologische Beschaffenheit des Enns-Thales', in *Jahrbuch der kaiserlich-königlichen Geologischen Reichsanstalt*, 3, 461–83.

63. Penck and Brückner (1909) *Die Alpen im Eiszeitalter*, vol. 3, 1064.

64. The survey map underwent five revisions before the turn of the century.

65. Wilhelm Haidinger (1850), 'Vorwort' in *Jahrbuch der kaiserlich-königlichen Geologischen Reichsanstalt*, 1.

5

Scientific Nationalism: A Historical Approach to Nature in Late Nineteenth-Century Hungary

Gábor Palló

This chapter approaches various occurrences of scientific nationalism in nineteenth-century Hungary; the term 'scientific nationalism' stands for both nationalism occurring within science and nationalism based on science. The latter form of nationalism, which used science directly for political purposes as an argument and justification for nationalist politics, had gradually become a widespread ideology in Hungary in the second part of the nineteenth century. It was not science, but politics, that was nationalist; however, science could not stay neutral on this central political issue. It developed a national character manifested both in its institutional structure and in its cognitive contents.

The chapter will focus on the nationalist character of Hungarian science, especially on its epistemic nationalism, in the period of the Austro-Hungarian Monarchy. In this case, the term 'epistemic nationalism' denotes nationalism that occurs in the cognitive content of science, and differs from political, institutional, cultural or emotional nationalism in science, a complex social entity. Nationalism can be seen as many things, including flag-waving emotion, a part of personal identity, culture or language, long-standing – almost tribal – traditions and ethnic solidarity. All belong to the notion of 'nationalism' and they all deserve analytical approaches. It could be an important investigation to pinpoint the forms of appearances of all these emotions in science. Some particular phenomena, such as priority debates based on nationality or boasts of national excellence or even superiority in particular fields, have occurred very often in the history of national sciences, as well as in the historiographies of sciences. Here, however, nationalism in the sciences will be analysed in a restricted political, non-emotional framework, as the first section will show.

The term 'science' will primarily be used in reference to the natural sciences, physics, chemistry and biology, disregarding the important historical changes in its meaning.

Epistemic nationalism can exist in science particularly in the form of natural history, which was a dominant – albeit decreasingly so – branch of the natural sciences in Hungary up to the First World War. After a brief introduction to the conceptual approach to scientific nationalism in general, the chapter will typify nationalism and then differentiate between nationalism based on science and non-epistemic scientific nationalism, using the Hungarian example to conclude with the epistemic version of nationalism in Hungarian science.

Nationalism in science

According to Ernest Gellner's widely quoted definition, 'nationalism is primarily a political principle, which holds that the political and the national unit should be congruent'.[1] In other words, the population of one nation, conceived as an ethnic group, should only live within the political boundaries of a nation-state. States should be nationally homogeneous; to give an example not cited by Gellner, Hungarians should live in Hungary not in Romania, Slovakia, etc. and Slovakians should live in Slovakia not in Hungary. The nationalist principle is violated, Gellner continues, if, first, a given state does not include all members of the nation; second, if it includes all members but also people belonging to other nations; and, finally, if it does not include all members of the nation, but includes members of a foreign nation as well. Gellner adds that nationalist sentiment is particularly sensitive in a situation where 'the rulers of the political unit belong to a nation other than that of the majority of the ruled'.[2]

The Hungarian political situation in the nineteenth century can easily be placed into this scheme. The nationalist principle was violated in the way described by Gellner's second point. Hungary included not only Magyars but also members of thirteen national minorities, and the country was ruled by the Habsburgs, who belonged to another nation. Gellner was right: Hungarian nationalist sentiment was very sensitive to this situation. Nationalist sentiment became a main feature of Hungarian high culture for a long time. Directly or indirectly, almost all aspects of culture, including poetry, literature, music, theatre and painting, were imbued with nationalism.

Hungarian nationalist sentiment had two faces. On the one hand, Hungarians had fought against the Habsburgs, their foreign rulers, and this fight was described in heroic, emotional terms. On the other hand, Hungarians ruled all the national minorities living inside their borders, which they despised. Hungarians felt superior to these national minorities, and wanted to assimilate them. National minorities were often represented in Hungarian high culture in an ironic and patronizing, if not cynical and hostile, way. Nationalism, however, in this chapter, is primarily considered to be a political principle, not a sentiment. Science, on the other hand, has traditionally been thought to be a system of knowledge containing universal laws. According to Mertonian norms, science is universalistic in the way that its truth is evaluated in terms of universal, impersonal criteria, and not on the basis of race, class, gender, religion or nationality.[3]

So does science have anything to do with nationalism? Gellner generated a typology of nationalism based on three factors: the existence of a centralized power, education, and shared culture (essentially high culture). Science is closely related to all of them. Disregarding now both Gellner's otherwise very relevant Austrian-type[4] and John Plamenatz's Eastern-type nationalism,[5] it seems sensible to look for nationalist features in Hungarian science in the same way as it is sensible to look for nationalist features in Hungarian literature or dance. As nationalism is a political principle, nationalist science should be considered as a political actor in realizing the goal of constructing a homogeneous high culture.

Nationalism based on science

Both faces of Hungarian nationalism, whether related to foreign rulers or to the national minorities, relied on various ideologies to legitimate the endeavour of constructing a homogeneous nation. Science proved to be instrumental, as was shown by the fast growing popularity of Darwinism after 1860. Indeed, Darwin's *The Origin of Species* was published in Britain in 1859, and just a few months later it was reviewed in a Hungarian journal.[6] The author of the review, Ferenc Jánosi, was a military officer, later a high-school teacher, journalist and secretary in the ministry of justice, who had studied law, theology and chemistry. He was everything but a naturalist, but enthusiastic about a breakthrough in his field. The first person to write a book on Darwinian evolution, in 1864, was Jácint Rónay, a Catholic priest, who, after the Revolution and War of Independence of 1848–49, fled from Hungary to Britain (in 1850) and returned seventeen years later.[7] He was not a naturalist, either; neither was Ágost Greguss, a professor of aesthetics, who was the first to speak against Darwinism at the Academy, in 1863.[8] Indeed, Darwinism seemed to be more popular among public intellectuals, and later politicians and sociologists, than among experts of natural history.

Two factors explain this phenomenon. First, from the 1850s, positivist philosophy grew to a dominant position in Hungary as compared with the earlier Hegelian influence. Parallel with the strengthening of positivism, science gained new appreciation in non-scientific circles. Second, Hungarian political, ideological and emotional nationalism gained a new vocabulary and new arguments from Darwinism that became widely known, in particular through the positivist philosopher Herbert Spencer's works.

Darwinism provided a scientific framework for proving Hungarians to be nationally superior to the ethnic minorities such as Slovaks, Romanians or Ruthenians, who were considered to be less successful groups in the struggle for survival than Hungarians. Many political writers used variants of this argument. For instance, Gusztáv Beksics, publicist, expert of constitutional law and member of parliament, argued for a policy of assimilation, that is, assimilating national minorities instead of facilitating their struggle for autonomy. Beksics claimed that the Hungarian race was superior to the national minorities because Hungarian is not a pure race, rather a mixture of several races, unlike the Romanians, Slovaks and others. To enhance their superiority, Hungarians should mix with these less developed races, resulting in an even stronger Hungarian race and the disappearance of the weak minorities. According to Beksics, assimilating to Hungarians was in the interest of all races living in the Carpathian basin.[9] This kind of logic almost automatically led to a vigorous eugenic movement starting in the first decade of the twentieth century in Hungary.[10]

Biological argumentation was widespread in Hungary in the period of the Austro-Hungarian Monarchy, as in other parts of Europe. Science served as an intellectual basis for nationalist emotions and nationalist politics in a country where the relationship between Magyars and non-Magyar minorities was a central political issue.

Institutional nationalism of science

Although science is generally considered to be an international, transnational, cosmopolitan or universalistic endeavour, it has been organized nationally at least since nation-states existed. Science can be called nationalist as a social entity in a non-epistemic sense if it is declared that the system of scientific institutions is set up and operated in order to promote the interest of the given nation either as a part of its high culture or as an entity that serves the technological and economical progress of the given nation.

To promote the fight against the Habsburg violation of the nationalist principle, a number of scientific institutions were founded in Hungary in the nineteenth century. The example of the Hungarian Academy of Sciences (Magyar Tudományos Akadémia) is telling.

The establishment of a Hungarian scientific society (Magyar Tudós Társaság) was decided by the Hungarian Diet in 1825 on the initiative of Count István Széchenyi, whose intention was to construct a modernized Hungary. Although the scientific society was intended to be financed by private donations without relying on the state budget, the establishment had to be endorsed by the Habsburg Emperor, who was also King of Hungary.[11] The Hungarian Academy of Sciences has always been a learned society, although, according to the historian James McClellan, learned societies were typically established in the eighteenth century, while the nineteenth century was the time when specialized scientific societies were formed. Such a scientific society was the Geological Society, founded in London in

1807, which aimed at bringing together the professionals working in the field of geology. A learned society, like the *Académie française*, established in the seventeenth century, covered diverse subjects, such as language, philosophy, literature, the fine arts, history, medicine, agriculture, economics and the sciences. It was a social and cultural enterprise incorporated by and within the Ancien Régime.[12]

Indeed, the Hungarian Academy of Sciences belonged to a group of national institutions established in the nineteenth century, such as the National Theatre or the National Museum. Similar institutions in Western Europe served to represent the central power of the state, sometimes embodied by an emperor: Napoleon, for instance. In Hungary, however, they represented the revolt against the central power, the Austrian rulers. In keeping with Gellner's nationalist principle, these institutions, by emphasizing their homogeneously national character, served as tools in the fight against foreign rulers. One of the most important battlefields was the demand to use the Hungarian language in all areas of Hungarian culture.

The Academy's statutes set it a double goal: the cultivation of the Hungarian language, and the cultivation and popularization of science. Its work began in six sections: linguistics, philosophy, history, mathematics, law, and natural sciences, which showed the learned-society character of the Academy. All its activity, which started in 1831, was intended to be carried out in Hungarian.[13]

Hungarian scientific language, however, was still under construction as part of a movement, called language renewal, aiming to form the so-called peasant language into a usable Hungarian language for the purpose of high culture. Accordingly, for instance, chemistry attempted to change all foreign words to new Hungarian ones that were constructed by rules based on the suggestions of various authors. Even common words such as 'material' (*anyag*), 'nitrogen' (*légeny*), 'oxygen' (*éleny*), 'mercury' (*higany*) and 'reaction' (*egyesülés*), along with expressions like 'sphere' (*gömb*) and 'gravitation' (*nehézkedés*), received Hungarian names. Some of them survived but most of them died out. Throughout the nineteenth century, the Academy resounded with loud debates about what was the 'correct' scientific language. Finally, the chemistry community compromised and agreed to use some Latin-based expressions such as 'oxygen', 'oxides' and 'reaction', mixed with Hungarian words.

The issue of language was crucial in nationalist struggles. Teaching of the Hungarian language was gradually introduced in Budapest University after 1830, when a law was issued in Vienna that permitted the official use of national languages in the Empire. A memorandum, dated 1841, allowed the introduction of Hungarian as a language of instruction at the university. In fact, some subjects had already been taught in Hungarian in the faculties of law and medicine, but in certain areas of law Hungarian words were still missing because the administration used Latin or German expressions.

The same problems occurred at the faculty of theology and philosophy. Physics, heraldry, philosophy and mathematics, for instance, were taught in Latin, while other subjects were taught in Hungarian. Under neo-absolutism, between 1849 and 1860, the most important subjects had to be taught in German, too, resulting in a mixed-language education. In 1860, Franz Joseph ordered the reintroduction of Hungarian at the University. After this, mixed-language instruction gradually changed to Hungarian.[14]

The complexity of the issue can be exemplified by the case of chemistry. The first chemistry lecture for university students was given by Károly Nendtvich, a member of the Academy, during the 1848 revolution. Nendtvich, originally a surgeon, later a naturalist and botanist, was a prominent member of a political and cultural reform group with fervent nationalist emotions. He published about the importance of science and fought for the reform of the Hungarian language, mainly the language of chemistry. He was appointed to full professor in 1848. After the revolution he was dismissed from the university, but he was allowed to return to his former position at the Polytechnic (Joseph Polytechnicum, then called Joseph Industrieschule) in 1850. Here he became full professor in 1857. After the Austro-Hungarian 'Compromise' ('*Ausgleich*') in 1867, the Polytechnic was reformed to become the Palatine Joseph Technical University (Királyi József Műegyetem) in 1871. The university got its name from Palatine Joseph (Joseph Anton Johann von Österreich/József nádor (1776–1847)), a Habsburg archduke. Nendtvich served as rector of the university in 1873/74 and retained his full professor tenure until his retirement in 1882. Nendtvich had travelled widely in various foreign countries, including Germany, France, Belgium, Britain and North America, about the last of which he published a travel book.[15] In his old age, in 1889, he took a long trip to Italy, Tunis, Algeria and Spain. He also produced important and successful chemistry textbooks. After his retirement from the university, he participated in the activities of the early Hungarian anti-Semitic movement and published a book about the 'Jewish problem'.[16]

Epistemic nationalism

The study of the nationalist principle in the institutional system of science, including scientific language, considers science to be a social entity, disregarding its function of producing scientific knowledge. An intricate question is whether the product, knowledge, can be nationalistic in the sense of the nationalist principle. In other words: whether scientific knowledge can serve the particular goal of building or constructing a homogeneous nation, instead of serving the whole of mankind as the universalistic attitude would require.

Hungarian science was dominated by a natural-historical approach until the 1920s, instead of the universalistic natural-philosophical approach.

The latter follows the Aristotelian pattern of explaining phenomena from first principles by logical means. The vast amount of historical and theoretical literature discusses the best examples of this tradition: cases of Kepler, Galileo, Newton, Maxwell, Einstein and others. Natural history has been less appreciated by theoretically inclined historians, although Michel Foucault in philosophy and the huge industry of historical studies of evolution seems to attract a growing interest in natural history.[17] Natural history does not seek demonstrative truths, but rather describes nature, collects objects, like plants, animals and minerals, makes pictures of them, maps their locations, catalogues them, and constructs systems that help the practitioners to arrange their objects into an order – once, but today perhaps no longer thought to simulate the order that was given to the nature by God.

With a few exceptions, the two traditions are treated today as two separate worlds: researchers follow either the natural-philosophical-oriented tradition or the natural-historical one. Peter Galison, a historian of physics and one of the few who sees the connection between the two traditions, writes that there was a 'split in science itself between an abstract, reductionist approach to the physical world and a natural historical approach that authors from Goethe to Maxwell had dubbed the "morphological" sciences.'[18] He adds that 'Opposing the "one-sided" working of abstract science lay another ideal of investigation, embodied in the morphological sciences.'[19] Galison noticed an intimate relationship between natural history and art: 'painters and poets tried to capture the power of storms and grand scale of forests, cliffs and waterfalls. And both artists and scientists recognized a tension between the rationalizing, lawlike image of nature proffered by the natural philosophers and the irreducible, often spiritual aspect of nature presented by their contemporaries in the arts.'[20]

In research of the typical morphological sciences like botany and zoology, Hungarians did not seek to establish new systems or notions of species, or develop theories about the fixity or transformation of species. They rather collected, described, named and pictured plants and animals that they could find in Hungary. Pál Kitaibel, Imre Frivaldszky and many others were proud of their collections. The same could be said about geologists, geographers and mineralogists in Hungary. In chemistry, the natural-historical line can be seen by the dominance of analytical chemistry in the nineteenth and early twentieth centuries over organic syntheses or general, physical chemistry. The analysis of minerals and mainly mineral waters, or agricultural products attracted the interest of chemistry professors like Károly Nendvich, also a botanist, who analysed various types of coal found in Hungary. Károly Than, Béla Lengyel and Lajos Ilosvay, emblematic professors in the late nineteenth century, were also excellent analysts. Seen in this way, chemistry in Hungary had inclinations similar to those of biology, because both were oriented towards natural history. In addition, many of the scientists mentioned above had an attraction to paintings, drawing and poetry, as if they exemplified Galison's description. Their epistemic nationalism was manifested in the subjects of their investigations: the description of nature in Hungary rather than the search for new universal laws.

This characterization of Hungarian science can be extended to physics, the modern embodiment of natural philosophy with its universalistic attitude. Loránd Eötvös, the leading personality of physics in Hungary at that time, was also highly skilled in drawing. He wrote poems, and later became an enthusiastic photographer. These attributes of a naturalist did not prevent Eötvös from becoming a physicist. After the mid-1890s, he published on gravitation, a crucial subject of Newtonian mechanics. Eötvös developed an extremely sensitive torsion pendulum, or torsion balance, by which gravitation could be studied with unprecedented precision. He showed, to a high degree of accuracy, that gravitational mass and inertial mass are equivalent, which is an essential postulate in both Newtonian and Einsteinian physics.[21] No scientific statement can be more universalistic. The equivalence of inertial and gravitational mass is supposed to be valid in all parts of the universe, in all countries, independent of nations, locality, politics, religion and other social and cultural factors.

Eötvös, however, was not solely motivated by realizing an exceedingly precise measurement to confirm the well known and widely accepted law. His instrument was so sensitive to changes in gravitation that it proved to be useful for research on geological strata below the surface of the Earth. In other words, the torsion balance was suitable for extending natural-historical description and mapping the unseen part of nature, like the microscope and telescope. In 1901, Eötvös, the president of the Hungarian Academy of Sciences at the time, said that his pendulum was 'simple like Hamlet's flute; one just has to know how to play it and the musician can draw delightful variations from it.'[22] With his instrument, he said, the physicist 'can read the smallest change in gravitation'.[23] He continued, 'Wherever I place my device, with my procedure, I can measure how much and in what direction gravitation changes.'[24] Finally, he explained his motives in a non-physicist rhetoric, a style as picturesque and poetic as the way Galison characterized natural history:

> Encircled by a wreath of mountains, the flatness of the Great Hungarian Plain lies here below our feet. Gravitation formed its surface to its liking. What kind of mountains did it bury, and what kind of hollows did it fill with soft material until the formation of the plain, which grows the golden ear of wheat to give life to the Hungarian nation? As long as I walk on it, as long as I eat the bread it provides, I would like to answer this question.[25]

Eötvös's intention was apparently to link the natural-historical approach to the natural-philosophical one. He wanted to follow a nationalist goal, to map Hungarian mineral resources under the surface of the earth, in addition to carrying out highly precise measurements with universal significance.

Conclusion

During the long nineteenth century, that is until the First World War, Hungarian science was distinctively nationalistic in its style. This nationalist character fulfilled the requirements of building up an independent nation-state, and reflected both the political situation and the high culture of the country. The cultural fertility of the Austro-Hungarian Dualist Monarchy extended to the natural sciences. In both parts of the Monarchy, a number of important results were achieved and influential scientists worked on them. However, the political position of the two parts, Austria and Hungary, was not symmetrical. This asymmetry was reflected by the continuation of Hungarian nationalism born in the late eighteenth and early nineteenth centuries. Nationalism was a characteristic feature of Hungarian culture, literature, music and science; compared with Austrian universalism, Hungarian scientific thinking was local, practical and historical.

This characterization does not mean that Hungarian science was insulated from other nations' sciences. Hungarian science worked on the periphery of German world-science, it was connected to Austrian and German science by thousands of strong personal, institutional, and intellectual ties and to other national sciences by less strong ties.[26] Some Hungarian scientists published in foreign languages, mostly in German, and participated in the activity of the international scientific community, by participating in conferences, establishing cooperation and the like. They worked for two knowledge markets, a national and an international market. Most of the professors who were active during the second part of the nineteenth century, including Eötvös and Nendtvich, studied and worked in Germany. Eötvös improved his gravitation measurement in the hope of winning the Beneke prize offered by the Royal Scientific Society of Göttingen in 1906, and he wrote up this work in German; the prize committee had invited applications for the best solution of gravitation measurement, well knowing that this was Eötvös' specialty. Nendtvich, who fought vehemently for the use of Hungarian in teaching, also wrote his first textbook in German, enjoying the advantage of being able to write and speak German, the lingua franca of science in the German-speaking world.[27] To foster the connections of Hungarian science with the German-speaking scientific community and to provide a wider scientific forum for results of local interest, the Academy of Sciences launched a scientific journal in German on Eötvös's initiative. The title was telling: *Mathematische und Naturwissenschaftliche Berichte aus Ungarn* (Mathematical and Scientific Reports from Hungary). This alone shows the local, national character of the publications. And yet this local, practical and historical knowledge born in Hungary aimed at contributing to international science.

Notes

1. Ernest Gellner (1996) *Nations and Nationalism* (Oxford: Blackwell Publishing), 1.
2. Ibid.
3. Robert Merton (1973) 'The Normative Structure of Science', in idem, *The Sociology of Science: Theoretical and Empirical Investigations* (Chicago: University of Chicago Press), 267–78.
4. In Gellner's Austrian-type nationalism both power and high culture are in the hands of the rulers, basically the Habsburgs. The powerless have no access to education but share folk cultures. They fight to elevate the folk culture to the rank of high culture. Gellner, *Nations and Nationalism*, 94–97.
5. Plamenatz analysed cultural nationalism instead of political nationalism. He distinguished between two types of European nationalism, a Western and an Eastern type on the basis of 'backwardness'. While Western nationalism was formed in nations (such as Germany and Italy) that were equipped with all modern cultural institutions, skills, behaviours, styles, expertise and values, Eastern nationalism was formed in nations that lacked them (such as Slavic nations). According to Plamenatz, in the process of adaptation to their surroundings, a special tension develops between the ancient cultural tradition and the new requirements. This tension is expressed in the nationalism of the so-called backward nations. John Plamenatz (1976) 'Two Types of Nationalism', in Eugene Kamenka (ed.), *Nationalism. The Nature and Evolution of an Idea* (London: Edward Arnold), 23–36.
6. Ferenc Jánosi (1860) 'Új természetrajzi elmélet – A nemek eredete', *Budapesti Szemle*, 1, 10, 383–418.
7. Jácint Rónay (1864) *Fajkeletkezés. Az embernek helye a természetben és régisége* (Pest: Demjén és Sebes). On Rónay's biography, see Lajos Pál (1976) *Rónay Jácint* (Budapest: Akadémiai).
8. Gregus Ágost (1863) 'Az ember helye a természetben', *Budapesti Szemle*, 18, 420. Greguss Ágost (1864) 'A haladás elvéről', in *Magyar Tudományos Akadémia Értesítő: A Philosophiai, Törvény- és Történettudományi Osztályok Közlönye* (Budapest: Magyar Tudományos Akadémia), 435.
9. See, for example, Gusztáv Beksics (1895) *A román kérdés és a fajok harca Európában és Magyarországon* (Budapest: Athenaeum); Gusztáv Beksics (1896) *A magyar faj terjeszkedése és nemzeti konszolidácziónk különös tekintettel a mezőgazdaságra, birtokviszonyokra és a népesedésre* (Budapest: Athenaeum).
10. I detailed the nationalist features of the nineteenth-century Darwin reception in Hungary in an article: Gábor Palló (2009) 'Darwin utazása Magyarországon' [Darwin's trip to Hungary], *Magyar Tudomány*, 6, 714–26, and in a paper: 'The adaptation potential of Darwinism: The unending reception in Hungary', delivered at the conference 'Darwin now: Darwin's Living Legacy', Bibliotheca Alexandria, Egypt 14–16 November 2009. In various contexts other authors have also published about the issue; see, for example, Marius Turda (2004) *The Idea of National Superiority in Central Europe, 1880–1918* (Lewiston, NY: Edwin Mellen Press); Katalin Mund, (2008) 'The reception of Darwin in nineteenth-century Hungarian society', in Eve-Marie Engels and Thomas F. Glick (eds.) *The Reception of Charles Darwin in Europe* (London, New York: Continuum), 441–62. Sándor Soós (2008), The Scientific Reception of Darwin's Work in Nineteenth-Century Hungary', in ibid., 430–40.
11. The standard source for the history of the Hungarian Academy of Sciences was published for the anniversary of the Academy. Zsigmond Pál Pach (ed.) (1975)

112   Gábor Palló

A *Magyar Tudományos Akadémia Másfél évszázada 1825–1975* [*One and a Half Centuries of the Hungarian Academy of Sciences*] (Budapest: Akadémiai Kiadó). A volume was published on the role played by the Academy in the field of natural sciences: László Vekerdi (1994) *'A Tudománynak háza vagyon': Reáliák a Régi Akadémia terveiben és működésében* [*'It is the House of Science': The Academy and the Natural Sciences*] (Piliscsaba-Budapest: Magyar Tudománytörténeti Intézet).

12. James McClellan (1985) *Science Reorganized: Scientific Societies in the Eighteenth Century* (New York: Columbia University Press), 3.

13. On the early work of the Academy, see Agnes R. Varkonyi (1975) 'A Magyar Tudományos Akadémia megalapítása 1825–1831' and 'A Magyar Tudós Társaságtól a Magyar Nemzeti Akadémiáig 1831–1849', in Pach (ed.) *A Magyar Tudományos Akadémia*, 11–27 and 31–51, here 23–49.

14. István Sinkovics (ed.) (1985) *Az Eötvös Loránd Tudományegyetem Története 1635–1985* [*The History of the Eötvös University*] (Budapest: ELTE), 144–46, 163–68, 189–90.

15. Károly Nendtvich (1858) *Amerikai utazásom. Egy földabroszszal és három kőrajzzal* [My Trip to America] (Pest: Heckenast).

16. There is no published biography of Nendtvich. His name is mentioned in some books and articles, e.g. Ferenc Szabadváry and Zoltán Szőkefalvi-Nagy (1972) *A kémia története Magyarországon* [The History of Chemistry in Hungary] (Budapest: Akadémiai Kiadó), 201–03. Károly Nendtvich (1884) *Die Judenfrage in Oesterreich-Ungarn. Eine kulturhistorische Studie* (Pest).

17. Michel Foucault (1973) *The Order of Things: an Archaeology of the Human Sciences* (New York: Vintage).

18. Peter Galison (1999) *Image and Logic: A Material Culture of Microphysics* (Chicago: Chicago University Press), 75.

19. Ibid., 79.

20. Ibid., 75.

21. Roland v. Eötvös (1890) 'Über die Anziehung der Erde auf verschiedene Substanzen', *Mathematische und Naturwissenschaftliche Berichte aus Ungarn*, 8, 65–68. The long series of gradually improving measurement techniques has been related in a posthumously published paper by Eötvös and his assistants: Roland Eötvös, Desiderius Pekár and Eugen Fekete (1922) 'Beiträge zum Gesetze der Proportionalität von Trägheit und Gravität', *Annalen der Physik*, 68, 11–66.

22. Eötvös Loránd (1901) 'A Föld alakjának kérdése. Elnöki megnyitó beszéd, 1901' [The problem of the shape of the Earth. Opening address], *Természettudományi Közlöny*, 33, 321–28 (accessible online at: http://mek.oszk.hu/03200/03286/html /eotvos1/foldalak.html).

23. Ibid.

24. Ibid.

25. Ibid.

26. I have detailed Hungarian–German scientific relations in Gábor Palló (1995) 'Deutsch-ungarische Beziehungen in den Naturwissenschaften im 20. Jahrhundert', in Holger Fischer and Ferenc Szabadváry (eds.) *Technologietransfer und Wissenschaftsaustausch zwischen Ungarn und Deutschland. Aspekte der histor-ischen Beziehungen in Naturwissenschaft und Technik* (Südosteuropäische Arbeiten 94) (Munich: Oldenbourg), 273–89.

27. Karoly Nendvich (1839) *Grundriss der Stöchiometrie nebst einem geschichtlichen Überblick derselben für angehende Chemiker und Pharmaceuten entworfen* (Budae).

6
Acts of Creation: The Eötvös Family and the Rise of Science Education in Hungary

Tibor Frank

When asked about the reasons for the appearance of so many excellent mathematicians in Hungary at the turn of the nineteenth and twentieth centuries and afterwards, Professor George Pólya of Stanford University answered: '[a] general reason is that mathematics is the cheapest science'.[1] This was, indeed, important in a relatively underdeveloped country. As to specific reasons, Pólya listed the *Középiskolai Mathematikai Lapok* (High School Papers in Mathematics), the Eötvös Competition, and the personality of the mathematician Lipót Fejér.[2] In order to understand and appreciate the significance of Baron Loránd Eötvös, the man mainly responsible for this achievement, we should first consider the nature of the Hungarian brand of creativity and its peculiarities.[3]

Hungarian creativity

Hungarian creativity is embedded in a complex tradition. Two aspects deserve particular emphasis: the almost constant entanglement with internal and international conflicts, wars and revolutions, and the long coexistence with German culture and civilization. Through many centuries of Habsburg rule and beyond, German philosophy, science, literature, education and music shaped and harnessed the intellectual energies and talents of one Hungarian intellectual generation after another. The social history of the Hungarian cast of mind – indeed of the way of thinking across much of East-Central Europe – is deeply rooted in war and conflict, abetted by a foe of an entirely different nature: poverty. Often in a cross-fertilizing way, both the German impact and the many international conflicts left a lasting imprint on the Hungarian mind, its ways of solving problems, creating new ideas and organizing thoughts.

Problem solving, almost a passion, permeated all aspects of life, from the mundane to the abstruse. Much of this came from the multiethnic, multicultural, multilingual nature of Hungarian and Austro-Hungarian society, which constantly threw up problems to be solved – economic, social, political and cultural. One may conjecture that this call for problem solvers offers a partial explanation for the country's longstanding abundance of brilliant mathematicians and scientists; János Bolyai, Lipót Fejér, John von Neumann, George Pólya and Paul Erdős are merely the best known in a seemingly unending succession of outstanding talents that spilled over, beyond the realm of pure mathematics, into physics, chemistry, engineering and many other fields.

A potent factor in maintaining this record of achievement lay in the way in which secondary education was reorganized after the Austro-Hungarian Compromise of 1867. In 1870–72, the educationist Mór Kármán, father of the aviation pioneer Theodore von Kármán, was commissioned by the minister of education, Baron József Eötvös (1813–71), to undertake a first-hand study of Germany's acclaimed high-school system with Professor Tuiskon Ziller at the University of Leipzig. Mór Kármán was convinced that 'the student has to pass through the same stages of intellectual development as those by which the culture of humankind reached its state today'.[4] Mór Kármán added that 'It is not the life of a whole people that can be explained through the stages of the life of an individual; on the contrary: it is the life of the nation that opens up the life of the individual.' As the basis of his educational philosophy and practice Mór Kármán declared that 'it is not expedient to take some division of the sciences for a system of our general knowledge rather than the content of culture as a whole. It is rendered even less practical by the experience that scientific thinking is a very late phenomenon in the development of human spirit.'[5] In Mór Kármán's high school, individual studies were not to be applied in some professional isolation, but in 'a well-connected harmony to prepare for the sciences, to found a general culture and, as their ultimate goal, to develop a noble moral character.'[6] Kármán's ideas were based mostly on those of nineteenth-century German thinkers such as Johann Friedrich Herbart, Georg Wilhelm Friedrich Hegel, Johann Wolfgang von Goethe, Tuiskon Ziller, Kurt Breysig, Wilhelm Dilthey, Karl Lamprecht, Bruno Hildebrand, Gustav Schmoller, Karl Bücher, Theodor Waitz and Moritz Wilhelm Drobisch. He called Germany 'the homeland of education'.[7] Kármán's work prepared the ground to make sure that the best Hungarian schools consistently had access to first-class teaching resources capable of encouraging students to attain standards that still compare favourably with many (and not just lower-tier) colleges in the United States.

Following the German model, the high school (*Gymnasium*) placed heavy emphasis on the classics, Hungarian language and literature, and universal knowledge, without neglecting mathematics and the natural sciences. These were unashamedly elitist institutions, with a student intake typically drawn from a rather narrow upper-middle section of Hungary's still relatively conservative, even feudalistic society. However, they could attract teachers of a very high calibre; many of these were recognized scholars in their fields, as reflected in their subsequent membership of the Hungarian Academy of Sciences and appointment to university professorships. As a result, the country's top schools, such as the Lutheran high school in Pest or the *Mintagimnázium* (The 'Model' High School) of the University of Budapest, succeeded for several decades in cultivating an astonishingly consistent succession of brilliant young minds, of whom John von Neumann, Eugene Wigner, Edward Teller and Theodore von Kármán were only a few of the more prominent.

The German influence was also more widely felt during this era, Hungary in many ways was an outpost of German culture, whose icons – from writers Goethe and Schiller and philosophers Kant and Schopenhauer, through composers Beethoven, Brahms and Wagner and painters von Kaulbach and von Piloty, to scientists Gauss, Haeckel and Brehm – were held in unparalleled esteem. Even news of the wider world outside the German universe was usually refracted through the medium of the German language and, inevitably, German cultural paradigms when it reached Hungarian aristocratic libraries or the coffeehouses and salons of Budapest's middle classes.

Of course, there was also a measure of animosity towards innovation: conservatism prevailed in much of the Austro-Hungarian Monarchy. Although Hungary was in several ways an ideal creative spawning ground, many of its achievements were made in the face of official Austrian and Hungarian disapproval. For the greater part of the nineteenth century the national tradition was conservative and the mentality hostile to innovation – not least thanks to the obverse side of German and Austrian culture, with its authoritarian insistence on strict and often antiquated rules and standards, established patterns of thinking and unalterable methods. The general ambiance favoured preserving the status quo rather than supporting new ideas, and accordingly the ruling conservative forces of the pre-1867 period ignored or spurned many reform-minded Hungarians, which more often than not drove the latter into exile (as in the cases of Governor-President Lajos Kossuth, Generals György Klapka, György Kmety, Richárd Guyon, János Czetz and Antal Vetter, cabinet ministers Baron József Eötvös, Lázár Mészáros, Sebő Vukovics, Count Kázmér Batthyány, scholars Ferenc Pulszky, Jácint Rónay), to the lunatic asylum (Prime Minister Bertalan Szemere) or to suicide (Count István Széchenyi, Count László Teleki). There is, of course, a contradiction here between conservatism and renewal. The best and most successful members of this generation, such as Baron József Eötvös, thought in terms of humankind rather than a narrowly conceived nation. As he wrote to his son in 1869, 'I am convinced that the duty of the individual is to work for the good of humankind according to his abilities, and I believe that if we work on the material and spiritual development of our fellow human beings in a certain circle called our homeland, we are in fact working on the development of the entire human race.'[8]

After the Austro-Hungarian Compromise of 1867, however, Franz Joseph I, Emperor of Austria (1848–1916) and Apostolic King of Hungary (1867–1916), reigned in a period of constant change during the half century of his Dual Monarchy. Innovative spirits flourished in many walks of life; great industrial enterprises sprang up and in their search for a competitive edge founded product-oriented experimental laboratories in fields as diverse as telephony, lighting, pharmaceuticals, armaments and electric locomotion, to name but a few. Later generations were to look back on these as the 'good old days' of peace and prosperity.

At the same time, the likes of Sigmund Freud, Franz Kafka, Robert Musil, Lajos Kossuth and Endre Ady, in their very different ways, were all highly critical of the Habsburg Monarchy and exposed its troubled nature. In many ways, as the poet Endre Ady said, official Hungary was a 'cemetery of souls', where new ideas were still doomed to failure. Unfortunately, the country was unable to sustain either full employment or a modicum of social services. Between the 1880s and the First World War, between one and a half and two million people left Hungary, mainly for the United States, as part of an international migration. Their exodus, in some ways part of a natural and recurrent historical pattern, was essentially economic – a search for work by, in many cases, uneducated, illiterate peasants who hoped that they would be able to remit part of their pay packet to their families back home or, having 'made it', return with whatever earnings they had managed to save. Indeed, the majority of these masses are more properly considered 'birds of passage' rather than true 'emigrants' in the sense that many of them intended only to make money; 'their ultimate goal was to save and return to the place of their birth with their savings'.[9] The majority never wanted to become American citizens and longed to return to their native land. Many, however, were stranded by the First World War and its aftermath, the collapse of the Monarchy and Hungary's subsequent dismemberment and dire economic plight leaving nothing worth returning to.

The economic advance that occurred under the Austro-Hungarian Monarchy did, however, spur the development of the Hungarian language, which was intended to furnish it with an adequate technical vocabulary as a vehicle for professional communication and understanding – again reflecting the willingness of the culture to adapt itself to the modern world. This was particularly noticeable at the universities, where Hungarian again gradually displaced German as the main language of tuition. This process began after the Parliament of 1860–61, as part of a return to the pre-1848 era. This became most noticeable at the Faculty of Law, but German was partly retained, for example, at the Music Academy; the transition proceeded slowly and was not completed until 1914.[10]

Latin had already ceased to be the language of parliamentary and government administration after 1844. Throughout the period of Hungarian political reforms (1825–48), the language of tuition in schools was to have become

Hungarian, though in fact it failed to do so. Neither the language of public administration, nor the language of military leadership had been changed to Hungarian by 1848.[11] Conceived by university professors Franz Exner (Prague) and Hermann Bonitz (Vienna), a new regulation entitled *Entwurf der Organisation der Gymnasien und Realschulen in Oesterreich* was submitted by the minister of religion and education Count Leo Thun-Hohenstein in 1849. Thun tried to make German the language of tuition in the upper classes of high schools, but his attempts lost momentum mainly on account of the opposition of the Hungarian schools. As of 1861, the language of high schools was made Hungarian again. Under the Habsburg Monarchy the majority language of the area was supposed to be the language of tuition. In most of Hungary this meant Hungarian, as far as high schools were concerned.[12]

Creating Budapest

In order to understand the social and cultural background to Baron Loránd Eötvös's career, we need to take a closer look at fin-de-siècle Budapest in all its splendour and squalour. The emergence of Budapest as the capital city of Hungary provided the setting for a rapidly growing, ethnically mixed élite, including the Eötvös family.

Soon after the creation of the Austro-Hungarian Monarchy (1867) and the unification of Buda, Pest and Óbuda into the representative and impressive Hungarian capital city of Budapest (1873), a new, complex and modern Hungarian intellectual élite emerged. Centred in the capital, this modernizing group came partly from the decaying aristocracy and landed gentry of feudal origins and partly from intellectually ambitious members of the assimilating (predominantly German and Jewish) middle class. While creating metropolitan Budapest in the intellectual sense, they constituted themselves as a group through what proved to be a completely new and unique social and psychological experience.

Several economic and social factors contributed to the emergence of this gifted and creative professional group at the time of the rise and fall of the Austro-Hungarian Monarchy (1867–1918). In a country where the gradual decay of feudalism had become visible and the political and social system based on huge landed estates had come under sharp attack, the beginnings of a new, capitalist society stimulated work in science, technology and the arts. The transformation of the Habsburg Monarchy and the dream of a 'Hungarian Empire' contributed to an economic prosperity that brought about a boom in building and transportation, advances in technology, and the appearance of a sophisticated financial system. The rise of a new urban middle class affected the school system.

Assimilation was the key word of the period: religious conversion, the dropping of German, Slavic and particularly Jewish family names, and even occasional ennoblement were standard practices.[13] The capital city played the role of a Hungarian melting pot throughout the four decades preceding the First World War. It attracted a vast number of migrant workers, professionals and intellectuals from every part of the kingdom of Hungary and beyond. It became an energized meeting ground for a multitude of ethnic and religious groups with varying social norms, modes of behaviour and mental patterns. The mixing and clashing, fusion and friction of such diverse values and codes of behaviour resulted in an unparalleled outburst of creativity, a veritable explosion of productive energy. In this exciting and excited ambiance, a spirit of intellectual competitiveness was born which favoured originality, novelty and experimentation. Budapest expected and produced excellence and became deeply interested in the secret of genius. Sensing the approaching demise of the Austro-Hungarian Monarchy seemed to have generated unusual amounts of sensitivity and creativity.[14]

Baron József Eötvös: liberal author and statesman

József Eötvös, born into a distinguished aristocratic family with conservative leanings, was to become one of the leading liberals of Hungary's pre-1848 reform generation. His rich personal library, which still survives in Budapest, shows him to have been a man of immense erudition and sophisticated culture, with an avid interest in world history.[15] Baron Eötvös read a wide variety of international literature, philosophy, history and politics in German, English, French, Italian, Latin and, of course, Hungarian.[16] His scientific knowledge was impressive. He read Alexander von Humboldt's *Cosmos*, as well as Jean Baptiste Joseph Delambre's *Histoire de l'Astronomie*. He was deeply interested in astronomy and geology; in his *The Village Notary*, however, he criticized the phrenological teachings of Franz Joseph Gall and the physiologist Johann Caspar Lavater.[17] A dedicated critic of the Hungarian nobility and of the feudal political system that it upheld, he wrote one of the most significant novels of the pre-March era, the three-volume *A falu jegyzője* (Pest, 1845) (English edition: *The Village Notary*, 1850). As Arthur J. Patterson, a contemporary British observer, put it, Baron József Eötvös was a witness who cannot be accused of partiality to this class since 'his novel, *The Village Notary*, was written to satirize the very institutions to which they were especially attached'.[18] In its own poetic way, describing the misery in which most people lived in early nineteenth-century Hungary, the book contributed to the preparation of the 1848 revolution.[19] The author said of one of his characters:

> He belonged to that section of the Hungarian nobility – alas! It is dying out every day – which in spite of its faults and weaknesses deserves our respect for at least one reason: because it has preserved to the present day the customs and the nationality of our forefathers; their want of culture,

but at the same time their good-hearted simplicity; many prejudices which in our days appear downright ridiculous, but along with them the firm conviction, without which our nation would have disappeared amid its hard struggles for existence – the conviction that *the Magyar shall never perish out of the world.*[20]

Eötvös's novel was a huge success in contemporary Britain, for it gave readers insight into the system of local autonomy which the author advocated.[21]

Eötvös's role in the pre-March era and afterwards was exemplified by his unique membership of both the revolutionary government of 1848, in which he became minister of religion and education under Prime Minister Count Lajos Batthyány, and in the government of the Austro-Hungarian Compromise of 1867, under Count Gyula Andrássy. This great liberal author and statesman considered in 1848 that the establishment of the Hungarian educational system was a duty of the state rather than of the Church – this made him distinctly unpopular with the Catholic Church, which considered the idea of the so-called 'shared schools' an attack against religion.[22] Eötvös did not approve of the radicalization of the Hungarian revolution and left Hungary after the lynching of Lieutenant-General Count Franz Philipp von Lamberg in late September 1848. He went through Vienna to stay in Munich, where he devoted all his energies to literature and philosophy.[23]

It was in 1848 that he started writing his pamphlet *'Über die Gleichberechtigung der Nationalitäten in Österreich'* (On the Equalization of the Nationalities in Austria) in German. This was perhaps the most highly valued of his political pamphlets. The essential feature of nationality, he wrote there, is its autonomous development, which is made possible by its own language.[24] Eötvös proved to be a representative of the idea of federalism. He considered the nationality question a purely legal issue and denied a racist interpretation of the Habsburg Empire. He added that language itself is not nationality, but only a tool to sustain it.[25] Eötvös tried to establish a balance between the existence of Austria and the gratification of the nationality requirements, bringing his ideas close to a balance between *Staatsnation* and *Kulturnation.*[26] Throughout the 1850s, his emphasis was on strengthening the Habsburg Empire – a political idea that ran counter to the national feelings of Hungarians for well over a decade after the lost revolution and war of independence.[27] Historian Gábor Gángó wittily labelled Eötvös in the early 1850s an 'imperial patriot',[28] adding that the most important contemporary critics of the Habsburg Empire all agreed that citizens of the Empire did not have a fatherland and that the Monarchy might easily be dissolved in the not too distant future. Baron József Eötvös, like his contemporaries František Palacký (1798–1876), Joseph Alexander von Helfert (1820–1910), Joseph Chowanetz (1814–88), Victor Franz Freiherr von Andrian-Werburg (1813–58), Franz Schuselka (1811–86), Carl Möring (1818–1900) and Joseph Mathias Graf von Thun-Hohenstein (1794–1863), discussed during and after 1848 the problems engendered by the lack of imperial patriotism and the likely case of the dissolution of Austria along national lines.[29] In his *Gleichberechtigung*, Eötvös made it clear that the ruling ideas of his era included freedom, equality and nationality. He suggested in no uncertain terms that 'where the concept of the state is not synonymous with that of nationality this must by necessity lead to the dissolution of the former'.[30]

Eötvös returned to Buda in 1853 and published a book on the warranties of Austria's power and unity (*Die Garantien der Macht und Einheit Oesterreichs*, 1859), which was interpreted in Hungary as an attempt to make concessions to Austria and was consequently most unpopular. In a separate treatise he discussed Hungary's standing from the point of view of a united Germany (*Die Sonderstellung Ungarns vom Standpunkte der Einheit Deutschlands*, 1859). As of 1861 he started to engage in politics again, became a member of parliament and was, after 1865, most active in preparing the Austro-Hungarian Compromise. His new political weekly, *Politikai Hetilap* (Political Weekly), discussed the issues that had led to the Compromise. He became the president of the Hungarian Academy of Sciences in 1866. He had been vice-president since 1855, so it was a natural for him to become president upon the death of the previous president, Count Emil Dessewffy, in January 1866.

The outbreak of the Austro–Prussian War brought his political views relating to Austria and Hungary closer and closer to the idea of a compromise. By 1866 he had come to the conclusion that 'He who fights the dualistic [i.e. Austro-Hungarian] transformation of the empire actually supports the policies of Bismarck.'[31] It was his vision 'to safeguard the standing of Austria in Germany and thereby grasp the initiative vis-à-vis the unity of Germany and to form this in such a way as to secure the hegemony of Austria'; as he noted in his diary on 18 July 1866:

> To organize the countries of the Crown of St Stephen [i.e. the Kingdom of Hungary] as a separate state, in temporary connection with the other provinces of the empire in terms of certain common affairs, ... to change the connection between us and Austria into a simple personal union – this was the policy which I have declared since the end of the [Austro-] Italian war in memoirs, in my pamphlet *Die Sonderstellung Ungarns aus dem Standpunkte der Einheit Deutschlands* and in all my speeches.[32]

It was his view that the Austro-Hungarian Compromise should not be concluded with the Habsburg Empire, but with 'the peoples of constitutional Austria'.[33] Yet, as I have written elsewhere,

> the Hungarian liberal leadership believed in the creation of a unitary Hungarian national state. They could not reconcile with this the territorial autonomy, collective national rights, and independent political institutions demanded by the minority nationalities. The conflict between the

unity of the state and the demands of a multinational population was a problem the liberal Hungarian nationalists tried to eliminate by equating the concepts of state and nation. The ideals of the liberal thinkers were not reflected in the actions of the Hungarian political leadership, which tried to transform the 'common homeland' into a Hungarian national state.[34]

As minister of religion and education after 1867, József Eötvös 'thought of pursuing politics as a science, considering the government a huge field of praxis, in which to try the correctness of my principles, and I daily make the most interesting experiments and collect the most accidental experiences; otherwise I would have turned long ago against my office with disgust.'[35] Under József Eötvös, one of the first new pieces of legislation was the emancipation of the Jews in 1867.[36] After 1867 he was instrumental in launching the law on elementary education and reforming higher education. In 1867–68 Eötvös was busy developing the technical university, which was given new buildings in Buda and a new type of comprehensive examination. The law academies were provided with a new curriculum, and the mining and forestry academy at Selmecbánya (now Banská Štiavnica in Slovakia) was taken under the supervision of the state. In 1868 Eötvös initiated the establishment of a new university at Kolozsvár (today Cluj, Romania), based on two existing institutions in the city. Eötvös asserted the 'royal', i.e. state, character of the University of Pest, without denying that the university property belonged partly to the Roman Catholic Church. It was under his guidance that the building of the new campus of the University of Pest was begun, to be completed only under the long tenure of his brother-in-law Ágoston Trefort. József Eötvös initiated university reforms in 1870, which led to heated discussions on the nature of the university and its autonomy. He was instrumental in regulating Catholic autonomy. The parliamentary debates on the autonomy of the university were heavily influenced by political discussions and made it difficult for Eötvös to carry through his university reform.[37] The minister established the Jewish Congress of 1868–69 to enable the Jewish population to deal with their own issues, particularly in education.[38] During the debate on the budget in 1870 he was viciously attacked by the opposition; his health broke down as a result and he died in 1871.

Father and son

In order to fully understand Loránd Eötvös's background we should look at the dominant patterns of family structure in fin-de-siècle Austria-Hungary and particularly in Hungary. Upper- and middle-class Hungarian families in the late nineteenth and early twentieth centuries were characterized by the dominant role of fathers, with mothers relegated to the role of preserving the German trinity of *Kinder, Küche, Kirche* (children, kitchen, church). Most families were supported by the single income of the father, who enjoyed ultimate authority. More often than not, fathers had the final word in serious matters such as the education of the children as well as decisions about their marriages and jobs. Indeed, fathers loomed so large in Hungarian as well as Austrian families that one of the most significant issues to be resolved for young people was their relationship with the paterfamilias. The Eötvös family was no exception, as indicated by the correspondence of József and Loránd Eötvös.[39]

Apparently, Sigmund Freud's concept of the dominating father figure was experienced not only in most middle-class families, but also in the aristocracy and the lower nobility. The problem was conceptualized by Freud's notion of the 'father complex'. In his 1899 *Die Traumdeutung* (*The Interpretation of Dreams*), Freud observed that

> even in our middle-class families, fathers are, as a rule, inclined to refuse their sons' independence and the means necessary to secure it, and thus to foster the growth of the germ of hostility which is inherent in their relation. A physician will often be in a position to notice how a son's grief at the loss of his father cannot suppress his satisfaction at having at length won his freedom. In our society today fathers are apt to cling desperately to what is left of a now sadly antiquated *potestas patris familias*; and an author who, like Ibsen, brings the immemorial struggle between fathers and sons into prominence in his writings, may be certain of producing his effect.[40]

As Claudio Magris adds in his 'The Habsburg Myth', the source of Freud's general assumption not only is a basic rule of psychology, but also is recognized today as a particular Austrian social and family structure based on the dominating figure of the father. The patriarchal institution of the family, Magris concludes, reflected the hierarchical order of the Habsburg system.[41]

How did the pattern discussed above relate to the two Barons Eötvös? The older Eötvös maintained regular contact and indeed a tender, loving relationship with his son Loránd while the latter studied in Germany. Written with affectionate care, the correspondence between father and son reflected their devotion to nineteenth-century ideals such as rationalism, progress, enlightened thinking and openness of mind. A touch of social responsibility permeates the letters, some of which were published.[42] 'From my point of view,' Loránd noted in a letter to his father from Heidelberg, dated 15 December 1868,

> political progress is also the result of the progress of the natural sciences. I do not want to mention the influence exerted, through industry, by the

natural sciences on the entirety of social relations, as no one could doubt this influence, which is based primarily on the improvement of the vehicles of communication and the possibilities combined with it, enabling people to live in small spaces in large groups and big cities. The most conspicuous influence is that which the principles of the natural sciences exert directly on the principles of politics, as these alone are capable of destabilizing the dogma. The human mind, which is growing more and more convinced of its power with every step taken by the natural sciences, will gradually acknowledge that what is old is not necessarily purposeful and once we crawl out of the historical tangle, social relations will be organized on the model of the nat[ural] sci[ences] by such laws that originate from these very relations. I think this will be progress.[43]

His father replied in characteristic fashion on 20 December 1868. He agreed with what his son intimated about the significance of the natural sciences but added, with a reference to Goethe's *Faust,*

The human mind and its development, as manifested in the transformation of ideas, is just as much a subject of science as the phenomenon of material nature, and if we abstract from the two-thousand-year history of the human species the laws of the intellectual evolution of people and use these when organizing bourgeois society, this procedure is not a jot less scientific than the one followed by the natural sciences.[44]

The relationship between father and son was different from the hierarchical order seen in many nineteenth-century families, though the father thought it his duty to constantly guide his son via his frequent letters. While the younger Eötvös was studying in Germany, his father found it necessary to educate his son in all walks of life. József cherished his son's personality and valued his many talents.

One of his letters sheds interesting light on his views of their own family: 'We Eötvös are of rather a secretive nature; this is our family fault, and therefore something that we should avoid.'[45] Though he made constant references to his financial difficulties, József noted: 'I would give anything that is needed to your scientific education...as far as I can, I am ready to renounce my own comfort only to enable you to live in a carefree and pleasant way – but I should request you also to renounce certain pleasures for our sake'.[46] József considered his son's doctorate of paramount importance: 'Once you have your doctorate, you may choose your vocation freely. I meddle in it only in so far as I can do so by means of my advice, and I will do anything to make your goals easier to attain, as far as I can with my abilities, but I sincerely ask you to complete your present career before you go any further and choose some other ways – I believe the time may come when you acknowledge the truth of my position.'[47] In a lengthy letter to Loránd, his father showed a measure of dissatisfaction with his son's lack of devotion to the subjects necessary to accomplish his plans for a journey to the Arctic: 'Did you deal enough with the particular branch of sciences necessary for such a trip, to declare yourself duly prepared? Am I unjust in demanding of you ... to be prepared?'[48] In a sense, Baron József Eötvös's resolutely parental voice reflects a mild version of the authoritarian family tradition of the nineteenth century. Even this highly creative and liberal father advised his most imaginative and scientifically most original son, who was to be thrice nominated for the Nobel Prize, not to allow his dreams to have more influence on his decisions than necessary and asked him to 'use all his power to defeat this inclination'.[49]

Hungary and the German cultural tradition

The influence of German culture and Germany as a civilization was so strong in Hungarian history that we must address it in a variety of contexts. His university experiences in Germany also had a huge impact on Loránd Eötvös throughout his life. Both as a language and as a culture, German was a natural accomplishment for upper- and middle-class Hungarians up to the Second World War. The lingua franca of the Habsburg Empire and of the Austro-Hungarian Monarchy, German was used at home, taught at school, spoken on the street, needed in the army.[50] This tradition was more than a century old: the links between Hungary and both Austrian and German culture went back to the eighteenth century. For a considerable period in the eighteenth and nineteenth centuries, Hungary (or large parts of it) lay in many ways on the fringes of the greater realm of German culture.[51] We should emphasize again that the average middle-class 'Hungarian' was typically of German origin (*Schwab*) or of Jewish origin and for him it was German culture and civilization that connected Hungary and the Austro-Hungarian Monarchy with Europe and the rest of the world. Middle-class sitting rooms in Austria, Hungary, Bohemia, Galicia and Croatia typically boasted the complete works of Goethe and Schiller, the poetry of Heine and Lenau, and the plays of Grillparzer and Schnitzler.[52] It was not only that German literary works and translations were read throughout the Empire: German was the language of the entire culture. When Baron József Eötvös visited his daughter in her castle in Földeák, near Makó, in eastern Hungary, he noted: 'What contrasts! I cross Szeged and Makó, then visit my daughter to find Kaulbach on the wall, Goethe on the bookshelf and Beethoven on the piano.'[53]

As already stated above, throughout the Austro-Hungarian Monarchy and beyond, Hungarians looked to Germany to import modern theories and establish modern practices. The study of the German school system had a great tradition throughout the nineteenth century. For generations of Hungarian lawmakers, German schools provided the finest example in

Europe. Two widely spaced examples are characteristic. Young Bertalan Szemere, a future Prime Minister of Hungary, went to Berlin in 1836 to study 'what was best in each country – schools in Germany, public life in France, and prisons in Britain'.[54] A generation later, in the early 1870s, the ideas and know-how of modern teacher training were studied in, and imported from, Germany by Mór Kármán, at the instigation of Baron József Eötvös (see above).

Efforts to study and imitate what was German were also natural because German was then the international language of science and literature: in the first eighteen years of the Nobel Prize, between 1901 and 1918, there were seven German Nobel laureates in chemistry, six in physics, four (and one Austro-Hungarian) in medicine, and four in literature.[55] Loránd Eötvös himself was three times nominated for the Nobel Prize, in 1911, 1914 and 1917.[56] Scholars and scientists read the *Beiträge*, the *Mitteilungen* or the *Jahrbücher* of their field of research or practice, published in some respectable German university town such as Giessen, Jena or Greifswald. The grand tour of a young intellectual, artist or professional would unmistakeably lead the budding scholar to Göttingen or Heidelberg, and increasingly Berlin. Artists typically went to Munich to study under the art professor Karl von Piloty.[57]

Founded in 1854, the authoritative *Pester Lloyd* of Budapest continued as one of the most appreciated and well read papers of the Budapest middle-class until 1944. German in language but committed to Hungarian culture,[58] this daily paper helped to bridge the gap between the two cultures. In much of the eighteenth and nineteenth centuries, German novels and poetry written and published in Hungary were just as integral a part of the Greater German (*Gesamtdeutsch*) literature as anything written in Königsberg or Prague.[59] The Jewish population of the Empire/Monarchy, particularly its educated urban middle class, embraced German primarily as a new, common language and contributed to making the Austrian realm a part and not just an outskirt of German civilization.[60] For Jewish families with social aspirations, German was the language of education and upward mobility.

There was also some influence in the other direction. In 1911, a German monthly on matters Hungarian, very similar to the Hungarian journal *Nyugat*, was launched in Berlin under the title of *Jung Ungarn*.[61] In 1916 a centre for Hungarian culture in Berlin, the Collegium Hungaricum, was founded. Robert Gragger went to teach Hungarian studies at the University of Berlin and became director of the Collegium. He also published the *Ungarische Jahrbücher*, a quality journal presenting Hungarian scholarship. Gragger's Collegium particularly attracted young Hungarians at the beginning of their careers.

Berlin in the pre-war era proved to be an irresistible magnet for the new Hungarian intellectual and professional classes. The Hungarian middle class felt at home in imperial Germany and sent their sons and daughters there to study. After completing their courses in Budapest before the First World War, Hungary's up-and-coming mathematicians saw Göttingen and Berlin as the most important places to study. As a very young man, the celebrated mathematician Lipót Fejér spent the academic year 1899–1900 in Berlin, where he attended the famous seminar of Hermann Amandus Schwarz. In 1902–03 he studied in Göttingen and in subsequent years returned to both universities.[62] A gifted student of Fejér, Gábor Szegő followed in his footsteps and went to study in pre-war Berlin, Göttingen and Vienna, later becoming professor of mathematics at Stanford.[63]

Men of letters followed a similar path in large numbers. The gifted Hungarian poet and future film theoretician Béla Balázs went to Berlin to study with Georg Simmel in 1906 and dedicated his doctoral dissertation '*Az öntudatról*' ('On Self Consciousness', later renamed '*Halálesztétika*', 'The Aesthetics of Death') to his German master.[64] The heroine of Balázs's first literary opus, *Doktor Szélpál Margit*, spent three years in Berlin as a student, a typical pattern in pre-war German-Hungarian relations.[65] Critic, author and art patron Baron Lajos Hatvany studied classics with the prestigious Ulrich von Wilamowitz-Moellendorff in Berlin – an experience he was later to denounce in his sarcastic *Die Wissenschaft des nicht Wissenswerten* (1908), first published in Leipzig, Germany.[66] His second book, *Ich und die Bücher*, was published simultaneously in German in Berlin and in Hungarian in Budapest, in 1910.[67] Others who left Hungary for Berlin included important businessmen such as stock exchange wizard Alfred Manovill, who joined the Berlin bank Mendelssohn and Co. well before the war at the age of twenty-four and acted as the honorary president of the Berliner Hungarian Club until the advent of Hitler.[68]

Baron Loránd Eötvös: physicist and science organizer

The key personality in late nineteenth-century Hungarian science and mathematics, introduced above, was Baron Loránd Eötvös (1848–1919). The younger Loránd was not only a major physicist in his own right, but also one of the truly great organizers of Hungarian science.[69] As mentioned above, his father sent him to study chemistry, physics and mathematics at the (Ruprecht-Karls) University of Heidelberg in 1867–70 under the chemist Robert Bunsen and the physicists Hermann von Helmholtz and Gustav Kirchhoff.[70] This was the heyday of the natural sciences in Heidelberg, a centre for liberal thought and open-mindedness.

Loránd benefited from the intellectual and political legacy of both his father and his uncle, Ágoston Trefort, who for sixteen years continued József Eötvös's great liberal work as Hungarian minister of religion and education (1872–88) and president of the Hungarian Academy of Sciences (1885–88).[71] As a minister, Trefort had a tremendous impact on Hungarian education. He was instrumental in developing the school system and introduced the new law regulating high schools. He focused especially on developing scientific and medical higher education and erected most of the new buildings (including the clinics, laboratories and library) of the University of Budapest[72] as well as the new buildings of the Technical University.[73] He introduced the institution of the seminar for German language and literature at the University of Budapest, where it had been explicitly rejected by Count Thun in 1852. Based on a proposal by Loránd Eötvös of 1878, Trefort connected the seminars to Hungarian teacher training, thereby renewing the latter in Hungary, from the academic year 1887/88 onwards.[74] He initiated the Music Academy of Hungary, inviting Franz Liszt to Budapest to preside over the new institution, which finally opened in 1875. It is indeed difficult to assess and sum up his vast contribution to Hungarian education and especially higher education.[75] His policies took a distinctly Hungarianizing direction without, however, some of the viciously nationalistic tendencies that prevailed in his aftermath.[76]

Loránd Eötvös himself became minister of religion and education, though for a very short time (1894–95), in addition to his distinguished service as president of the Hungarian Academy of Sciences (1889–1905). He paid the utmost attention to every detail in Hungarian higher education. A good example is his note dated 31 January 1878, well before he was appointed to any of his high offices, which shows his great interest in the role of teacher training in the structure of scientifically based Hungarian higher education. The note also reveals Eötvös as an astute observer of Hungarian academia and teacher training.[77] Loránd Eötvös questioned the existing structure of teacher training and asked the faculty of philosophy to find the most suitable way to incorporate it into academia. On the basis of Eötvös's suggestion, the Faculty submitted a proposal to Minister Trefort, making it clear that 'the future of our scientific life and our high schools urgently demands that our university students … finish the university equipped with the principles, method and practice of research and scientific thinking and experience.' It is no exaggeration to claim that the Eötvös-Trefort family dominated Hungarian science and education policies throughout the period between 1866 and 1905.

Aside from his brief time in government, Loránd Eötvös generally stayed away from politics as much as he could. But he joined in the patriotic rhetoric that swept across the country on the death of the national leader of the Hungarian Revolution of 1848–49, Lajos Kossuth (a member of the same Hungarian national government as József Eötvös in 1848), in 1894 – a year that was very near to a major turning point in Hungarian history. The relative stability that had reigned in Hungary after the Compromise with Austria in 1867 was coming to an end towards the close of the century.[78] The growing nationalistic spirit of public opinion was considerably strengthened by the national mourning that followed the death of Kossuth after 45 years in exile and his 1894 funeral in Budapest. At the time of the funeral, the Habsburg court feared a resurrection of the revolutionary spirit of 1848–49.[79] Since Ferenc Deák and József Eötvös, however, nobody had tried to gain the loyalty of the non-Hungarian population of Hungary and much had been done to alienate them.[80] For younger people of non-Hungarian nationalities, it was especially difficult to find their way towards higher education and especially to research.[81] Most of the select few came from the Hungarian gentry, as well as the rising German and Jewish middle class.

In this situation, it is noteworthy that the younger Eötvös abstained from passing a hasty political judgement on the national hero in an increasingly nationalistic Hungary in a Habsburg realm. As President of the Hungarian Academy of Sciences, he characteristically declared on one occasion:

> We stand very far from the passionate struggles of politics... But we cannot become indifferent towards a pain which permeates the hearts of thousands, nay, millions of our nation. ... We are the historians of the Hungarian nation and thus it will be our duty to pass a judgement through the unbiased research of history about his great, immortal figure and to shed some light on those lofty virtues of patriotism which would make his name cherished in the thoughts of every Hungarian without partisanship.[82]

As a scientist, Loránd Eötvös is best remembered today for the Eötvös Rule concerning surface tension and for his gravitational torsion balance measurements, which opened up pioneering ways to identify new sources of energy and particularly of natural gas.[83] His educational achievements are, however, just as important. With his German educational background and inspiration, Eötvös created a small, private Mathematics Circle in Budapest, in the autumn of 1885, to build an informal network among university professors and high-school teachers and their best students.[84] The idea came from his student years in Heidelberg, where 'I learnt the better things I know from the scientific discussions with people of my sort, as sooner or later a discussion emerged and one was forced to gather one's strength and survey his subject independently.' Answering the warnings of his father against joining a German *Burschenschaft* (student fraternity) and going to the German *Kneipe* (tavern), he stated that he had already said a number of times and was completely convinced that German science would not exist without the *Kneipe*.[85] From 1891 onwards, this circle continued as the Mathematikai és Physikai Társulat (Society for Mathematics and Physics) with some three hundred members (including three women). Loránd Eötvös served as the first president of the Társulat, which launched *Mathematikai és Physikai Lapok* (Mathematical and Physical Papers). In his inaugural address, Eötvös expressed his hope that 'we will do great service to the general cultural development of the country, because undoubtedly, the success of teaching in both higher and secondary schools depends above all on the scientific preparation of the teachers.'[86] The special emphasis on the training of mathematics and physics teachers and on the achievements of secondary-school students in Hungary can thus be traced back to Loránd Eötvös. Hungarian-born Mathematics Professor Peter Lax of the Courant Institute of Mathematical Sciences at New York University remembered Eötvös as a professor of his parents, who were joined by a host of students in 'the lecture room just to be able to hear him lecture'.[87] Between 1886 and 1919, all students of medicine in Budapest were taught 'experimental natural sciences' by Eötvös.[88]

In his 1891 inaugural address as the newly installed Rector of Budapest (now Eötvös Loránd) University, Eötvös emphatically argued for independent research and independent thinking: 'I do not call a scientist someone who knows a lot, but someone who does research in science. ... Independence in thinking can only be fostered by a teacher who thinks independently himself, and it is this very independence that is most necessary for the scientist as a man of practice.'[89] His ideas were obviously based on his experiences at some of the best German universities (mainly Heidelberg) in his student years.

When Loránd Eötvös became minister of education in 1894,[90] the appointment was looked upon as opening the way for great scientific opportunities. The time was ripe to launch a new, practical and successful Hungary also in the realm of the sciences. With the approaching national and nationalist celebration of 1896 to commemorate 1000 years of the state of Hungary, it was the moment to impress the world with the country's achievements. Accordingly, Continental Europe's first subway system and largest parliament building were constructed in Budapest, the Royal Palace was rebuilt, three times its previous size, and a host of public buildings, theatres, museums and universities were created, all of them tributes to Hungary's architectural and building skills, innovative spirit in engineering and entrepreneurial excellence.

As students were expected to be challenged in regular national interschool competitions in mathematics and science, the Mathematikai és Physikai Társulat honoured Eötvös in 1894 by naming after him a new annual mathematics and physics contest 'in order to discover those who are exceptional in these fields'.[91] The first and second prizes (the Eötvös Prizes) were awarded to the best secondary-school graduates. As only secondary-school material was included in the test, no additional study was necessary for the exam. Results were reported directly to the Ministry of Religion and Education, along with the names of the teachers of the winners, and were duly published in the *Mathematikai és Physikai Lapok*.

To support preparations for future competitions, 1894 also saw the inauguration of *Középiskolai Mathematikai Lapok* (High School Papers in Mathematics), edited by Dániel Arany (1863–1945), an outstanding high-school mathematics teacher from the West-Hungarian city of Győr.

Between 1896 and 1914, László Rátz (1863–1930), the future mathematics teacher of John von Neumann and Eugene Wigner, continued Arany's editorial work. The problems to be solved covered a wide variety of fields including algebra, calculus, combinatorics, geometry, number theory and trigonometry, and the solutions always required creative thinking. Esteem rather than money was the reward of the best students. The way in which these competitions were organized, along with the related new publications, provided a well structured and carefully regulated framework of preparation for the professional challenges, mostly in Hungary, that these students would face. The language of these journals and competitions was, of course, Hungarian, by then the dominant language of Hungarian education.

The idea of founding awards and contests was not restricted to the Eötvös Prize. For example, it was at the personal instigation of Loránd Eötvös that Andor Semsey, a rich patron of the Hungarian National Museum, established fellowships and prizes in various scholarly and scientific fields to inspire talented young people to conduct research.[92] Also, upon the death of the well known local high-school mathematics teacher Adolf Prilisauer (1859–1913), his city of Kaposvár in Western Hungary, along with his former teaching colleagues, established a prize for the best student (later, students) in mathematics.[93]

The *Középiskolai Matematikai és Fizikai Lapok* (High School Papers in Mathematics and Physics), the *Eötvös Loránd fizikai verseny* (Eötvös Loránd Competition in Physics) and the *Arany Dániel országos matematika verseny* (Arany Dániel National Competition in Mathematics) have survived until today and maintain the tradition of excellent mathematics and science education based on early training, a competitive spirit and the recognition of talent.[94] They also constitute a living tribute to the scientific, teaching and organizing abilities of Loránd Eötvös – and his family.

Conclusion

This aristocratic father and son played an outstanding part in the civic transformation of Hungary from feudalism to capitalism. They had a significant role in building an educated middle class that came partly from the Hungarian gentry and partly from the aspiring and rapidly assimilating German and Jewish bourgeoisie. Members of the liberal elite, the Eötvös family saw the rise and then the decline of liberalism and the building of nationalism in Hungary and tried to reverse the latter two tendencies in their own ways. The Eötvös family, including Ágoston Trefort, were, in Romain Rolland's elevated expression, *au-dessus de la mêlée*,[95] above the crowd, all of them hugely influenced by the German cultural tradition, which introduced science and scientific thought into a largely uncultured and uneducated Hungary.

The Hungarian political elite, especially in the first part of the nineteenth century, included several other outstanding representatives of a liberal-minded aristocracy: Count István Széchenyi, Baron Miklós Wesselényi, Counts Lajos and Kázmér Batthyány, Count Gyula Andrássy, Count László Teleki, Count István Károlyi, Baron Zsigmond Kemény and Baron Miklós Jósika, to mention only a few. The Eötvös, father and son, together with József Eötvös's brother-in-law, Loránd's uncle Ágoston Trefort, recognized the fact that there is no duality or opposition between universal science and national science. Universalism and national growth may only develop together, József Eötvös suggested.[96] The Eötvös family did their very best to create the infrastructure of modernity through educational reform, to develop higher education and to contribute to international science. On top of this, they provided a lasting example of how the elite may use its international experience and unique knowledge to enhance the level of culture and education in a country that was underdeveloped and essentially feudal in the midst of a modernizing Europe.

Notes

1. George Pólya interviewed by Agnes Arvai Wieschenberg (1984) 'Identification and Development of the Mathematically Talented – The Hungarian Experience', PhD dissertation, Columbia University, 86–87.
2. George Pólya (1961) 'Leopold Fejér', *Journal of the London Mathematical Society*, 36, 501.
3. Parts of this chapter have been taken from my book, Tibor Frank (2009) *Double Exile: Migrations of Jewish-Hungarian Professionals through Germany to the United States, 1919–1945* (Oxford: Peter Lang). For specific references, see this work.
4. Mór Kármán (1880) 'A történeti fejlődés útja' [The Way of Historical Development], in Szilárd Faludi (ed.), *Mór Kármán válogatott pedagógiai művei* [Selected Educational Works by Mór Kármán] (http://mek.niif.hu/07600/07677/07677.pdf, downloaded 3 June 2011), 3–19. This and all subsequent translations are by Tibor Frank, unless otherwise stated.
5. Mór Kármán, *Az oktatás (értelmi nevelés) feladatai* [The Tasks of Teaching (Intellectual Education)], in *Mór Kármán válogatott pedagógiai művei*, 38–45, here 39.
6. Mór Kármán (1908 [1880]) *Általános utasítás a gimnáziumi tanítás tervéhez* [General Instruction for the Study Plan of the Gymnasium], in *Mór Kármán válogatott pedagógiai művei*, 70–87, here 71.
7. Mór Kármán (1896) *A pedagógia helye a tudományok sorában* [The Place of Education among Sciences], in *Mór Kármán válogatott pedagógiai művei*, 20–27, here 20.
8. József Eötvös to Loránd Eötvös, 2 February 1869, in Elek Környei (ed.) (1964) *Eötvös Loránd. A tudós és művelődéspolitikus írásaiból* [Loránd Eötvös. From the Writings of the Scientist and the Cultural Politician] (Budapest: Gondolat Kiadó), 36.
9. Julianna Puskás (2000) *Ties That Bind, Ties That Divide: One Hundred Years of Hungarian Experience in the United States* (New York, London: Holmes & Meier), 304. For the financial implications see Neil Larry Shumsky (1995) '"Damned if You Do and Damned if You Don't": American Hostility to Return Migration, 1890–1924,' in Daniela Rossini (ed.) *From Theodore Roosevelt to FDR: Internationalism*

and Isolationism in American Foreign Policy (Ryburn Publishing, Keele University Press), 93–112.

10. József Kardos, Elemér Kelemen and László Szögi (2000) A magyar felsőoktatás évszázadai [Centuries of Hungarian Higher Education] (Budapest: Nemzeti Tankönyvkiadó), 74. On teaching in German at the Music Academy see Antal Molnár (1958) Eszmények. értékek, emlékek [Ideals, Values, Memories] (Budapest: Zeneműkiadó, 1981), 91.

11. Ezerszáz év a magyar történelemben [One Thousand and One Hundred Years in Hungarian History], Reformkor – a magyar nyelv ügye [The Age of Reform – The Issue of the Hungarian Language] (http://ezerszazev.uw.hu/1848_magyarnyelvugye. html, downloaded 3 June 2011).

12. Péter Tibor Nagy, A műveltség demokratizálása a magyarországi elitoktatásban (1867–1918) [The Democratization of Culture in Hungarian Elite Education, 1867–1918] (www.hier.iif.hu/en/letoltes.php?fid=kutatasok/94, downloaded 3 June 2011).

13. Cf. József Gerő (ed.) (1940) A királyi könyvek. Az I. Ferenc József és IV. Károly által 1867-től 1918-ig adományozott nemességek, főnemességek, előnevek és címerek jegyzéke [Royal Books. A List of Persons Who Received Noble and Aristocratic Ranks, Titles, and Coats of Arms by Kings Ferenc József I and Károly IV between 1867 and 1918] (Budapest).

14. László Mátrai (1976) Alapját vesztett felépítmény [Superstructure Without Base] (Budapest: Magvető); Kristóf Nyíri (1980) A Monarchia szellemi életéről [The Intellectual Life of the Monarchy] (Budapest: Gondolat); J.C. Nyíri (1988) Am Rande Europas. Studien zur österreichisch-ungarischen Philosophiegeschichte (Vienna: Böhlau); Péter Hanák (1998) The Garden and the Workshop. Essays on the Cultural History of Vienna and Budapest (Princeton, NJ: Princeton University Press); Károly Vörös (ed.) (1978) Budapest története [The History of Budapest], vol. IV (Budapest: Akadémiai Kiadó), 321–723; John Lukacs (1988) Budapest 1900: A Historical Portrait of a City and Its Culture (New York: Weidenfeld and Nicolson); Mary Gluck (1985) Georg Lukács and His Generation 1900–1918 (Cambridge, Mass., London: Harvard University Press); István Hargittai (2006) The Martians of Science: Five Physicists Who Changed the Twentieth Century (New York: Oxford University Press), 3–31.

15. Gábor Gángó (ed.) (1995), Die Bibliothek von Joseph Eötvös (Budapest: Argumentum). The library of József Eötvös is in the custody of the Institute of Literature of the Hungarian Academy of Sciences.

16. Miklós Bényei (1972), Eötvös József olvasmányai [The Readings of József Eötvös] (Budapest: Akadémiai Kiadó).

17. Miklós Bényei (1978) 'Eötvös József természettudományos műveltsége' [The scientific culture of József Eötvös], in Imre Dankó (ed.) Árkádia. Antológia a Déri Múzeum Baráti Köre fennállása 50. évfordulójára [Arcadia. An Anthology on the Occasion of the 50th Anniversary of the Circle of Friends of the Déri Museum] (Debrecen: Déri Múzeum), 116–24.

18. Arthur J. Patterson (2004) The Magyars: Their Country and Institutions (London: Smith, Elder & Co., 1869), vol. I, 271. Cf. Tibor Frank (2004) 'Arthur J. Patterson and the Austro-Hungarian Settlement of 1867', in László Péter and Martin Rady (eds.) British–Hungarian Relations Since 1848 (London: Hungarian Cultural Centre – School of Slavonic and East European Studies, University College), 47–62.

19. István Sőtér, Eötvös József (2nd ed. Budapest: Akadémiai Kiadó), 169.

20. József Eötvös, The Village Notary, quoted by Arthur J. Patterson in The Magyars, vol. I, 271. Cf. István Sőtér, Eötvös József, 137–69. Emphasis in the original.

21. Tibor Frank (2005) *Picturing Austria-Hungary: The British Perception of the Austro-Hungarian Monarchy 1865–1870* (New York: Columbia University Press), 132, 375.
22. Béla Köpeczi (1993) 'Eötvös József és a magyar művelődés' [József Eötvös and Hungarian culture], in *Eötvös József a művelődéspolitikus* [József Eötvös the Cultural Politician] (Budapest, Ercsi: Országos Pedagógiai Könyvtár és Múzeum, Ercsi Önkormányzat), 5–9.
23. József Eötvös (1851, 1854) *A XIX. század uralkodó eszméinek befolyása az álladalomra*, vols. 1–2 (Bécs és Pest) [*Der Einfluß der herrschenden Ideen des 19. Jahrhunderts auf den Staat* (Leipzig: F.A. Brockhaus, 1854]). See Gábor Gángó (1999), *Eötvös József az emigrációban* [József Eötvös in Exile] (Debrecen: Kossuth Egyetemi Kiadó).
24. [Joseph Eötvös] (1850) *Über die Gleichberechtigung der Nationalitäten in Österreich* (Pest: C.A. Hartleben), 8, quoted by Gábor Gángó, *Eötvös József*, 147.
25. Gábor Gángó, *Eötvös József*, 121–47.
26. Ibid., 177. For further discussion of the distinction between *Staatsnation* and *Kulturnation*, see the chapter by Johannes Feichtinger in this volume.
27. For many years after 1945, dominant Hungarian literary criticism was not kind to Eötvös the politician as opposed to Eötvös the novelist. See István Sőtér (1965) 'A kritikai realizmus első klasszikusa: Eötvös József' [The first classic of critical realism: József Eötvös], in Pál Pándi (ed.) *A magyar irodalom története 1772-től 1849-ig* [The History of Hungarian Literature 1772–1849], vol. III of István Sőtér, gen. ed., *A magyar irodalom története* [The History of Hungarian Literature] (Budapest: Akadémiai Kiadó), 629–31.
28. Gábor Gángó, *Eötvös József*, 121.
29. Ibid., 121–29.
30. Joseph Eötvös, *Über die Gleichberechtigung der Nationalitäten in Österreich*, 36, quoted in Gángó, *Eötvös József*, 164.
31. József Eötvös to Miksa Falk, 17 June 1866, quoted in Tibor Frank (2005) 'Eötvös és Carneri: Ismeretlen osztrák-magyar véleménycsere a dualista berendezkedésről' [Eötvös and Carneri: An unknown Austro-Hungarian discussion of the dualist establishment], *Aetas* 20, 3, 150.
32. József Eötvös, *Naplójegyzetek – Gondolatok 1864–1868* [Diary Notes – Thoughts, 1864–1868], ed. Imre Lukinich (Budapest: MTA, 1941), 181, quoted by Tibor Frank in 'Eötvös és Carneri', 150.
33. György Szabad (1971) 'Eötvös József a politika útjain' [József Eötvös on the roads of politics], *Századok*, 105/3–4, 667.
34. Tibor Frank (1994) 'Hungary and the Dual Monarchy, 1867–1890', in Peter F. Sugar, Péter Hanák and Tibor Frank (eds.) *A History of Hungary* (Bloomington and Indianapolis: Indiana University Press), 255.
35. József Eötvös to Loránd Eötvös, 2 February 1869, in Elek Környei (ed.) (1964) *Eötvös Loránd*, 37.
36. Tibor Frank (ed.) (2006) *Honszeretet és felekezeti hűség. Wahrmann Mór 1831–1892* [Love of the Homeland and Loyalty to the Denomination: Mór Wahrmann 1831–1892] (Budapest: Argumentum), 58, 73, 111.
37. Kardos, Kelemen and Szögi, *A magyar felsőoktatás évszázadai*, 77.
38. Brigitta Eszter Gantner (2006) 'Az egyetemes izraelita kongresszus 1868–69-ben. Küzdelem az asszimilációért és ellene' [The Universal Jewish Congress in 1868–69. Struggle for and against assimilation], in Tibor Frank (ed.) *Honszeretet és felekezeti hűség*, 91–110.
39. Mihály Benedek (ed.) (1988) *Eötvös József levelei fiához, Eötvös Lorándhoz* [The Letters of József Eötvös to his Son, Loránd Eötvös] (Budapest: Szépirodalmi Könyvkiadó).

40. Sigmund Freud (1965) *The Interpretation of Dreams*, trans. James Strachey (New York: Avon Books), 290. One could easily add several other literary examples to Freud's reference to Ibsen, from German, Austrian and Hungarian literature.

41. Claudio Magris (1963) *Il mito absburgico nella letteratura austriaca moderna* (Torino: Einaudi); Hungarian ed. *A Habsburg-mítosz az osztrák irodalomban* (Budapest: Európa, 1988), 91.

42. Benedek (ed.) *Eötvös József levelei fiához, Eötvös Lorándhoz*. I have also studied Loránd's letters, originally in the custody of Patrona Hungariae Gyakorló Általános Iskola, Gimnázium, Diákotthon, Zene- és Szakiskola, Budapest, copies of which are today in the University Archive of Eötvös Loránd University, Budapest. I am grateful to Dr. László Szögi, Director of the University Library and Archive of Eötvös Loránd University for kindly providing access to the Eötvös documents in the custody of the Archive.

43. Loránd Eötvös to József Eötvös, Heidelberg, 15 December 1868. University Archive of Eötvös Loránd University, Budapest. Cf. note 33; in his presidential inaugural speech at the 1886 general assembly of the Hungarian Academy of Sciences, Loránd's uncle Ágoston Trefort used closely related arguments: 'Among sciences today the greatest power is natural science, which completely transformed human thinking [and] extinguished prejudices and superstition. ... Natural science is an absolutely necessary organ of culture. The history of natural science is the history of mankind', quoted in Miklós Mann (1982) *Trefort Ágoston élete és működése* [The Life and Work of Ágoston Trefort] (Budapest: Akadémiai Kiadó), 187.

44. József Eötvös to Loránd Eötvös, 20 December 1868, in Benedek (ed.) *Eötvös József levelei fiához*, 43–44.

45. József Eötvös to Loránd Eötvös, 20 November 1869, ibid., 79.

46. József Eötvös to Loránd Eötvös, 29 November 1869, ibid., 82–83.

47. József Eötvös to Loránd Eötvös, 25 December 1869, ibid., 91.

48. József Eötvös to Loránd Eötvös, December 1869, fragment, ibid., 97.

49. József Eötvös to Loránd Eötvös, 25 December 1869, ibid., 90.

50. István Deák (1990) *Beyond Nationalism: A Social and Political History of the Habsburg Officer Corps, 1848–1918* (New York, Oxford: Oxford University Press), 83, 89, 99–102.

51. Zoltán Farkas (1914), *A biedermeier* [The Biedermeier] (Budapest: Singer és Wolfner), 5.

52. Cf. Gyula Illyés (1938) *Magyarok. Naplójegyzetek* [Hungarians. Notes of a Diary], 3rd ed. (Budapest: Nyugat, n.d.), vol. II, 239.

53. István Sőtér (1967) *Eötvös József*, 314.

54. Journal entry from Berlin, 31 October 1836. Cf. Bertalan Szemere (1983) *Utazás külföldön* [Travelling Abroad] (Budapest: Helikon), 59.

55. *The World Almanac and Book of Facts 2011* (New York: World Almanac Books), 266–68.

56. Mihály Beck (1995) 'A Nobel-díj és a magyar Nobel-díjasok' [The Nobel Prize and the Hungarian Nobel Laureates] (http://www.kfki.hu/chemonet/hun/teazo/nobel/nobeldij.html, downloaded 6 May 2009); cf. Mihály Beck (1995) 'A Díj és a magyarok' [The Prize and the Hungarians], *Természet Világa*, 126, 531–35.

57. Károly Lyka (1982) *Magyar művészélet Münchenben* [Hungarian Artist-Life in Munich], 2nd ed. (Budapest: Corvina); László Balogh (1988) *Die ungarische Facette der Münchner Schule* (Mainburg: Pinsker-Verlag).

58. József Kiss (1969) 'Petőfi in der deutschsprachigen Presse Ungarns vor der Märzrevolution', in *Studien zur Geschichte der deutsch-ungarischen literarischen Beziehungen* (Berlin), 275–97.

59. László Tarnói (1998) *Parallelen, Kontakte und Kontraste. Die deutsche Lyrik um 1800 und ihre Beziehungen zur ungarischen Dichtung in den ersten Jahrzehnten des 19. Jahrhunderts* (Budapest: ELTE Germanisztikai Intézet), 203–322; Ulrik R. Monsberger (1931) *A hazai német naptárirodalom története 18 21-ig* [A History of German Calendar Literature in Hungary to 1821] (Budapest).

60. György Szalai (1974) 'A hazai zsidóság magyarosodása 1849-ig' [The Magyarization of the Hungarian Jewry to 1849], *Világosság* 15, 216–23; Róza Osztern (1930) *Zsidó újságírók és szépírók a magyarországi német nyelvű időszaki sajtóban, a 'Pester Lloyd' megalapításáig, 1854-ig* [Jewish Journalists and Authors in the German Periodical Press of Hungary, up to the Foundation of the Pester Lloyd in 1854] (Budapest).

61. Széchenyi Ágnes (2009) 'A *Nyugat* német mása: *Jung Ungarn*, Berlin 1911' [The German version of *Nyugat*: *Jung Ungarn*, Berlin, 1911], in *Nyugat népe. Tanulmányok a Nyugatról és koráról* [The People of the West. Studies on the Nyugat and its Era] (Budapest: Petőfi Irodalmi Múzeum), 178–99.

62. Gábor Szegő (1960) 'Leopold Fejér: In Memoriam, 1880–1959', *Bulletin of the American Mathematical Society*, 66/5 (September), 346–47. On the 'peregrination' of Hungarian students to German universities, see László Szögi (2006) 'Studenten aus Ungarn und Siebenbürgen an den deutschen Universitäten, 1789–1919', in Márta Fata, Gyula Kurucz and Anton Schindling (eds.) *Peregrinatio Hungarica. Studenten aus Ungarn an deutschen und österreichischen Hochschulen vom 16. bis zum 20. Jahrhundert* (Contubernium. Tübinger Beiträge zur Universitäts- und Wissenschaftsgeschichte, 64) (Munich: Franz Steiner Verlag).

63. [Gábor Szegő] 'Lebenslauf', Gábor Szegő Papers, SC 323, Boxes 85–036. Department of Special Collections and University Archives, Stanford University Libraries, Stanford, CA.

64. Herbert Bauer (1908) *Az öntudatról* [= Béla Balázs, *Halálesztétika*] (Budapest: Deutsch Zsigmond).

65. Béla Balázs (1909) *Doktor Szélpál Margit* (Budapest: Nyugat), 10. Cf. Péter Kozák (2011), *Doktor Szélpál Margit. Az első tudósnő magyar színpadon* [Doktor Szélpál Margit. The first woman scientist on the Hungarian stage] (Budapest: Gabbiano Print).

66. Ludwig Hatvany (1914) *Die Wissenschaft des nicht Wissenswerten* (Leipzig: Julius Zeitler, 1908; 2. Auflage, Munich: Georg Müller).

67. Ludwig Hatvany (1910) *Ich und die Bücher (Selbstvorwürfe des Kritikers)* (Berlin: Paul Cassirer); in Hungarian: Lajos Hatvany, *Én és a könyvek* (Budapest: Nyugat).

68. 'Alfred Manovill 50 Jahre' [(German) manuscript of a newspaper article in the Michael Polanyi Papers], Box 20, Folder 2, Department of Special Collections, University of Chicago Library, Chicago, Ill.

69. For further discussion of this topic, see the chapter by Gábor Palló in this volume.

70. Promotionsakten von Roland von Eötvös, Ruprechts-Karl-Universität, Universitätsarchiv, Heidelberg, UAH H-IV-102/72; copies in the custody of the University Archive, Eötvös Loránd University, Budapest. Cf. István Rosta (2008) *Eötvös Loránd* (Budapest: Elektra), 11.

71. Mann, *Trefort Ágoston*.

72. László Szögi (2010) *The Illustrated History of the Eötvös Loránd University* (ELTE) (Budapest: Eötvös Loránd Tudományegyetem), 73.

73. Csaba Borsodi and Anna Tüskés (2010) *Az Eötvös Loránd Tudományegyetem Bölcsészettudományi Karának története képekben 1635–2010* [An Illustrated History of the Faculty of Humanities of Eötvös Loránd University, 1635-2010] (Budapest: Eötvös Loránd Tudományegyetem Bölcsészettudományi Kara), 110.

74. Imre Szentpétery (1935) *A Bölcsészettudományi Kar története 1635–1935* [The History of the Faculty of Humanities, 1635–1935] (Budapest: Királyi Magyar Egyetemi Nyomda), 410, 503–14.

75. Ibid., 497–552.

76. Mann, *Trefort Ágoston*. Trefort, however, made every effort to preserve Magyar leadership within the borders of the Kingdom of Hungary. This is clearly shown by an unpublished document he sent to the Lutheran superintendent Lajos Geduly on 5 June 1875 banning a Slovak Lutheran school textbook that 'denied the sovereignty of our homeland and contained false doctrines directed against the constitution' (No. 10418), Tibor Frank private collection, Budapest.

77. Loránd Eötvös, Note from Budapest, 31 January 1878, Tibor Frank private collection, Budapest. Partially published by Imre Szentpétery (1935), with additional documents, op. cit., 506–07.

78. Géza Jeszenszky (1994) 'Hungary through World War I and the End of the Dual Monarchy', in Sugar, Hanák and Frank (eds.) *A History of Hungary*, op. cit., 267.

79. Ibid., 269.

80. Ibid., 270. See also note 75.

81. Ibid., 287–88.

82. 'Megemlékezés Kossuth Lajos haláláról, a Magyar Tudományos Akadémia rendkívüli összes ülésén, 1894. március 28' [Commemorating the death of Lajos Kossuth at the extraordinary plenary session of the Hungarian Academy of Sciences, 28 March 1894]. Published by Elek Környei (ed.) (1964) *Eötvös Loránd*, 215–16.

83. Loránd Eötvös (1912) 'Bericht über Arbeiten mit der Drehwage ausgeführt im Auftrage der kön. ung. Regierung in den Jahren 1909–1911', *Verhandlungen der XVII. allgemeinen Konferenz der internationalen Erdmessung in Hamburg*, vol. I, 427–38. Cf. Gy. Tóth, L. Völgyesi (2003) 'Importance of Eötvös torsion balance measurements and their geodetic applications', *Geophysical Research Abstracts*, vol. 5, 13217. For the collected works of Eötvös, see Loránd Eötvös (1953) *Gesammelte Arbeiten*, ed. P[ál] Selényi (Budapest: Akadémiai Kiadó), LXXX + 384 p. On the ever-growing Eötvös literature see Sándor Mikola and János Renner (eds.) (1930) 'Irodalom' [Bibliography], in Izidor Fröhlich (ed.) (1930) *Báró Eötvös Loránd Emlékkönyv* [Baron Loránd Eötvös Memorial Volume] (Budapest: Magyar Tudományos Akadémia), 287–317; Elek Környei (ed.) (1964) *Eötvös Loránd*, 379–421.

84. Eötvös also surrounded himself with a circle of fellow Hungarian physicists in Heidelberg. See Gyula Radnai (1991) 'Az Eötvös-korszak' [The Eötvös era], *Fizikai Szemle*, 10, 349.

85. Fröhlich (ed.) *Báró Eötvös Loránd*, 29, 44–45.

86. Loránd Eötvös (1892) 'Szaktársainkhoz' [To our colleagues], *Mathematikai és Physikai Lapok*, 1/1, quoted by Agnes Arvai Wieschenberg (n. 1), 23.

87. Interview of Peter Lax, 3 May 1983, quoted by Ágnes Árvai Wieschenberg (Note 1), 56.

88. See the annotations of Eötvös in the course book of the present author's grandfather, Herman Flesch, Royal Hungarian University of Budapest, 1897–98, 5, 7. Tibor Frank private collection, Budapest.

89. Loránd Eötvös (1964) inaugural speech as Rector of the University of Budapest, 15 September 1891, in Környei, *Eötvös Loránd*, 201–02.
90. Emperor Franz Joseph I to Baron Loránd Eötvös, 10 June 1894, Prime Minister Sándor Wekerle to Baron Loránd Eötvös, 10 June 1894, Archive of Eötvös Loránd University, Budapest. Cf. note 33.
91. 'Értesítő a Mathematikai és Physikai Társulat választmányának f. é. Június hó 22-ikén tartott üléséről' [Minutes of the 22 June meeting of the Mathematical and Physical Society], *Mathematikai és Physikai Lapok*, 3, 1894, 197–98, quoted by Ágnes Árvai Wieschenberg (Note 1), 26.
92. Izidor Fröhlich, 'Báró Eötvös Loránd emlékezete' [The Memory of Baron Loránd Eötvös], in Fröhlich (ed.), *Báró Eötvös Loránd Emlékkönyv*, 47–48.
93. Gyula Kovács-Sebestény and Károly Pongrácz (1913) 'Felhívás' [Appeal], Kaposvár, June 1913. *A kaposvári Magyar Királyi Állami Főgimnázium Emlékkönyve 1812–1912* [Centenary Memorial of the Hungarian Royal State High School at Kaposvár] (Kaposvár: Szabó Lipót Könyvsajtója), 177–78.
94. János Gordon Győri, Mária Halmos, Katalin Munkácsy and Józsefné Pálfalvi (eds.) (2007) *A matematikatanítás mestersége. Metertanárok a matematikatanításról* [The Arts and Crafts of Mathematics Education. Master Teachers on Teaching Mathematics] (Budapest: Gondolat Kiadó).
95. Romain Rolland (1914) 'Au-dessus de la mêlée', *Journal de Genève*, supplement, 22 September.
96. Péter Hanák (1993) 'A tudomány kettős kötődése Eötvös József eszméiben' [The double bind of science in the ideas of József Eötvös], in Ferenc Glatz (ed.) *A tudomány szolgálatában. Emlékkönyv Benda Kálmán 80. születésnapjára* [Serving Scholarship: *Festschrift* for the 80th Birthday of Kálmán Benda] (Budapest: MTA Történettudományi Intézet), 348–51.

7
Patriotism, Nationalism and Internationalism in Czech Science: Chemists in the Czech National Revival

Soňa Štrbáňová

Introduction

In the multiethnic and multicultural space of the Habsburg Monarchy, the interrelations between patriotism, nationalism and internationalism in science expressed themselves in a very complex way and were sources of profound tensions. Tensions included inconsistencies between the inherent international character of science, on the one hand, and science operating as 'national science' in its social and political embodiment within scientific institutions defined by usage of the national language and using 'national' communication systems, on the other. The same applies to the fact that science has always been an international enterprise, but individual scientists have identified themselves with an ethnic group or a nation, which meant that they were anchored in nationally defined institutions where they used the language of their ethnic group for teaching as well as writing articles and books.

Carol Harrison and Ann Johnson have pointed out that 'research on nationalism has largely ignored the nexus between science and national identity', and that although 'scholars have lavished attention on the historical project of nation-building, calling attention to its complexity', they have been 'remarkably reticent about the place of science and technology in the construction of national identities'.[1] In accordance with this statement, historians have habitually interpreted the Czech National Revival in the nineteenth century as a matter of cultivating the Czech language and elevating it to the language of intellectuals, thus identifying the movement primarily with progress in education, literature and the fine arts. It is also necessary to point out that even in cases where historians have dealt with the role of science and scholarship in the creation of national identity during the Czech National Revival, they have usually had in mind the humanities – especially history, linguistics, philology and related fields – while the natural sciences have been neglected. Jan Janko and I have tried to fill this gap in our book about the role of science in the Czech National Revival up to the 1870s.[2] There we demonstrated how the cultural, social and political emancipation of the modern Czech nation was closely linked to scientific and technological development in all its representations, including research, communication, education and institutionalization. This emancipation meant above all drawing a demarcation line between the Czechs and the Germans, who originally formed a socially, politically and economically stronger layer of society in the Czech Lands.[3]

Language did indeed become the leading symbol of national movements all over Europe in the nineteenth century, including the Czech National Revival, and this symbol also asserted itself in science; one of the most important items on the agenda of the Czech National Revival was the creation of a modern Czech scientific language.[4] This effort had political as well as practical goals. The Revivalists aimed to show that the Czech language was able to express even the most complex scientific ideas and at the same time serve as a tool of teaching and communication. In reality this meant pursuing linguistically Czech science and establishing Czech scholarly institutions in which the language of teaching and publishing would be Czech.[5] Another goal of the Revival was the education of broad strata of the Czech population and the popularization of scientific achievements. Also at the practical level, we should keep in mind that the nineteenth century, especially its second half, was the century of the advancement of industry and trade, with its demand for specialists at all levels of education. The common people, however, did not speak German; therefore Czech terminology was needed to open to them the road to schooling, and Czech-language terminology in the educated professions became an indispensable tool for this purpose.

In the nineteenth-century Czech Lands not only language but also political symbols and ideas, especially the ideas of nation and patriotism, found their reflection in science. Patriotism took on multiple forms, including 'provincial', 'state', 'national' and 'Slavic', with differing accentuation in different periods. In the second half of the nineteenth century, patriotism became allied with nationalism in both the Czech- and German-speaking environments. The narrow-minded national antagonism that developed in the 1880s infiltrated all layers of society in the Czech Lands, including the Czech and German communities. In this article, I will focus on the attitude of the Czech scientific community towards this ever-increasing Czech–German antagonism and militant nationalism in Czech society in the last third of the nineteenth century, examining especially the position of chemists.

Ethnic division in the chemical community in the Czech Lands in the second half of the nineteenth century

In the second half of the nineteenth century chemistry was more developed in the Czech Lands than in other parts of the Habsburg Empire, whether we consider the level of university education or research, [6] and the Empire's chemical industry was sited mainly in the Czech Lands (Bohemia and Moravia).[7] According to some sources, 75 per cent of the Austro-Hungarian chemical industry was located in the territory of the future Czechoslovakia,[8] mostly in Bohemia. In 1880, the Czech Lands' share of Austria-Hungary's total production of food was 65.5 per cent, of sugar 95 per cent and of chemicals 37.7 per cent.[9] In Bohemia the share of total production in the textile industry, which was closely related to the chemical industry, was 42 per cent,[10] while, as Jan Havránek writes, the 'second strongest branch was ... the food industry, which in 1880 represented 33 per cent of the total value of all industrial production in Bohemia and was important in Moravia, as well.'[11] While industries connected with agriculture (food, fermentation and sugar industries) had mostly Czech proprietors, the production of inorganic and organic chemicals and of textiles was predominantly in German hands. According to Havránek, 'The sugar industry became as important a source for the accumulation of Czech capital as textile manufacturing was for the accumulation of German capital in Bohemia.'[12]

Likewise, the Czech chemical community gradually established an institutional base separate from the German one. In the 1860s and 1870s, when the fall of the Bach absolutist government enabled a revitalization of political, social and public life, Czech chemists founded their first chemical societies. The earliest one, the Isis Association for Natural Sciences,[13] started in 1866 as the predecessor of the Society of the Czech Chemists,[14] formed in 1872, which still exists today.[15] It had an almost exclusively Czech membership[16] and was mostly sponsored by wealthy owners or directors of the sugar factories (so-called founding members);[17] hence the journal of the Society, *Listy chemické*, served until 1883 not only scientific objectives but also the needs of the sugar-refining community as its publication and information base [18] At the same time, German chemists did not have a specialized professional chemical association[19] in the Czech Lands, although the scientific association Lotos,[20] founded in 1848,[21] also included chemists. University education in chemistry in the Czech language took place at the separated Czech Polytechnics[22] from 1869 and at the Czech University, Prague, which came into existence in 1882 when the Karl-Ferdinands-University divided into independent Czech and German counterparts.[23] In Brno (Brünn)[24] Czech chemical education began in 1899, when the Czech Technical University was founded.[25] All these actions brought forth a differentiation and gradually a complete separation of the Czech and German chemical communities. The establishment of the Czech Academy of Sciences, Letters and Arts[26] in 1890[27] is considered the culmination of the efforts of the Czech National Revival to create a network of Czech educational and scientific institutions consisting of Czech secondary schools, scientific societies and journals, and universities.[28] Although the Academy had rich financial resources for supporting research and publishing, its possibilities were limited as most research took place at universities and industrial laboratories. Another serious constraint was the conservatism of the Academy, which was reflected in its programme, structure and composition. The division of the Academy into four classes reflected a preference for the humanities and arts: only the Second Class (to which the chemists belonged) was dedicated to mathematical, natural and medical sciences, a perplexing disproportion at a time of vigorous development of the sciences in Europe.[29]

The consolidation of the institutional base of linguistically Czech science in the 1860s–1880s, including chemistry, also led to growing tensions in the Czech scientific community. On the one hand, it came under the increasing pressure due to nationalistic attitudes in all strata of Czech society, but on the other it endeavoured to become an integral part of the international scientific community. Most Czech scientists were loyal to the Monarchy[30] but feared the increasing influence of Germany on Austria and therefore nurtured the Slavic idea as a counterweight.[31] In the 1890s patriotism in science evolved in some circles into open nationalism, chauvinism and extreme isolation from German science, as mentioned above. Official contacts between Czech and German universities and other institutions in the Czech Lands practically ceased to exist, and personal links between Czech and German academics became rare. Determined nationalism was typical especially for the Czech Academy, which was intentionally founded as an exclusively Czech learned society,[32] accepting only Czechs as regular members, using only the Czech language for communication and aimed especially at cultivating Czech literature and art. In this way the Czech Academy demarcated itself from the traditionally bilingual Royal Czech Society of Sciences (founded in 1784) and other nationally undefined or liberal scientific societies. Bilingualism was considered inadmissible not only by Josef Hlávka,[33] but also some scientists, like the pharmacologist Karel Chodounský.[34] Anti-German sentiments manifested themselves, for instance, in the election of the foreign, so-called non-resident members of the Second Class of the Academy, as can be seen in Table 7.1.

As the table shows, in the years 1891–1908 the Second Class elected 14 so-called non-resident or foreign members. The selection 'process', mostly arbitrary, favoured candidates related or known to the members of the Academy; with a few exceptions they were not prominent scholars. In accordance with the Academy's nationalistic attitudes, we can see among the foreign members only two Germans – citizens of the German Empire: Karl Josef Küpper, former professor of descriptive geometry at the Prague German Technical University,[35] and the mineralogist Paul Groth. The case

Table 7.1   Non-resident (foreign) members of the second class of the Czech Academy of Arts and Sciences before 1914

1891	Dmitri Ivanovich Mendeleev, chemist (St Petersburg)
1891	Karol Olszewski, chemist (Cracow)
1892	Charles Hermite, mathematician (Paris)
1894	Adalbert Carl von Waltenhofen, physicist (Vienna)
1895	Charles A. Oliver, ophthalmologist (Philadelphia, USA)
1896	William Ramsay, chemist (London)
1897	Karl J. Küpper (Bonn), former Prof. of descriptive geometry (Technical University, Prague)
1901	André Victor Cornil, physician (Paris)
1901	Tadeusz Browicz, pathologist (Cracow)
1903	Emil Christian Ch. Hansen, chemist (Carlsberg-Copenhagen)
1908	Eduard Suess, geologist (Vienna)
1908	Nikolay Nikolayevich Beketov, chemist (St. Petersburg)
1908	Paul Groth, mineralogist (Munich)
1908	Marie Curie (Paris)

of the prominent German chemist Friedrich August Kekulé demonstrates the firmness of the opposition to German scientists in the Czech Academy. His desire to become a member was so strong that in 1893 he even asked his Czech pupil Bohuslav Raýman to intervene on his behalf, but in spite of this he was never elected.[36] A certain lack of interest in the election of foreign members is also indicated by the fact that their total number was only fourteen, although the statutes of the Academy enabled each class to elect twenty members.[37]

To be a good son of one's nation or to get involved in a supranational scientific network?

Czech scholars involved in the natural and exact sciences had been asking themselves this crucial question since the middle of the nineteenth century, when so-called 'provincial' patriotism started to be replaced by Czech nationalism, but the question became poignant especially from the 1880s, when pronounced nationalism and Czech–German antagonism dominated not only the Czech Academy but also Czech society as a whole. Czech chemists faced the controversial issues associated with this question quite early, for the first time in connection with the creation of Czech chemical terminology. In the 1820s practically no chemical expression existed in the Czech language, so Czech chemical terminology had to be invented from scratch.[38] This process soon acquired political overtones, because some influential revivalists considered it to be non-patriotic to introduce terms of German or even other non-Slavic origin. This circumstance led to the evolution of two streams in the formation of chemical nomenclature.

The so-called 'purists' followed the unrealizable 'patriotic' requirement of eliminating all terms of 'foreign' origin, and eventually arrived in the 1850s at a totally incomprehensible, useless and even absurd chemical language ridiculed by most chemists and mocked by other writers. The more reasonable, yet still 'patriotic', path was to devise modern Czech chemical terminology as a compromise between 'purist' requirements and the international chemical terminology that was evolving at the same time. However, in order to avoid the strong influence of German and Latin, new terms of Slavic (especially Polish) origin were created.[39]

By the 1850s, despite various twists and turns, the fundamentals of modern Czech scientific (and chemical) terminology had been established as one of the essential achievements of the National Revival. Its inauguration was a semi-official act, for in 1853 the authorized *German–Czech Dictionary of Scientific Nomenclature* was published at the instigation of the ministry of culture and education.[40] The dictionary assembled all known Czech terms used in the sciences; the chemist Vojtěch Šafařík,[41] then only 24 years old, used his philological talent to compile and modify the chemical entries, eliminate 'purist' oddities and adapt the whole system of chemical nomenclature to make it usable for teaching and writing scientific literature. The existence of a modern Czech chemical nomenclature conditioned the creation of a viable network of chemical institutions, which in turn stimulated after the 1860s the independent development of chemistry in Czech, representatives of which attempted to catch up with international developments in their field.[42] In the main, Czech chemists were well aware of the danger of contemporary chauvinism, which threatened to abuse science for political goals. They understood that erecting nationalistic barriers would decrease the quality of domestic research and education, obstruct participation in international networks and inevitably lead to international isolation.

On the domestic scene, Czech chemists and their associations were expected to maintain the demarcation line between themselves and their German colleagues. Nonetheless, a detailed investigation reveals some leaks even in this seemingly impervious barrier, especially in the domain of industry and applied chemistry, where the economic advantages of cooperation between the Czechs and Germans overpowered patriotic appeals. The two strongest professional groups, the brewers and sugar industrialists, established bilingual (although Czech dominated) associations closely cooperating with the Society of Czech Chemists: the Spolek pro průmysl pivovarský v království Českém (Association for the Brewing Industry in the Czech Kingdom) in 1873 and the Spolek pro průmysl cukrovarnický v Čechách (Association for the Sugar Industry in Bohemia) in 1876. Both societies published their journals in Czech and German versions. We should not omit the powerful all-Austrian Österreichische Gesellschaft zur Förderung der chemischen Industrie (Austrian Society for Support of the Chemical

Industry), founded in Prague in 1878,[43] the headquarters of which remained in Prague and offered a base where Austrian, Czech and German industrial chemists could join forces. Its first president was Wilhelm Friedrich Gintl,[44] professor at the German Technical Academy (Deutsche Technische Hochschule) in Prague; among its Czech initiators was the chemist František Šebor,[45] a leading sugar specialist.

A prerequisite of the international partnerships in which Czech chemists engaged from the 1870s was publication in internationally comprehensible languages. Therefore, most of them did not obey the appeal of 'patriots' for exclusive publication in Czech and submitted their important papers either to German, French or English periodicals only or to these and concurrently to Czech journals, especially the *Listy chemické* (*Chemical Letters*). Many papers by Czech chemists appeared in *Liebig's Annalen*, the *Bulletin de la Société de chimie*, the *Journal of the Chemical Society*, the *Sitzungsberichte der Kaiserlichen Akademie der Wissenschaften* in Vienna and other leading journals. Czech chemists did not hesitate to contribute to specialized handbooks issued by German chemists; the foremost inorganic chemist, Bohuslav Brauner,[46] for instance, was the author of critical chapters on the atomic weights of 60 elements in Abegg's *Handbuch der anorganischen Chemie*.[47]

Personal contacts, especially correspondence, between Czech chemists and German, British, French and other European scholars, were also common. Young Czech chemists frequently studied or conducted research at German, French or British universities, and the bonds with their foreign tutors and colleagues often persisted for the rest of their lives. Brauner's correspondence, for example, contains letters from Ernst Mach, Adolf Lieben, Richard Abegg, Henry Roscoe, William Ramsay, Zdenko Skraup and many other German and British scientific personalities.[48] In Bohuslav Raýman's papers[49] are letters from his teachers F.A. Kekulé, Charles Friedel and Charles Adolphe Wurtz[50] and other leading chemists. However, it is necessary to remark that contacts with British or French chemists were socially more acceptable than relations with German ones.

The Society of Czech Chemists, which had a richly structured membership[51] including some members from abroad,[52] also attempted to cultivate international contacts. The reasons were obvious. In the 1880s and 1890s the rapid development of chemistry and chemical industries in Germany, England, France and other European countries called for international cooperation that was vital for the quality of scientific research and economic development. Science and industry were both becoming international enterprises, and those who did not participate were falling hopelessly behind. International cooperation meant sharing or at least communicating new findings, laboratory methods and industrial processes, attending international meetings, exchanging students, teachers and literature, reading international journals, participating in circulation of knowledge. These were understood to be prerequisites for the scientific progress and economic prosperity of one's nation; therefore, being patriotic also meant being international. Although the programme of the Society had all this in mind, its possibilities were somewhat limited by the increasing nationalism in Czech society. Among the Society's international activities was the exchange of journals, including periodicals in German, French, English, Danish, Polish, Russian, Croatian and other languages.[53] Nonetheless, the general public mood reflected itself here as well, specifically in the elections of foreign honorary members, acts that should have been the expressions of the highest esteem for outstanding contemporary chemists (see Table 7.2).

If we look at the list of elected foreign personalities, we see a trend quite similar to that in the Czech Academy (see Table 7.1): prevalent among the honorary members were internationally little-known Slav chemists (including one physician), and one searches in vain for the names of prominent German, French and British scientists, who dominated contemporary chemistry. Such elections of foreign members by the Society stopped in 1896, earlier than in the Czech Academy.

Table 7.2 Foreign honorary members of the Society of Czech Chemists and the Society for Chemical Industry in the Czech Kingdom*

Honorary Member	Elected to Society of Czech Chemists	Elected to Society for Chemical Industry
Aleksandr Mikhailovich Butleroff (St Petersburg, Russia)	1880	
Dmitri Ivanovich Mendeleev (St Petersburg, Russia)	1880	1894
Nikolai Aleksandrovich Menshutkin (St Petersburg, Russia)	1880	
Bronisław Radziszewski (Lviv, now Ukraine)	1880	1894
Louis Pasteur (Paris, France)		1893
Emil Christian Hansen (Carlsberg, Denmark)		1893
Karl Lintner (Weihenstephan, Germany)		1894
Maximilian Maercker (Halle/Saale, Germany)		1894
Vilém Dušan Lambl** (Warsaw, Poland)		1895
Emilio Nölting (Mühlhausen, Germany)		1896

Notes: *These societies existed for some time simultaneously due to a temporary split in 1893.

**The foreign honorary membership of V.D. Lambl (1824–95) is quite curious as Lambl was not a chemist but a Czech physician, from 1871 professor of therapy at the Medical School of the Imperial University of Warsaw. Lambl is best known for his discovery of the protozoan parazite *Lamblia intestinalis*.

The re-orientation of Czech chemists toward their Slavic colleagues culminated in the years 1880–1914, when five Congresses of Czech Scientists and Physicians (1880, 1882, 1901, 1908 and 1914) took place and became explicit platforms of Slavic scientific cooperation. Chemists and physicians were eager to establish permanent bonds between Czech scholars and their associations on the one side and Poles and Russians on the other. The idea behind this initiative was to use these Slavic linkages as a political tool to reinforce the positions of Slavic scholars and their national institutions within Austria-Hungary.[54] This new focus on Slavic science and Slavic patriotism in science was intended to replace the former reliance on German science and even included attempts to devise a pan-Slavic scientific nomenclature.[55] The leading role in this effort was played by Czech scholars who believed that a common Slavic scientific language could bring together the Slavic scientific community and maintain political equilibrium against the strong international German-speaking community.

This inclination toward Slavic science also found its reflection in the activities of Bohuslav Brauner, who became an enthusiastic propagator of Dmitri Mendeleev's Periodic Table of Elements and helped to ensure its acceptance by chemists all over the world.[56] Brauner obviously considered his cooperation with Mendeleev not only a scientific but also a patriotic task, associated with the deflection from German science to Slavic science. This is made explicit in Brauner's obituary for Mendeleev:

> We Czechs are often censured because as cultivators of science we only feed off the leftovers from the table of great German science. ... Although I truly respect German science, and especially the achievements of Germans in chemistry, I will always be proud that I never had to eat at the table of German science, and that the guiding star of my life was Slavic science established by Dimitrii Ivanovich Mendeleev.[57]

At the same time, though, Brauner well understood the necessity for balance between international and national science:

> It has been often discussed in scientific circles in smaller nations whether science is national or international. ... Those who consider science a purely international matter are wrong, but mistaken are also those who dream about pursuing purely national science... and consider such pursuit of a purely patriotic science to be a God knows what 'patriotic' act. World science is a huge current that continuously rushes forward. The true scientist must swim in this current with the others... Those who want to take a higher place on the ladder of world science... must master the main world languages... Those who can must aim at maintaining contact, personal or literary, with the main representatives of world science; otherwise their activities will become mere patriotic provincialism... This is true

especially of scientists who are university teachers. Those who do not act this way will easily become biased and fall behind.[58]

These two quotations from the same essay reflect the schizophrenic trap into which Czech chemists and natural scientists in general were pushed by the social and political situation in the Czech Lands. It sounds alarming that even Brauner, cosmopolitan by background, upbringing[59] and behaviour, was led to give way to some kind of nationalism. In his case, however, this was an inoffensive and unrealistically romantic Slavic nationalism, in which some Czech scholars found support at the turn of the twentieth century.

Attempts to soften nationalism within the Czech Academy of Sciences and Arts

The organic chemist and biochemist Bohuslav Raýman, a friend and contemporary of Brauner and a representative of the younger generation of Czech scholars, chose a different path (for biographical information see Table 7.3). In 1890 Raýman became Secretary of the Second Class of the newly founded

Table 7.3 Biographical data of Bohuslav Raýman

1852	Born in Sobotka, Northern Bohemia
1872–74	Chemistry studies, Czech Technical University, Prague with V. Šafařík
1874–76	Chemistry studies, doctorate at Bonn University with F.A. Kekulé
1876/77	Winter Term – Faculté de médecine, Paris, with C.A. Wurtz
1877	Summer Term – École nationale des mines, Paris, with C. Friedel
1877	Study trips to Belgium, Netherlands, England and Scotland, visiting also Roscoe's and Schorlemmer's laboratories
1877–82	Assistant at the Czech Technical University, Prague
1878	Habilitation based on thesis on aromatic compounds at the Czech Technical University, Prague
1879–80	President, Society of Czech Chemists
1882–1907	*Dozent* for organic chemistry, Czech Technical University
1885–91	Co-editor of the journal *Listy chemické* (*Chemical Letters*)
1886	Doctorate in philosophy, Czech University, Prague
1890	Associate professor (*Dozent*) of organic chemistry, Czech University, Prague
1890	Associate member of the Czech Royal Society of Sciences
1890	Full Member, Czech Academy of Sciences and Arts, Secretary of its 2nd Class
1893	Founder of the journal *Bulletin international*
1896	Professor of organic chemistry, Czech University, Prague
1899	Secretary General of the Czech Academy of Sciences and Arts
1903	Honorary Member of the Society of Czech Chemists
1910	Death in Prague from stroke, aged 58

Czech Academy, and his voice was from the very beginning one of the few within the Academy that argued for the elimination of the stimulating influence of the Academy on nationalism in scientific circles.

Raýman's activities in the years prior to his entrance to the Academy prepared him well for the role he would play in the future. He had many years' experience at universities and laboratories in Germany, France and Britain, where he worked under the eminent chemists Friedrich August Kekulé, Charles Friedel, Charles Adolphe Wurtz and Henry Roscoe. His correspondence shows ample international contacts with German, British, Polish, Russian and even Indian chemists. As a teacher at both Czech universities in Prague, Raýman trained several generations of Czech chemists, among whom he disseminated the ideas developed by the Justus Liebig school, with which he aligned himself. Before he joined the Czech Academy of Sciences, he was also member of the bilingual Royal Czech Society of Sciences[60] and the Société chimique de France, and he was a key functionary of the Society of Czech Chemists. Raýman's contacts went far beyond chemistry. We find in his papers[61] letters from the Swedish Slavist Alfred Jensen, who had a high position in the Nobel Institute, and the English Slavist William Richard Morfill. Czech artists, poets and writers also belonged to his circle of friends.[62] His endeavour to raise Czech science to top European levels was anchored in his authentic patriotism, which was neither declarative nor chauvinistic.

Raýman's main concerns in his position as Secretary of the Second Class of the Czech Academy stemmed from the Academy's statutes,[63] which in several paragraphs strongly accentuated the Czech character of the Academy. The Czech historian of science Jiří Beran accurately summarizes this peculiarity of the statutes:

> Detailed study of the Statutes shows that the Academy understood the 'Czechness' and national character of science in the philological and geographical sense[s]. Science pursued exclusively in the Czech language, research into the local historical monuments and natural environment, these were the main aspects accentuated by Hlávka.[64] Only foreign members had the right to publish in foreign languages, while Czech scholars were only allowed to append summaries in a foreign language to their publications. In this way he retarded the active participation of Czech science and art in world culture, and at the same time enforced passive sharing of foreign results, so that he evoked disagreement especially among scientists.[65] Hlávka's concept of national science created to a certain extent barriers to its further development.[66]

How bizarre such useless demonstrations of patriotism could be is shown by the example of the Russian physical chemist Nikolay Beketov. A foreign member of the Czech Academy, he was praised in the Academy's official biography in 1910 for publishing only in Russian (with minor French exceptions).[67] Although the policy of publishing only in Czech pertained only to books and journals issued by the Academy itself, such principles affected to a certain extent the whole Czech intellectual community and prompted discussions about the essence of patriotism in science.

As stated above, nationalism mingled with conservatism manifested itself from the very beginning in the Czech Academy. While nationalism intervened in the sphere of language and communication, conservatism also interfered with the activities and routines of the Academy. Conservatives always joined forces, especially in the election of new members and Academy officers. In 1890, in consequence, some outstanding Czech scholars, such as the geologist, zoologist and palaeontologist Antonín Frič, the chemist František Štolba and the zoologist František Vejdovský, were not elected to the Academy.[68] The foremost poet Jan Neruda also never became a member of the Academy, which is attributed by some historians to his political engagement in the radical wing of the Young Czech Party.[69]

Eventually, Raýman was the man who succeeded in finding at least a partial solution. Scientific works of the members of the Academy were published in the *Rozpravy ČAVU* (*Debates of the Czech Academy*), issued by each class separately. In accordance with the statutes, most works published by Second Class students had a foreign-language summary, usually in French. In 1892 Raýman proposed[70] a relocation of these summaries into a separate periodical. For this purpose, the Second Class of the Czech Academy launched a new journal, the *Bulletin international. Classe des Sciences mathématiques, naturelles et de la médecine – Académie des Sciences de l'Empereur François Joseph I.* Gradually, the summaries became longer and tables and illustrations were added, until the abstracts became full papers and the *Bulletin* evolved into a fully fledged periodical. This was the first foreign-language journal published by a Czech scientific institution, in an attempt to compensate for the handicap of publishing scientific papers in a language that almost nobody except Czechs understood. Several years later, in 1901, in his article 'Patriotism and Science' published in the popular journal *Živa*, Raýman supported the necessity for the *Bulletin international* in the following words:

> The main task of our universities is to cultivate science, *science in every sense of the word...* Our scientific worker must share the results of his research into natural phenomena with foreign workers. I suppose that there may exist a great historian who would write his works only in Czech, but I deny that there could exist a physicist, chemist, mathematician, biologist who would not take into account the world research to which he would contribute with his own truth for the progress of science. It is nonsense to expect that the great nations will come by themselves to search out our results; it is in our patriotic interest, and private, too, to make them acquainted with our work.[71]

Final remarks

In the course of the Czech National Revival in the nineteenth-century Czech Lands, patriotism played an important role in promoting the creation of a Czech scientific terminology, which in turn was a prerequisite for the formation of Czech research and educational institutions and the communication base of Czech science and scholarship after the 1860s. 'Slavic patriotism' as an offshoot of this Czech patriotism also gained ground in the Czech Lands after the 1880s as the starting point of cooperation with Slavic scientists, especially Poles and Russians. This cooperation reinforced the central European version of the Slavic idea conceived by Czech, Russian and Bulgarian Slavic philologists in the 1830s and 1840s.[72] Despite the romantic features and idealized image linked with Pan-Slavism, this trend also strengthened communication channels of science across Europe in the second half of the century.[73]

In the 1860s patriotism in science in the Czech Lands started to shift into nationalism and chauvinism, marked by strong demarcation from German science. These tendencies were especially evident in the 1890s within the Czech Academy of Sciences and Arts, which pleaded for the almost exclusive use of the Czech language in communication and showed a tendency to exploit science for political goals. Some scholars, especially chemists, understood that such trend set the tone for the whole domestic scientific scene, and would inevitably lead to international isolation and decrease the quality of domestic research and education. Although these chemists were a minority in the Second Class[74] of the Academy, they played a decisive role in the struggle against nationalistic excesses and in finding new solutions for international cooperation in the years before the First World War.

The reasons why chemists had such a central position in the Czech scientific community are quite complex, and only some of them can be mentioned here. First, they were backed by the powerful chemical industry, especially the sugar and fermentation industries, where progress and profitability were unthinkable without international cooperation. Second, at the turn of the 19th century industrial chemists became a massive social force in Europe, as well as the Czech Lands.[75] Third, Czech chemists were well prepared for international cooperation, since many of them had studied or carried out research abroad, mostly at German universities, or attended German universities in the Czech Lands. Last, but not least, chemists were well aware of the potentially disastrous consequences of nationalistically motivated isolation for the development of Czech science and technology, so they nurtured their international ties despite strong political pressure, and their senior representatives were powerful enough to influence scholars from other sciences. We must add, however, that although chemists attempted to overcome the divisiveness of nationalism, on the surface the mutual isolation of Czech and German science in the Czech Lands appeared to be irreversible.

However, strategies for coping with nationalism in the sciences in the Czech Lands in the second half of the nineteenth century appear to have differed considerably from those employed in the humanities.[76]

Notes

Research for this chapter was supported by the Grant Agency of the Academy of Sciences of the Czech Republic (Grant No. IAAX00630801). The author is indebted to Dr. Vlasta Mádlová for suggestions that helped to improve the chapter.

1. Carol E. Harrison and Ann Johnson (2009) 'Introduction: Science and National Identity' in C.E. Harrison and A. Johnson (eds.) *National Identity: The Role of Science and Technology* (*Osiris*, Second Series, 24) (Chicago: University of Chicago Press), 1–14, here 4.
2. Jan Janko, Soňa Štrbáňová (1988) *Věda Purkyňovy doby* [*Science in Purkinje's Time*] (Prague: Academia).
3. For a detailed study on ethnicity in the Czech Lands see Jiří Kořalka (1996) *Češi v habsburské říši a v Evropě 1815–1914* [*Czechs in the Habsburg Empire and Europe 1815–1914*] (Prague: Argo).
4. On the creation of Czech professional and scientific terminology see Janko and Štrbáňová, *Věda Purkyňovy doby*, especially 107–14, 215–26; Soňa Štrbáňová and Jan Janko (2003) 'Uplatnění nového českého přírodovědného názvosloví na českých vysokých školách v průběhu 19. století' [Assertion of the new Czech scientific nomenclature at the Czech universities in the nineteenth century], in Harald Binder, Barbora Křivohlavá and Luboš Velek (eds.) *Position of National Languages in the Education, Educational System and Science of the Habsburg Monarchy 1867–1918* (Prague: Výzkumné centrum pro dějiny vědy), 297–311, abstract in English 732–33; Ludmila Hlaváčková (2003) 'Čeština v medicíně a na pražské lékařské fakultě (1784–1918)' [Tschechisch in der Medizin und an der Prager Medizinischen Fakultät (1784–1918)], in Binder, Křivohlavá and Velek (eds.) *Position of National Languages*, 327–44, abstract in German 733–34.
5. We should keep in mind a terminological problem: in texts written in English, the terms 'nation' and 'national' have different meanings from their Czech translations '*národ*' and '*národní*'. While the term 'nation' and words derived from it refer to membership of a nation or citizenship in a state, their Czech equivalents indicate ethnicity. For this reason, we prefer to use the expression 'linguistically Czech science' rather than 'Czech national science'.
6. There is abundant literature on the development of chemistry in the Czech Lands, but the particulars are mostly scattered in various books and articles. Important references can be found, for instance, in Soňa Štrbáňová (2008) 'Czech Lands: Chemical Societies as Multifunctional Social Elements in the Czech Lands, 1866–1919', in Anita Kildebaek Nielsen and Soňa Štrbáňová (eds.) *Creating Networks in Chemistry; The Founding and Early History of Chemical Societies in Europe* (Cambridge: RSC Publishing), 43–74. See also Robert Rosner (2004) *Chemie in Österreich 1740–1914* (Vienna, Cologne, Weimar: Böhlau Verlag), 181–213, 289–329.
7. Jan Havránek (1967) 'The Development of Czech Nationalism', in *The Nationality Problem in the Habsburg Monarchy in the Nineteenth Century: A Critical Appraisal, Part II, The National Minorities, Austrian History Yearbook* 3, Part 2 (Houston: Rice University Press), 223–60, here 228; Rosner, *Chemie in Österreich*, 298.
8. Redakce (1928) '1918–1928', *Chemický obzor*, 3, 325–28.

9. Bohumil Hájek, Ladislav Niklíček and Irena Manová (1981) 'Profesor Vojtěch Šafařík – jeden ze zakladatelů české chemie, 2. Část [Prof. V. Šafařík – one of the founders of the Czech chemistry, part 2], *Sborník Vysoké školy chemicko-technologické v Praze*, A, 23, 50–108, here 59.

10. Havránek, 'Development of Czech Nationalism', 228.

11. Havránek, 'Development of Czech Nationalism', 229. As Havránek emphasizes, the decreasing importance of industries dominated by German capital (glass, porcelain, textiles) resulted also in the decline of political influence of Bohemian Germans at the beginning of the twentieth century

12. Havránek, 'Development of Czech Nationalism', 230.

13. Přírodovědecký spolek Isis.

14. Spolek chemiků českých.

15. Today it is called Česká společnost chemická (Czech Chemical Society). A detailed account of the Society's early history is given in Štrbáňová, 'Czech Lands: Chemical Societies', where the numerous changes in the title and structure of the Society are also described.

16. Probably the only German member of the Society, for a few years at the turn of the nineteenth century, was Carl Zulkowski (1833–1907), professor of chemical technology of the Deutsche Technische Hochschule in Prague (1887–1904).

17. Štrbáňová, 'Czech Lands: Chemical Societies', 51.

18. Štrbáňová, 'Czech Lands: Chemical Societies', 55.

19. The all-Austrian Österreichische Gesellschaft zur Förderung der chemischen Industrie (1878) was not a typical chemical society and will be mentioned further on.

20. Naturhistorisches Verein Lotos.

21. Emilie Těšínská (2003) 'Vznik a působnost přírodovědného spolku "Lotos" v českých zemích' [Die Entstehung und das Wirken des naturwissenschaftlichen Vereins 'Lotos' in den böhmischen Ländern] in Binder, Křivohlavá and Velek (eds.) *Position of National Languages*, 327–44, abstract in German 735–36.

22. Ständisches Polytechnisches Institut, founded in 1806, from 1876 independent Czech and German Technical Universities.

23. Česká universita Karlo-Ferdinandova and Deutsche Karl-Ferdinands Universität.

24. K.k. Technische Hochschule in Brünn.

25. From the extensive literature on the history of chemical education at Czech universities see, for example, Otakar Quadrat (1966) *Nástin historického vývoje Vysoké školy chemicko-technologické v Praze (do roku 1945)* [Outline of the historical development of the Chemical Technological University in Prague until 1945] (Praha: SPN); Otakar Franěk (1969) *Dějiny České vysoké školy technické v Brně, vol. 1* [History of the Czech Technical University in Brünn, vol. 1] (Brno: VUT); relevant chapters in Václav Lomič and Pavla Horská (1979) *Dějiny Českého vysokého učení technického v Praze* [History of the Czech Technical University in Prague], part 1, vol. 2. (Prague: SNTL); Jan Havránek (ed.) (1997) *Dějiny Univerzity Karlovy III, 1802–1918* [History of the Charles University, vol. III, 1802–1918] (Praha: Karolinum); and Miroslav Schätz (2002) *Historie výuky chemie* [History of Chemistry Instruction] (Prague: Vysoká škola chemicko-technologická).

26. Česká akademie císaře Františka Josefa I. pro vědy, slovesnost a umění (ČAVU). In this article the expression 'Czech Academy' or the abbreviation ČAVU will be used.

27. Sources sometimes indicate different years for its constitution, from 1888 to 1891, e.g. Jiří Beran (1971) 'II. třída ČAVU v letech 1891–1914' [The Second Class

of the Czech Academy in the years 1891–1914], *Dějiny věd a techniky*, 4, 193–208, here 193. 1888 is the year when the founder of the Academy and its patron, the architect Josef Hlávka, donated the money for its financing.

28. Ibid., 193.

29. Jiří Beran, for instance, draws attention to these characteristics of the Czech Academy in (1973) 'Vznik a hlavní tendence ve vývoji České akademie věd a umění' [The origins and main tendencies in the development of the Czech Academy of Sciences and Arts] *Práce z dějin přírodních věd*, 4, 91–100. Beran notices that at the end of 1891, the Czech Academy had only one chemist and one physicist, but five jurists and even more philologists among its acting members (p. 194). The Czech Academy was divided into the following classes: I. philosophy, political sciences, jurisprudence, social sciences, history and archaeology, II. mathematical, natural and medical sciences, III. philology, IV. literature, fine arts and music.

30. Their loyalty is well documented by the several volumes of a memorial issued by the Czech Academy of Sciences, Letters and Arts on the occasion of the 50th anniversary of the accession of Emperor Franz Joseph I: (1898) *Památník na oslavu padesátiletého panovnického jubilea Jeho veličenstva císaře a krále Františka Josefa I: vědecký a umělecký rozvoj v národě českém: 1848–1898* (Prague: Nákladem České akademie císaře Františka Josefa pro vědy, slovesnost a umění).

31. Havránek, 'The Development of Czech Nationalism', 237.

32. Documents related to the establishment and statutes of the Czech Academy are published in Jiří Beran (1989) *Vznik České akademie věd a umění v dokumentech* [The Creation of the Czech Academy of Sciences and Arts in Documents], Práce z dějin ČSAV, B, vol. 2. (Prague: Ústřední archiv ČSAV). A survey of all members with ample additional biographical and other information is published in Alena Šlechtová and Jiří Levora (1898) *Členové České akademie věd a umění 1890–1952* [The members of the Czech Academy of Sciences and Arts 1890–1952] Práce z dějin ČSAV B, vol. 3. (Prague: Ústřední archiv ČSAV).

33. Beran, *Vznik České akademie*, 27.

34. Karel Chodounský (1888) 'Česká akademie věd' [Czech Academy of Sciences], *Časopis lékařů českých*, 27, 385–87.

35. Karl Josef Küpper (1828–1900) was professor of geometry in Prague until 1898 and left for Bonn only after his retirement. His assistant was the Czech mathematician Eduard Weyr (1852–1903), a member of the Czech Academy, and Küpper was known for his friendliness to the Czechs. These facts can explain his election. See Franz Stark, Wilhelm Gintl and Anton Grünwald (eds.) (1906) *Die k. k. Deutsche Technische Hochschule in Prag 1806–1906* (Prague: Selbstverlag), 358–59; Jindřich Bečvář (1995) 'Eduard Weyr', in Jindřich Bečvář (ed.) *Eduard Weyr, 1852–1903* (Prague: Prometheus), 35–66.

36. For details see Soňa Štrbáňová and Jan Janko (1993) 'Kekulé s character in the light of his ennoblement', in John Wotiz (ed.) *The Kekulé Riddle. A Challenge for Chemists and Psychologists* (Clearwater, FL; Vienna, IL: Cache River Press), 195–210.

37. (1890) 'Stanovy České Akademie císaře Františka Josefa pro vědy slovesnost a umění' [Statutes of the Czech Academy], *Almanach ČAVU*, 1, 72–141, § 11, 79.

38. For more details see Štrbáňová and Janko, works cited in Note 5.

39. Due to this compromise and discrepancies between Czech and international chemical terminologies, contemporary Czech chemists still sometimes have trouble correlating Czech and international terminology.

40. (1853) *Německo český slovník vědeckého názvosloví pro gymnasia a reálné školy. Od komise k ustanovení vědeckého názvosloví pro gymnasia a reálné školy. Deutsch-böhmische wissenschaftlice Terminologie* [German–Czech Dictionary of Scientific Terminology for Gymnasien and Realschulen. By the Commission for the Establishment of Scientific Terminology for Gymnasien and Realschulen] (Prague: Kalvéské knihkupectví Bedřich Tempský).

41. This official scientific dictionary may be considered evidence of efforts by some of the revivalists to disentangle themselves from the pressure of narrow-minded 'patrioteering'. The foreword was written by the foremost Slavist P.J. Šafařík (1795–1861), father of the chemist Vojtěch Šafařík (1829–1902), who, in turn, in 1860 wrote the first Czech university textbook on chemistry and in 1869 became professor of chemistry at the Czech Polytechnic.

42. For more details see, for instance, Soňa Štrbáňová (1986) 'Vztah české a světové chemie' [The relationship between Czech and world chemistry], *Sborník Vysoké školy chemicko-technologické v Praze*, A, 31, 9–24.

43. Its establishment is described in (1879) 'Geschichte der Bildung der Gesellschaft', *Berichte der Österreichischen Gesellschaft zur Förderung der chemischen Industrie*, nos. 1 and 2, 1–2; see also Rosner, *Chemie in Österreich*, 298.

44. Wilhelm Friedrich Gintl (1843–1908) was professor at the German Technical University (1870) and president of the Verein für chemische und metallurgische Produktion in Aussig (1890), the largest Austro-Hungarian chemical cartel. Rosner, *Chemie in Österreich*, 269–70; Stark, Gintl and Grünwald, *Die k. k. Deutsche Technische Hochschule*, 379.

45. František Šebor (1838–1904) was a leading Czech chemical production engineer, president of the Society of Czech Chemists (1892–97) and an honorary member (1893); see Oldřich Hanč (ed.) (1996) *100 let Československé společnosti chemické její dějiny a vývoj 1866–1966* [100 Years of the Czechoslovak Chemical Society, its History and Development 1866–1966] (Prague: Academia), 62.

46. Bohuslav Brauner (1855–1935), professor of the Czech University (1897), is probably the most widely known Czech chemist. He specialized in research into certain groups of elements in the Periodic System. His biographic references include: Gerald Druce (1944) *Two Czech Chemists* (London: The New Europe Publishing Co.), 5–44; Jaroslav Heyrovský (1935) 'Professor Bohuslav Brauner died February 15th 1935', *Collection of Czechoslovak Chemical Communications*, 7, 51–56; Jan S. Štěrba-Böhm (1935) *Bohuslav Brauner* (Prague: Česká akademie věd a umění), with bibliography of Brauner; S.G. Schacher (1973) 'Brauner, Bohuslav', in *Dictionary of Scientific Biography*, vol. 2 (New York: Scribner), 428–30; Soňa Štrbáňová (2003) 'Brauner, Bohuslav', in Dieter Hoffmann, Hubert Laitko, Staffan Müller-Wille and Ilse Jahn (eds.) *Lexikon der bedeutenden Naturwissenschaftler*, vol. 1 (Heidelberg and Berlin: Spektrum Akademischer Verlag), 249–51.

47. Richard Abegg und Friedrich Auerbach (eds.) (1905–13) *Handbuch der anorganischen Chemie in vier Bänden* (Leipzig: Hirzel). Brauner's chapters are in vol. II, 1, 1908; vol. II, 2, 1905; vol. III, 1, 1906; vol. III, 2, 1909; vol. III, 3, 1907; vol. IV, 2, 1913.

48. Brauner's extensive correspondence is kept at the Archives of the Museum of Czech Literature (PNP-Památník národního písemnictví). PNP, Brauner Bohuslav, Personal Papers, Correspondence.

49. Bohuslav Raýman (1852–1910) specialized in organic chemistry and biochemistry, and was *dozent* at the Czech Technical University (1978) and professor at the Czech University (1890).

50. Only a fragment of Raýman's correspondence has been preserved; it is kept in the Archives of the Museum of Czech Literature (PNP), PNP, Raýman Bohuslav, Personal Papers. On Raýman's correspondence, see Soňa Štrbáňová (2011), 'Raýmanova osobnost ve světle jeho fondu v Literárním archivu Památníku národního písemnictví' [Raýman's personality as reflected in his papers kept at the Literary Archive of the Museum of Czech Literature], *Práce z dějin Akademie věd*, 3, 161–182, abstract in English, 173–174; His correspondence with Emil Fischer is described in Soňa Štrbáňová (1986) 'Tschechische Beiträge zur Entwicklung der Biochemie: B. Raýman und K. Kruis', *Philosophische, historische und wissenschaftstheoretische Probleme in Chemie und Technik*, Geschichte u. Organisation der Wissenschaft, Kolloquien Heft 57 (Berlin: Akademie der Wissenschaften der DDR), 147–58. For his correspondence with Kekulé, Friedel and Wurtz see, for example, Soňa Štrbáňová and Jan Janko (1991) 'Die Umstände der Nobilitierung F.A. Kekules,' *Chemie in unserer Zeit*, 25, 208–13; Štrbáňová and Janko, 'Kekulé s character'; Soňa Štrbáňová (2005) 'Correspondence strengthening the network of a scientific school: unknown letters of the French chemists C. Friedel and C.A. Wurtz to the Czech chemist B. Raýman', in Horst Kant and Annette Vogt (eds.) *Aus Wissenschaftsgeschichte und -theorie* (Berlin: Verlag für Wissenschafts- und Regionalgeschichte Dr. Michael Engel), 257–76. These papers also contain references to Raýman's biography.
51. In 1907 the unified Česká společnost chemická pro vědu a průmysl (Czech Chemical Society for Science and Industry) had 630 members. See Štrbáňová, 'Czech Lands: Chemical Societies', 65.
52. Beside foreign honorary members, the Society had foreign corresponding members like Gustav Janeček from Zagreb, Sima Lozanić from Belgrade, Bronisław Pawlewski from Lviv and Jule Tourtel from Tantonville near Nancy. Tourtel, unlike the three others, was not a chemist, but a brewer, co-owner (with his brother Prosper) of the famous brewery at Tantonville in the Lorraine, where Pasteur also carried out some studies.
53. Štrbáňová, 'Vztah české a světové chemie', 21.
54. More on this in Soňa Štrbáňová (1989) 'Congresses of Czech naturalists and physicians in the years 1880–1914 and Czech–Polish scientific collaboration', *Acta historiae rerum naturalium necnon technicarum*, 21, 79–122.
55. For more details see Štrbáňová, 'Congresses'; Hlaváčková, 'Čeština v medicíně a na pražské lékařské fakultě'.
56. Soňa Štrbáňová (2009) 'The role of Czech chemists in the reception and dissemination of the periodic system in Europe', unpublished paper from the conference Consumers and Experts: The Uses of Chemistry (and Alchemy), Sopron, Hungary.
57. Quoted from (1952) *Dopisy Dimitrije Ivanoviče Mendělejeva českému chemiku Bohuslavu Braunerovi* [Letters of D.I. Mendeleev to the Czech Chemist B. Brauner] (Prague: Technicko-vědecké vydavatelství), 69–70.
58. *Dopisy Dimitrije Ivanoviče Mendělejeva*, 68–69.
59. Noteworthy is Brauner's mixed parentage, which endowed him with all-embracing capabilities. The carrier of his 'chemical genes' was his German mother, Augusta Braunerová (1817–1890), whose father was Karl A. Neumann (1771–1866), the first professor of chemistry at the Prague Polytechnic, and whose grand-uncle was Caspar Neumann (1683–1737), one of the great figures in the European history of chemistry, professor of chemistry in Berlin and friend of G.E. Stahl. The father of Bohuslav Brauner was František August Brauner (1810–1880), a lawyer who became one of the most influential Czech politicians after 1848. Brauner's

sister Zdenka Braunerová (1858–1934) was a recognized modern artist, and his other sister, Anna, was married to a well known French writer, Elémir Bourges (1852–1925), winner of the Goncourt Prize.

60. Královská česká společnost nauk.
61. PNP, Raýman Bohuslav, Personal Papers.
62. A close friend was the most prominent Czech poet of his generation, Jaroslav Vrchlický. Raýman's correspondence kept at the PNP and the Archives of the Academy of Sciences of the Czech Republic discloses that he tried in vain for several years to obtain the Nobel Prize for Vrchlický.
63. Stanovy České Akademie (cit. Note 39).
64. Josef Hlávka (1831–1908) Czech architect and benefactor, founder and first President of the Czech Academy; see above.
65. Beran, *Vznik České akademie*, 42–43.
66. Beran, *Vznik České akademie*, 79.
67. (1910) 'Životopisy nových členů' [Biographies of new members] *Almanach ČAVU*, 20, 46–47.
68. Beran, *Vznik České akademie*, 54. The paragraphs that most accentuate the 'Czech character' of the Academy and the restriction in its communications to the Czech language are § 1 and § 5; Stanovy České Akademie (1890), 73, 76–77.
69. I am indebted for this information to Dr. Martin Franc. Hlávka was a member of the Old-Czech Party, but within the Academy no political tensions were noted.
70. Vlasta Mádlová (2011), Bohuslav Raýman a Česká akademie věd a umění [Bohuslav Raýman and the Czech Academy of Sciences and Arts], *Práce z dějin Akademie věd* 3, 197–208, English summary 197.
71. Bohuslav Raýman (1901) 'Vlastenectví a věda' [Patriotism and science], *Živa*, 11, 80–82, here 81. Emphasis in the original.
72. On this topic compare, for example, Antoaneta Balcheva (2010) 'Utopias in Search of National Identity', in Richard Vašek and Jan Rychlík (eds.) *Formování moderních národů ve střední a východní Evropě v 19. a 20. století* [Shaping of the modern nations in central and eastern Europe in the 19th and 20th centuries] (Prague, Sofia: Masarykův ústav a Archiv AV ČR), 167–72.
73. For instance, Bohuslav Brauner reported regularly about the Russian chemical publications in the *Journal of the Chemical Society*.
74. These chemists were Bohuslav Raýman, Bohuslav Brauner, Emil Votoček (1872–1950) and, as a corresponding member, Otakar Šulc (1869–1901), who was included for just one year because of his premature death. Raýman, Brauner and Votoček were a strong formation due to their scientific, teaching and organizing positions in the Czech scientific community and their international reputations.
75. In 1907 the total membership of the Chemical Society was 630, of which 60 per cent were industrialists and industrial corporations.
76. For brief discussion of this issue, see Soňa Štrbáňová and Antonín Kostlán, 'To be a good son of one's nation or get involved in supranational scientific network? Cases of chemists and historians in the Czech National Revival', in 3rd International Conference of the European Society for the History of Science, Vienna, 10–12 September 2008, *Abstracts*, 120–21; Antonín Kostlán, 'To be a good son of one's nation ... Czech historians between national program and scientific style', in XXIII International Congress of History of Science and Technology, Ideas and Instruments in Social Context, 28 July–2 August 2009, Budapest, Hungary, *Book of Abstracts*, 628.

8
Fault Lines and Borderlands: Earthquake Science in Imperial Austria

Deborah R. Coen

Our knowledge of a natural phenomenon, say of an earthquake, is as complete as possible when our thoughts so marshal before the eye of the mind all the relevant sense-given facts of the case that they may be regarded almost as a substitute for the latter, and the facts appear to us as old familiar figures, having no power to occasion surprise. When, in imagination, we hear the subterranean thunders, feel the oscillation of the earth, figure to ourselves the sensation produced by the rising and sinking of the ground, the rocking of the walls, the falling of the plaster, the movement of the furniture and the pictures, the stopping of the clocks, the rattling and smashing of windows, the wrenching of the door-posts, the jamming of the doors; when we see in mind the oncoming undulation passing over a forest as lightly as a gust of wind over a field of grain, breaking the branches of the trees; when we see the town enveloped in a cloud of dust, hear the bells begin to ring in the towers; further, when the subterranean processes, which are at present unknown to us, shall stand out in full sensuous reality before our eyes, so that we shall see the earthquake advancing as we see a wagon approaching in the distance till finally we feel the earth shaking beneath our feet, – then more insight than this we cannot have, and more we do not require.[1]

In *Contributions to the Analysis of Sensations* (1886), Ernst Mach famously argued that the goal of physics is to build up a description of the world out of the most basic components of human experience. The scientific observation of an earthquake involved all the forms of sensation explored earlier in Mach's treatise: movement, sight, time, even tone. In this way, seismology could serve as a test of the capacity of Mach's psycho-physical programme to produce practical knowledge in real environments, beyond the laboratory, as his evolutionary epistemology demanded. In other respects, however, Mach's choice of example is confounding. Having spent most of his life in Bohemia, he was unlikely to have experienced a major earthquake first hand. Moreover, how could he expect anyone, in the midst of catastrophe, to record all these details, and manage to survive?

I argue that Mach's choice of example reflected the ambitions of earthquake research in imperial Austria. The observation of earthquakes was more central to Austrian scientific culture in the 1880s than one might suppose. Historical seismologists believe that the seismicity of the territory of Austria-Hungary was increasing in the late nineteenth century.[2] As early as 1869, Mach may well have heard a lecture at the Lotos natural-historical society in Prague, of which he was a member, by one Rudolf Falb, soon to become notorious worldwide as an earthquake 'prophet'.[3] Later, in the wake of the earthquakes in Belluno in 1873, Lower Austria in 1876 and Zagreb (German: Agram) in 1880, Austrian newspapers spread vivid reports. As Karl Kraus later caricatured the culture of earthquake observing, 'No impact is perceived, no fluids spilled, without making known in the newspapers the very next day from what direction this came or in which direction it went. ... Day in, day out, today, tomorrow, for ever. Until the house of the world really comes crashing down'.[4]

Mach would also have known that earthquakes were being explained in new ways in Central Europe at this time. Until the 1870s they had been attributed variously to meteor strikes, the sinking deltas of lakes, and the collapse of subterranean hollow strata. Then in 1873 the Austrian geologist Eduard Suess proposed his 'tectonic' hypothesis: that most earthquakes result from movement along fractures in the earth's crust.[5] The defence of this hypothesis would rest on correlating the location and direction of earthquakes with features of local geology. A new approach to seismology arose, termed 'monographic'. It drew on the testimony of eyewitnesses, on intimate familiarity with the lie of the land, and on historical evidence of past quakes culled from local archives. It served both basic research and the practical purpose of mapping the distribution of seismic hazard.

In an argument significant for Mach's epistemology, Austrian geologists of the late nineteenth century claimed that complete knowledge of a seismic event was possible only by combining the observations of numerous individuals, few of whom were likely to have any scientific training: 'Likely in no other field is the researcher so completely dependent on the help of the non-geologist, and nowhere is the observation of each individual of such high value as with earthquakes... Only through the cooperation of all can a satisfying result be delivered.'[6]

Imperial science

The Habsburg Monarchy was a territory of remarkable geological as well as human diversity. The contrasts between mountains and plains made it a rich field for geologists, but its cultural diversity posed practical problems of scientific communication and collaboration. In the Habsburg world, the challenge of extracting 'scientific' observations from ordinary people caught in potentially lethal circumstances was compounded by that of organizing communication among scientists and lay observers across barriers of language and culture.[7]

Yet no nineteenth-century empire could ignore the threat of earthquakes. On 1 November 1755, when an earthquake, tsunami and fire swept away one quarter of the population of the magnificent city of Lisbon, it seemed that a global empire had been destroyed in a single morning. Throughout the age of empire, earthquakes mocked the self-confidence of European settlers in Asia and the New World by levelling their stone houses while leaving native structures unscathed. Scientific explorers paid tribute to the fortitude of South American natives in the face of seismic power, a fortitude to which Europeans could only aspire. When Charles Darwin witnessed his first earthquake in Chile, he was struck by a vision of imperial decline – not over the course of centuries but in a matter of seconds.[8]

The threat that earthquakes posed to the Habsburg Monarchy went beyond material damage and social disorder. The danger was implicit in the wary reports in Viennese papers about camps pitched by earthquake victims far from the imperial capital. The spectre of homelessness, particularly in the Empire's south-eastern corner, raised fears of irredentist nationalism. Suess himself observed that the damage wrought by an earthquake was social as well as material: 'the individual, who is bound to hearth and family by a thousand threads, suddenly...sees these threads broken, like a plant torn up by its roots.'[9] Disasters likewise exposed political uncertainties: which level of the Monarchy's intricate web of governance would be called on to respond?[10] In multiple ways, then, an earthquake could call into question the political framework that tied the Monarchy's fringes to its two capitals.

On the other hand, disasters called forth a rhetoric of humanitarianism, which some hoped would push national strife back into the shadows. This hope ran high in the wake of the severe 1895 earthquake in Ljubljana (German: Laibach), at a moment when nationalism and socialism seemed to be splintering Austrian politics into ever more hostile factions. In Vienna, several dozen writers and artists joined forces to produce a volume of engravings and verse to benefit the stricken city. Through bucolic landscapes, quaint caricatures and lilting rhymes, the contributors conjured up the Monarchy as a 'brotherhood' and Vienna as its 'golden heart' (see Figure 8.1).[11]

The liberal press took a similar line – for instance, in this hopeful ditty that appeared in the satirical, anti-clerical journal *Der Floh*:

Wo gab's Slovenen da, wo Deutsche,
Wo Sprachenzwist und Kampf um Macht?

Figure 8.1   Cover by Alfred Roller of *Für Laibach: zum besten der durch die Erdbeben-Katastrophe im Frühjahre 1895 schwer betroffenen Einwohner von Laibach und Umgebung,* ed. Genossenschaft der bildenden Künstler Wiens, 1895

Nur Menschen gab es, bange, bleiche,
Nur Menschen, zitternd, angstverwirrt,
Nur Brüder, Einige und Gleiche –
Die nun das Unglück – coalirt.
(Where were Slovenes, where were Germans,
Where was language strife and power thirst?
Only humans, pale and fearful,
Only humans, trembling, terrified,
Only brothers, equal one and all –
Whom now misfortune – unified.)[12]

It was not until after the Ljubljana disaster of 1895 that Vienna took charge of earthquake research throughout the Austrian half of the Monarchy. Mach himself, as secretary of the Vienna Academy of Sciences, presided over the formation of the Austrian Earthquake Commission. At its height this network encompassed over 1700 observers reporting from all sixteen crownlands. Its success reflected a culture of earthquake observation that took root in Austria in the last quarter of the nineteenth century.

A perfect earthquake

On the evening of 9 November 1880, readers of the *Neue Freie Presse* learned that, between 8:30 and 9:00 that morning, an earthquake had shaken much of the Balkan peninsula. Tremors were also reported from western Hungary, Carinthia and lower Austria, and even from Vienna itself. The quake had killed at least two residents of Zagreb, wounded 30 others, and damaged more than 3000 houses (see Figure 8.2).[13]

This was, in short, the perfect object of study for the observational macro-seismology of the day. It was a 'moderate' earthquake, neither so weak as to go unnoticed nor so strong as to leave observers in what was typically described as a state of 'senseless panic or utter despair'.[14] In addition, its impact was geographically extensive, meaning that observations could be collected throughout central Austria-Hungary.

Figure 8.2 An unattributed contemporary illustration of a Zagreb street brings into focus the conflicting responses to the earthquake of 1880. On one hand, the artist offered a very sober observation of the quake's worst damage; on the other, the human figures in the scene are terror struck, frozen in gestures of helplessness. The image conveys the disparity between the rational attitude of the artist after the fact and the irrational panic of the quake's victims. It seems to suggest the impossibility of the scientific observation of an earthquake in progress. (Jan Kozák Earthquake Images Archive)

The scientific response to the Zagreb quake embodied the dualist (potentially trialist) nature of power in the Monarchy after 1867. Separate investigations were launched in the imperial capitals of Vienna and Budapest, as well as in the South-Slav centre of Zagreb. The imperial Geological Institute in Vienna solicited reports from eyewitnesses, while the imperial Academy of Sciences sent Franz Wähner, an assistant at the Geological Institute in Vienna, to speak with witnesses in Zagreb and collect evidence of the damage to buildings. Independently, the director of the Geological Institute in Budapest was sent by the Hungarian Ministry of Agriculture, Industry and Trade to inspect the area, but his inquiries were almost entirely restricted to assessments of damage to buildings.[15] Meanwhile, the South-Slav Academy of Arts and Sciences in Zagreb commissioned two researchers to report on the quake.[16] From mid-November to late December, the researchers from the Vienna and Zagreb academies covered approximately 90 square miles, interviewing witnesses and inspecting damage to buildings. The scale of the investigation dwarfed that of any previous earthquake in Austria-Hungary.

The Vienna geologist Melchior Neumayr evoked the magnitude of this effort when reflecting on Wähner's investigation:

> To get an idea of the quantity of observations that are necessary for a correct assessment of an earthquake, we may consider the materials on the basis of which a recently published work by Wähner on the Agram earthquake has been composed. The author himself spent five weeks on location, occupied exclusively with this matter; several other geologists from Vienna, Pest, and Agram were similarly occupied, and it was possible to use their observations as well. Through the intervention of a few railway authorities, the reports of well over a hundred railway stations were furnished, the maritime authorities transmitted [reports] from all the port captains and lighthouse keepers of the entire stretch from Cattaro to the Italian border; in addition a great many private communications and newspaper reports arrived, such that observations were brought forth from approximately 750 different locales. Certainly there were still many holes, but overall on this basis an accurate insight into the nature of this earthquake was possible through the united effort of more than 1000 different observers, whose results were united in one hand.[17]

Still, the question remained: having collected such detailed reports, what was a scientist to do with them? How would lay testimony become scientific evidence? 'The analysis of the collected reports, the assessment of their reliability, the separation of the worthwhile from the unuseable, finally the synthesis and application of the information obtained, demanded a significant investment of time and effort.'[18] It would ultimately take Franz Wähner three years to accomplish these tasks.

'To discern the phenomenon in its physical elements'

Wähner's study of the Zagreb quake became a classic of its genre. In his introduction, however, Wähner stressed the modesty of his ambition:

> In my view, given the current state of the discipline, it cannot be the task of a monograph on a large earthquake to investigate the final telluric or even cosmic causes of [the event]. It will be necessary above all to discern the phenomenon itself in its physical elements [*Elementen*], before it can be permitted to discuss causes hidden from observation, and I have convinced myself that in this first area we still have a great deal to learn.

He formulated his goal thus: 'to unite the individual observations, made and collected with the greatest objectivity, into a total picture and to seek to discover the law that lies therein.'[19] Wähner's sense of his task rested on an implicit phenomenalism: to identify the 'physical elements' and unite them into a 'total picture'.

Wähner's study exemplified the monographic method of earthquake investigation on the model of Eduard Suess.[20] He engaged intimately with the testimony of eyewitnesses and insisted on reproducing observers' reports as completely and accurately as possible. He even analysed the word choice of his witnesses. As a native of northern Bohemia, however, he depended on the Croatian colleagues with whom he travelled for translations. For instance, when discussing reports of earthquake sounds, he inserted the original Croatian next to German translations. In one case, alerting his readers that the German '*Getöse*' corresponded to Serbo-Croat '*tutnjava*',[21] he quoted a Croatian observer's seemingly awkward description: 'The dreadful roar/boom [*Getöse/tutnjava*] began to shake more and more strongly with the earth and the houses.' He commented: 'Despite the unusual manner of expression, very indicative of the character of the ground motion.'

Wähner's conclusions reflected his attention to the descriptive language used by the observers themselves. In all the reports from the area of greatest destruction, he noted

> [W]e look...in vain for the mention of an instantaneous strong movement such as would be termed, in everyday life as well as by a physicist, an 'impact'. The ground motion was, to the contrary, generally perceived as a long-lasting, continuous movement...It seems to have been slower, gentler movements, though movements of great intensity, than would have resulted from a brief to-and-fro movement of the individual soil particles in a horizontal or diagonal direction.[22]

Thus, in his final assessment of the nature of the Zagreb earthquake, Wähner concluded that it took the form of a transverse wave that shook the particles of the earth's crust nearly vertically, and which spread out not from a central point but rather from an extended area. He thus discredited the claim that earthquakes were longitudinal waves that propagated radially. In this way, by means of the language of 'everyday life', he lent support to Suess's tectonic theory.

Seismology as *Landeskunde*

In the terminology of the day, seismology's monographic method was an exercise in *Landeskunde*, the term for regional history, both natural and human. In fact, the most comprehensive statement of this method was published under the title *Erdbebenkunde*.[23] As Lynn Nyhart points out, the suffix *'kunde'* in the nineteenth-century German-speaking world indicated a form of knowledge marked by three characteristics: it was comprehensive even at the expense of intellectual coherence; it derived from personal, sensory experience; and it was open to popular participation.[24] Suess's methods were distinctively *'kundig'*. He drew on meticulous first-hand observation of the geology of the region, gleaned from months of hiking and sketching. He had also culled information from countless local informers, and had immersed himself in local historical archives, familiarizing himself not only with the region's general history but also with the quirks of historical documents. This style of research thus blended expert and popular knowledge: as the author of a study of the 1858 earthquake in Žilina (Slovakia, German: Sillein) had argued, in the study of earthquakes it was 'very difficult ... to specify the limit where someone starts or stops being an expert.'[25]

As the geologist Rudolf Hoernes explained at the outset of a study of earthquakes in the crownland of Styria: 'one might criticize me for undertaking first a discussion of a large area with political, that is more or less artificial, borders.' This was a typical quandary for earth scientists in the Habsburg world: whether to define the region of study according to physical or linguistic borders. In response to his hypothetical critic, Hoernes noted that he had, in fact, been collecting seismic observations from the entire Eastern Alps; but the 'critical inspection' of this material had 'for many reasons' to be conducted piecemeal, 'not according to period but according to place'. Thus the evidence itself called for a *landeskundlich* approach; as Hoernes put it, 'The largest portion of the sources to be used for the critical annotation of the earthquake reports bear the character of contributions to the *Landeskunde* of the affected province.'[26]

Historians of nineteenth-century Germany typically think of *Landeskunde* as a distinctively German pursuit.[27] However, many practitioners of the monographic method in imperial Austria did not identify themselves culturally or nationally as Germans. While Hoernes and Canaval were organizing their observational networks in the Alps, the South-Slavic Academy of Sciences and Arts in Zagreb commissioned a permanent committee for the observation of seismic phenomena. The Zagreb Academy, officially sanctioned by Emperor Franz Joseph in 1866, was a product of the Croatian national revival and the movement for South-Slavic unity, at a time when the notion of a South-Slavic identity was widely disputed. Under the direction of M. Kispatic, the first PhD in the natural sciences at the University of Zagreb, the seismological committee collected observations of earthquakes throughout Croatia and Slavonia, as well as in Dalmatia and Istria (in the Austrian half of the Monarchy) and in occupied Bosnia. The earthquake catalogues appeared regularly in the proceedings of the Academy. Kispatic, the author of popular works on the geology and 'natural-cultural' history of the South-Slavic Lands, also compiled a historical catalogue of earthquakes in the region, published in 1891–92. For this task Kispatic was supplied with medieval and modern documents by a circle of prominent intellectuals in Zagreb, who were in the midst of penning the first contributions to a Croatian national history.[28] The committee gained significantly in prestige in 1893 when it was joined by Andrija Mohorovičić, a successful meteorologist soon to emerge as a groundbreaking seismologist. According to Kispatic, Mohorovičić 'found quite a number of patriots in Croatia and Slavonia who send in, on an almost daily basis,' reports of seismological as well as meteorological phenomena.[29] In this sense, the South-Slavic seismological network, like those in Styria and Carinthia, relied on the association between seismology as *Erdbebenkunde* and regional 'patriotism'. One might even argue that the seismological committee, by coordinating earthquake observation across the South-Slavic Lands, helped to naturalize the region's claim to nationhood.

Wähner's study of the Zagreb quake stood in many ways as a model of the 'monographic' method. Yet it also raised the fraught question of how *Erdbebenkunde*, a quintessential example of local knowledge, might be internationalized. As Wähner acknowledged of his Croatian collaborators, 'I was greatly aided by their knowledge of the land and people.'[30]

Elements of observation

In order to make the best use of the public as volunteer observers, Austrian scientists built on the model of the Swiss Earthquake Commission, founded in 1878. Canaval adopted the Swiss questionnaires and Hoernes adapted their instructions for volunteer observers.[31] The instructions sought to inculcate in the observer an appropriately critical attitude:

> The sense organs should be kept alert in the correct state of tension, without exciting them to a state that would exaggerate sensations. If the reporter does not feel himself entirely certain of an observation, he should not withhold from us the uncertain observation, but rather mark

it in his report as uncertain ... Therefore: observe conscientiously, but have no fearfulness with respect to us in the transmission of observations. All that is genuinely observed is welcome, even what is perceived in uncertainty, as long as it is marked as such.[32]

As observers learned to calibrate the uncertainty of their own observations, they would also acquire something of the epistemic foundations of empirical science.

These small-scale observing networks were remarkably inclusive. The Styrian network formed by Hoernes in 1880 consisted primarily of members of the professional classes, judging by their titles: teachers, physicians, pharmacists and civil servants. There were also, however, participants without titles. Moreover, the individuals named in Hoernes' reports drew their testimony from a far broader segment of the local population. In one case, Hoernes received, via an academic colleague, a letter written by an untitled resident of Kappel, in the mountains of present-day Slovenia, which contained (in Hoernes' judgement) 'very interesting data on the perception of the earthquake of November ninth in the Sulzbacher Alps'. The writer reported that two farmers with whom he had spoken had felt a single impact, though they were unsure of its direction. He continued: 'I sent my boy to Sulzbach in order to conduct a survey, since I had had no success by writing. He made enquiries of several farmers, who had felt the quake in Sulzbach strongly – in two impacts. The time and direction of the impacts could not be determined, since the parish priest in Sulzbach, from whom I hoped for accurate data, had not felt the quake at all.'[33] Despite expectations, then, the priest proved a less valuable informant than the farmers, even if the reports of the latter were insufficiently 'precise'. Surprisingly, seismological research was being conducted by a juvenile and some peasants, at three removes from a scientific expert.

Earthquake researchers also collected oral testimony from witnesses as they made their way through an affected region.[34] The Carinthian geologist Richard Canaval gave a vivid picture of his own tactics in a report on his investigation of the Gmünd earthquake of November 1881. His interactions with eyewitnesses were far more intimate and reciprocal than written correspondence allowed:

> I asked the observers with whom I was able to speak to lead me to the place where they had felt the earthquake, to place themselves in the position they had been in when it had occurred, and to show me the direction in which they had perceived the impact. With the help of a good compass the direction was then determined. It was possible in this way, with the aid of suitably applied questions (for example, which wall appeared to be shaken first) and other data (direction in which objects swung, rolled away or fell over, etc.) to correct many erroneous statements.[35]

In cases like this, observers were taught to disaggregate their experience of an earthquake into basic factual statements of time, direction and other physical 'elements'. Seismologists were then in a position to put these pieces together into an approximately complete picture of the physical phenomenon. It was only through the combination of the partial perspectives of individual observers that such a picture could emerge.

Ljubljana, 1895

Not until fifteen years after the Zagreb quake would scientists begin to organize earthquake observation across the entire Austrian half of the Monarchy. This apparent delay had several causes. The first involved disciplinary dynamics: in 1880, the *landeskundlich* approach seemed perfectly suited to answering geologists' pressing questions about crustal tectonics and orogenesis; it was not until the development of more sensitive seismographs over the following two decades that earthquake research would be reoriented around questions of the earth's internal structure, requiring larger-scale coordination of instrumental observations. Second, the infrastructure that would make it possible to standardize observations on such a large scale was still under construction in 1880. Railroad time had not yet been coordinated, and many parts of the empire did not yet receive telegraphic time signals. As Hoernes found to his shock in 1880, even clocks at neighbouring rail stations sometimes differed by several minutes.[36] Last, political resistance to scientific centralization should not be underestimated.

The event that raised the call for action in Vienna occurred at the end of Easter Sunday, 1895. In Ljubljana, the capital of the crownland of Carniola, most residents were snug in bed after the holiday when an unusual 'buzzing' disturbed their rest. This was quickly followed by 'powerful droning, rattling, thundering and rumbling'; chimneys crashed to the ground and church towers wavered. Over the course of the night, 30–40 further shocks ensued. The population fled the city and did not return until morning, when they found the city in ruins – barely an exterior wall without cracks, and interiors littered with debris. Only two deaths were attributed directly to the earthquake, but 10 per cent of the city's buildings were condemned to demolition.[37]

Because of the holiday, most of the Monarchy learned of the disaster in Ljubljana only from Tuesday's newspapers. The *Neue Freie Presse* reported 'panic' as far from the epicentre as Triest, Fiume and Abbazia. Eduard Suess's son, Franz Eduard, obtained a commission from the education ministry to investigate, and set off on Tuesday evening. Meanwhile, Edmund von Mojsisovics, the vice-director of the Geological Institute in Vienna, circulated questionnaires and persuaded the newspapers to print requests for information. In all, the Geological Institute would collect more than 1300 personal observations from more than 900 locations, and more than 200 negative reports. Franz Eduard Suess's final report would mine all this testimony alongside over 500 published sources. Remarkably, the younger Suess published all these sources as an appendix to his final monograph on the earthquake, which appeared in 1897 and ran to 590 pages. Two hundred pages alone were devoted to the observer reports received by the Geological Institute, which were necessarily 'mostly reduced to keywords; only particularly elaborate and typical descriptions are reproduced word-for-word.'[38]

For the city of Ljubljana, the earthquake was an occasion to modernize. With respect to architecture, as Andrew Herscher has shown, the reconstruction effort laid bare a conflict between local and imperial direction.[39] A similar tension was manifest in the scientific response to the earthquake. With the city still reeling from aftershocks, the Carniolan provincial government expressed an interest in installing a seismograph at the Staats-Oberrealschule in the capital. There was, however, no one willing to supervise its operation until 1896, when the school hired Albin Belar, then an assistant at the Imperial Marine Academy in Fiume. Belar had studied with Eduard Suess in Vienna. He spoke German at home but also knew Slovenian, Croatian and Italian.[40] In 1897, Belar drew up a petition to found, in Ljubljana, Austria's first seismological observatory. He later claimed to have coined the German word for such an institution: *Erdbebenwarte*. It would seem to be a miraculous stroke of fortune by which Belar's petition found its way to the direction of the Krainer Sparkasse, which promised the necessary financial support. But luck had less to do with it than connections. The bank's director was Joseph Supan (or Suppan), a member of the Reichsgericht and one of the foremost German nationalists in Krain.[41] Indeed, the bank itself would become the target of Slovenian nationalist fury over the following years, with Supan even facing accusations of embezzlement. Supan's brother Alexander, meanwhile, was one of Austria's leading physical geographers and the editor of the influential journal *Petermanns Mittheilungen*, in which he often discussed seismological questions. Apparently, an intervention from Alexander Supan had something to do with the continuing support of the Krainer Sparkasse for Belar's observatory.[42] Over the next few years, Alexander Supan himself would become an opponent of what he saw as the centrifugal tendencies of earthquake research in imperial Austria. Conspicuously, Belar's observatory also received the personal support of several Habsburg archdukes.[43] From the start, then, the Ljubljana observatory was implicated in the debates over imperial authority in the South-Slavic Lands.[44]

Belar's interest in seismology lay in instrument design and the interpretation of seismographic records – in his terms, 'the technical aspects of the science'.[45] Yet much of his work involved comparing the instrumental records of his observatory with the human records of the imperial commission.[46] 'It has become an urgent necessity to bridge the large gap that unfortunately still exists today between the earthquake observer using instruments and the observer who relies exclusively on human perceptions, his own or those of others.'[47] Belar sought to correlate the characteristic curves of the seismograph with an array of human descriptions beyond the term 'earthquake': 'earth shock, ground vibration, ground shaking, earth movement', etc.[48] Crucially, 'bridging the gap' between instrumental and human observations depended on communication between seismologists and the public. To this end, in 1901 Belar founded the monthly journal *Die Erdbebenwarte*, a unique forum that addressed both expert and popular audiences. It covered the latest research in seismological geophysics and geology as well as issues of seismic safety. He also lectured frequently to popular audiences in Ljubljana and elsewhere in the monarchy, and was a correspondent for newspapers in Vienna, Berlin, London and New York.[49] Belar helped to forge a seismological language that bridged scientists, citizens and instruments.

The Imperial Earthquake Commission

Just ten days after the Ljubljana earthquake, the Academy of Sciences in Vienna voted to establish a commission for the 'more intensive study of seismic phenomena in the Austrian Lands'. The initial form of the Earthquake Commission remained significantly decentralized: each crownland had one reporter responsible for recruiting observers, for mailing, collecting and compiling questionnaires and for investigating significant earthquakes within his province. There was thus no direct contact between the commissioners in Vienna and the volunteer observers.[50] In practice, this scheme was further decentralized by the provision of one reporter each for the German- and Czech-speaking regions of Bohemia and the German- and Italian-speaking regions of the Tyrol. Paradoxically, these were geographical rather than social divisions, meaning that each reporter was still required to be bilingual. The Commission's second task, that of collecting historical information on past quakes, was to be organized according to 'appropriate regional sections, so for example the Alps, the Sudetenland, and the Karst region'.[51] This, too, proved tricky to implement. The problem was, at core, the familiar conflict between the territorial and linguistic divisions of the Monarchy. Earthquake observing was uniquely susceptible to this tension, dealing as it did directly with both the land and with people's perceptions of the land.

For instance, the Czech-Polish physicist Václav/Waclaw/Wenzel Laska found that the preparation of a treatise on 'The Earthquakes of Poland' was problematic. 'I have called my project "The Earthquakes of Poland", but more accurate would be "The Earthquakes in the Polish Historical Sources", because in this I did not by any means think of political, but rather of historical borders, and those not so much of the land as of the sources.' Poland, of course, was not a political entity in 1900, having been swallowed by its neighbouring states over the previous century and a half. Although the Austrian Kaiser was far more tolerant of publications in Polish than the Russian Tsar, and Laska benefited from the collection of the Lemberg (L'viv) Ossolineum, his task proved frustrating: 'The acquisition of the necessary literature created great difficulties for me, because so many fundamental texts could not be found in the Austrian libraries. It seems superfluous to add another request for the kind loan of documents necessary to me, which in the present paper are marked as inaccessible to me.'[52]

The recruitment of observers – at no pay – also proved a challenge. A free copy of any volume in which one's observations appeared was all the Commission offered. As the director Mojsisovics wrote five years into the Commission's work:

> The collection of observational data from the observers scattered across the province is far more difficult than it might appear to those further removed from the situation. It requires the unlimited attention and the constantly renewed initiative of the reporters, to ensure that the individual observers send in their observations. We require more than any other branch of natural science the participation and cooperation of a broad class of the population. We must therefore make an effort to cultivate the awakening interest of the public; thus we make clear, by publishing the essential contents of the observations received and by listing the names of our collaborators, how valuable and important to us the prompt cooperation of the public is to the fulfilment of our task. As soon as the institution of earthquake-reporting has become more habitual [*mehr eingelebt haben wird*], all participants will be eager to serve the cause of our efforts, conscious of supporting a purely scientific enterprise and making contributions to its progressive development.[53]

In its recruitment efforts, the Earthquake Commission was repeatedly forced to revisit the question of the languages in which its network would operate. Originally, the questionnaires and instructions were printed only in German. In 1898, they were translated into Czech, Slovenian, Croatian and Italian (though not Polish or Ruthenian, because of the low seismicity of Galicia and Bukovina). In 1901 Belar requested that the portion of the annual *Chronicle* concerning Dalmatia also be printed in Croatian. According to his own reports to the Commission, 382 of his observers wrote in Croatian, compared with only 26 in German and 15 in Italian.[54] The Commission responded that it would be impossible to translate this section of the *Chronicle*, because to do so would unleash analogous demands from the other nationalities, and the cost of satisfying them all would be prohibitive. The Commission recommended that Belar have the chronicle printed instead in a local newspaper.[55] A month later Belar complained to the Commission that his efforts to recruit observers by means of articles in Croatian and Italian newspapers, translated by a professor at the commercial academy, had so far been fruitless. On the other hand, the district school inspectors had promised that they would 'with all their energy … make the negligent observers from the teaching profession mindful of their duties with respect to the reporters.'[56]

As this suggests, the reliance on observers from the teaching profession was also problematic. Belar explained that the difficulty of expanding the observing network in Dalmatia was due in part to 'relocations' [*Versetzungen*] of teachers.[57] As historians have recently noted, teachers in late imperial Austria often played the role of nationalist missionaries. Teachers were prominent in the burgeoning nationalist organizations, which went to great lengths to establish schools in which children would be educated in their putative 'mother tongue'. This was particularly true in the areas of linguistic and ethnic mixing that nationalists referred to as the monarchy's 'borderlands'. New teachers arrived ever more frequently in these often rural areas, seeing themselves as agents of both modernization and nationalization. Often they did not stay put for long – presumably, a projected school might not materialize, a school might close for lack of students, or another school elsewhere might beckon.[58] Thus the reporter for the Italian Tyrol lamented in 1901:

> The observations from the side of the school administrations in the individual counties have proved of little value. Relatively few questionnaires have been received about the shocks that have been generally perceived. The reason probably lies above all in the situation that the teaching staff in this region is hardly stable or fixed, and therefore frequent switches occur. For that reason the questionnaires lying with the individual school administrations easily sink into oblivion. In addition many of the teaching staff leave their place of employment during the summer months in order to return to their homelands [*um sich in die Heimat zu begeben*].[59]

Given the patriotic connotations of naturalist activities in the Habsburg lands, the recruitment of earthquake observers had something in common with forms of nationalist recruitment. Historians of imperial Austria have recently described the lengths to which nationalists went to overcome what now appears to have been widespread indifference to their cause. In the Bohemian town of Budweis in the 1880s and 1890s, for instance, Czech and German nationalists used a variety of tactics to polarize the population: they split the town's gymnastics and singing societies in two, manipulated the census data by intimidating neighbours into identifying with one language or the other, and urged their neighbours to patronize only businesses on their own 'side'. Numbers mattered profoundly in this contest. Because of the peculiarities of Austrian electoral laws, nationalists viewed the census as a zero-sum battle between Czechs and Germans.[60]

A similar sense of competition seems to have infused the recruitment of earthquake observers in Bohemia, where the areas of highest seismicity coincided with predominantly German-speaking regions. In 1901 the reporter for Czech-speaking Bohemia recounted his efforts to expand his network by sending instructions and questionnaires to 250 school directors, 200 postal servants, 100 railway officials and 50 non-civil servants, resulting in 200 new volunteers. Anticipating that some might find he had spent too much time on this effort, he insisted that 'this method seems to be the only one that accomplishes the purpose quickly.' By the end of 1901 German Bohemia had 608 registered observers," while the Czech region had 550. The reporter calculated that the density of observers was still twice as high in the German region: three square-kilometers per observer, versus six square-kilometers per observer in the Czech region. [61] At one point, this reporter demanded a Czech version of one of his publications from the Commission, because he supposedly could not distribute the German version to his 'Czech' observers. For their part, the Commission ordered that he rewrite the first version of this article, in part because the observations were supposed to be arranged 'according to other than linguistic viewpoints'.[62] Once again, we see how seismology's human-centred methods could be seen as requiring an organization according to either cultural or physical geography. Given the work that went into building these separate yet parallel networks, it was a fitting irony that the first reporter for German-speaking Bohemia was named Dr. Tzeckh.

The question of centralization

The fate of the imperial Earthquake Commission would play out, like imperial politics overall, as a struggle between proponents of centralization and decentralization. The debate paralleled that over the organization of weather and climate research in the Monarchy.[63] The arguments in favour of centralization stressed, first, the need for a speedy and concerted response to major seismic disasters; second, the value of sharing data between crownlands for the thorough study of mid-size earthquakes; and third, the importance of communication among the new generation of seismological observatories, whose sensitive instruments, it was hoped, would contribute to knowledge of the deep structure of the earth as a whole. Against this current, however, some members of the Earthquake Commission argued for the continuing value of a decentralized approach to seismic research. They insisted that seismicity could be studied only region by region, by scientists fluent in the local language, versed in local history, both human and natural, and familiar to the local community of observers.

In this way, the spatial scale on which seismicity would be studied became, in part, a matter of imperial politics. The movement to centralize the Austrian Earthquake Commission opened with a diatribe by Alexander

Supan in *Petermanns' Mittheilungen*, the geographical journal he edited. Supan strongly supported the project of building an earthquake-observing network, but he was acidly critical of the form the network and its reports had taken. He had hoped to find in the annual *Chronicle* 'a comprehensive picture of the seismic activity of Austria. But in this respect we are completely disappointed.'[64] This desire for a synthetic perspective on the Monarchy was one of Supan's defining traits as a geographer. His 1889 *Austria-Hungary* argued that the monarchy was, 'despite the diversity of its orographic components, nonetheless a geographic unity'.[65] From this perspective, the Chronicle's organization by crownland was 'mistaken in principle'. Thus, for instance, even though the neighbouring crownlands of Krain and Görz fell to a single reporter, this individual treated each crownland separately. 'Whoever wants to be informed about the distribution of a seismic phenomenon must compile the material themselves', Supan complained. In one case, the *Chronicle* had printed a report on a tremor in Bohemia without any mention that shaking had also been observed in Upper Austria. 'All this suggests that in the Austrian organization there is more than one weak point.'[66]

Supan hammered out its inconsistencies. The 'seismically peaceful' provinces of Moravia, Silesia, Galicia and Bukovina proved 'entirely apathetic' to seismology. Inexplicably, Ljubljana, the 'seismic capital' of Carniola, had been assigned to the reporter for Görz instead of to Belar, who was instead responsible for Dalmatia. Supan might have added here that Belar was himself fighting with Eduard Mazelle, the director of the marine observatory in Trieste, over access to the reports of observers along the Adriatic coast; Mazelle insisted on collecting the reports of the coastal area himself.[67] Supan hinted that this irrational system reflected the political tendencies of the Monarchy in recent years. 'The tiresome nationalities dispute plays a role even in this purely scientific question, in that Bohemia and Tyrol were each divided into two linguistic regions with separate reporters.'[68] The solution to this chaos, in Supan's view, could only come from an authoritative central office in Vienna. Only Viennese oversight could ensure that provincial reporters responded promptly and thoroughly to tremors and that information flowed smoothly across the borders of the crownlands.

Supan's polemic seems to have had a hidden target. Repeatedly, his criticism focused on the reporter for Carniola and Görz-Gradiska, the one he castigated for publishing on the two crownlands separately. It was 'characteristic', Supan added, 'that the reporter for Carniola makes do with private reporters in Ljubljana and takes not the least notice of the seismic observatory'. Again, Supan judged that this reporter's article on the Ljubljana aftershocks was 'seriously diminished in value' by the lack of a map. Finally, Supan counted eighteen shocks in Krain that had been reported only in a single location. 'That is surely highly conspicuous. Is the reporter immediately informed of every event? Has he made an effort immediately to collect reports in the neighboring areas?'[69]

Who was to be blamed for these failings? Who was the reporter for Krain and Görz-Gradiska in 1900? The answer is Ferdinand Seidl, a teacher at the Realschule in the capital city, Görz/Gorica/Gorizia. Like his work on the climatology of Carniola, which he described as a contribution to the *Landeskunde* of the crownland, Seidl's approach to seismology was distinctly *landeskundlich*. Seidl went on to author the first popular work on geology in Slovenian. During the First World War he published a pamphlet on the question of the South-Slavic borders; after the war, he directed the meteorological observatory in Ljubljana, in the Republic of Yugoslavia. As a reporter for the Earthquake Commission, Seidl worked primarily with Slovene-language reports, along with a small number in Italian and German. His reports made clear the value he attached to lay observations: 'The reports were made in many cases with the utmost care, and people were visibly at pains to leave no question of the questionnaire unanswered, in order to furnish a broadly useful report for the needs of scientific research.'[70] He was fastidious in his respect for the observers' own phrasings: 'In the preparation of this final report, as in previous years, particular care was taken not only to be faithful to the original reports, even if reducing them to keywords, but also that characteristic personal expressions and descriptions would be conveyed as literally and with as much fidelity to their meaning as possible.'[71] Seidl clearly pursued earthquake research as a form of patriotic *Landeskunde*. Against Supan's call for centralization, Seidl would likely have argued that seismic investigations relied on local knowledge and therefore required local control. As it happens, Seidl did send the Earthquake Commission a reply to Supan's article soon after it was published, and he requested that it be printed in the Academy's proceedings. His request was denied, however, due to the 'polemical nature' of his paper, although the Commission welcomed its publication elsewhere.[72] No trace of it remains.

The next critic to weigh in on the question of centralization was Giovanni Agamennone, the director of the Earthquake Service in Rome. In the newly unified Italy of the 1870s the coordination of both seismology and meteorology had been carried out explicitly as an act of secular nation-building.[73] Hence Agamennone took evident pride in contrasting Italian centralization with Austrian decentralization. The Italian network was under the direct control of the bureau in Rome, which could thus always locate the approximate epicentres when earthquakes occurred. The central bureau could, 'in the interest of science, and without losing time, seek further information at these important places, but also reassure the population in the case of stronger earthquakes.' This was impossible in Austria, since the observers communicated only with the provincial reporters, not with a central bureau. In the case of an earthquake affecting several districts at once, 'no adequate idea of the strength and extent of the various seismological phenomena' was possible, at least not until the annual reports from the provinces were collected in Vienna several months later. Agemennone also judged that the publications of the Italian service were better organized, since they ordered earthquakes chronologically rather than by district. 'Altogether, as far as the seismological service goes, in Italy the concept of centralization has prevailed, while in our neighbouring empire decentralization is favoured.'[74]

Belar next offered a balanced response to Agamennone. 'An organization according to the Italian model is not so easily realizable here in Austria,' he admitted, 'because of the well known linguistic diversity of the individual observers; it would thus be necessary at the respective central office for a whole system of linguistically skilled functionaries to analyse the incoming reports, which would in any case involve great expense.' Yet Belar made clear that the Austrian service could not do without an authoritative central bureau that would organize the observations 'without consideration of political divisions' and maintain contact with foreign institutes concerning events on the borders. To emphasize his point, he recounted a visit he had made to Agamennone in Rome: 'On the large table a large-scale military map lay spread out, and Prof. Agamennone was laying differently coloured pieces of paper on the individual places, in the manner that newspaper readers indicate the movements of two mobilized armies for the sake of orientation. In this way, in Agamennone's opinion, one could most quickly gain an overview of the location of the epicentrum and of the entire extension of the particular territory struck by the earthquake.'[75] The military metaphor was a reminder that earthquake-reporting in Austria was a matter of public safety and hence of imperial responsibility. Supan had likewise argued that a cartographic representation of the distribution of earthquakes was indispensible for the sake of an imperial overview: 'From our perspective it is necessary that not only the reporters but also the Vienna central bureau be immediately informed of every event and that each observation be immediately recorded cartographically.'[76] Like warfare, disaster response seemed to require central oversight.

Yet the military analogy did not convince everyone. Supan's own journal published a dissenting opinion from the reporter for Upper Austria, Johann Commenda. As the director of the Realschule in Linz, Commenda published on earthquakes as well as on *Landeskunde* in the local popular press. Against the arguments for centralization, Commenda argued for the indispensibility of local knowledge to the tasks of the Earthquake Commission:

> Certainly, though, in smaller crownlands the intermediary of the earthquake reporter must be called on, with the help of the local historical and natural scientific societies and of the local museums, and by means of contact with all suitable organizations, to search in the local papers, chronicles and other published and unpublished historical sources for

recorded information on earlier earthquakes, to sort them and to compile them. Only when this is appropriately carried out will a scientific analysis of the material be recommendable, which then naturally can and must ignore all linguistic and political borders.[77]

In the end, the balance between centralization and decentralization was preserved. The administration of the expanding network was transferred from the private Academy of Sciences to the state-run Central Institute for Meteorology and Geodynamics (ZAMG). While the ZAMG provided the necessary degree of central organization, it also managed to keep in place the lively intercourse between local scientists and the interested public. Its success can be measured, for instance, by the over 2000 reports collected in the course of two large earthquake swarms in Bohemia in 1900 and 1903.[78]

This finely balanced system collapsed in 1918. For Albin Belar, the defeat of Austria-Hungary meant the loss of his *'Erdbebenwarte'*, the perch from which, for twenty years, he had surveyed each spasm of the planet. The new Yugoslav government, suspecting him of German nationalism, seized the contents of his observatory, dispossessed him of his apartment, and sent most of his instruments to Belgrade. While his children emigrated to the US, Belar moved what remained of his instruments to the Triglav valley in the Julian Alps, where he owned a villa designed by his friend Max Fabiani, the architect of Ljubljana's post-disaster modernization. The Triglav valley was one of the 'natural monuments' for which Belar had first sought state protection in 1903 – one of the earliest proposals for a nature preserve in European history.[79] The Triglav National Park was finally established in 1924, with no acknowledgment of Belar's contribution. On the door to Belar's villa hung a sign in Slovenian: 'Silence, silence, for here resides a solitary man who uses his instruments to listen to the sound of earthquakes being born in the heart of the Earth.'[80] It was a poignant reminder of the difficulty of finding a suitable local window onto global physics.

Conclusion

If we return now to the *Analysis of Sensations*, Mach's use of the earthquake should seem less puzzling. As he would have known – whether from discussions in the Lotos society, from Austrian scientific journals or even from the *Neue Freie Presse* – the monographic method of earthquake study was being held up as a model of empiricism and an antidote to popular unreason. Moreover, this method embodied Mach's principle of analysing experience into its most basic elements. Only through the combination of the partial perspectives of individual observers could a complete picture emerge. Tellingly, the subject of Mach's passage on the earthquake in the original German is not the generic *man* ('one') but the first-person plural *wir* ('we'). Such a panoramic view of a seismic event could not be the property of a single psyche. According to Mach, Wähner and any number of their colleagues, there was no individual human perspective from which it was possible to gain complete knowledge of an earthquake. Seismology was, for this reason, an ideal instantiation of Mach's principle that physics would do well to abandon the fiction of the ego. For Mach, 'The primary fact is not the I, the ego, but the elements (sensations).'[81]

Mach's denunciation of 'self-centred views [*egoistische Anschauungen*]' in *Analysis of Sensations* was as much a political critique as an epistemological objection. He likened the metaphysical notion of the bounded self to the social phenomena of 'class bias' and 'national pride', and called on the 'broad-minded inquirer' to renounce them. Behind this appeal lay bitter personal experience. Just as Suess resigned his rectorship in 1889, so had Mach in 1883, both apparently worn down by the politics of nationalism.[82] Both longed to escape the 'egoism' of modern politics for a collaborative, supranational science.[83] Beyond egoism lay the possibility of 'complete knowledge'. To Mach's Austrian colleagues, his seismic example of complete knowledge in the *Analysis* would have evoked a thriving culture of earth-quake observation – a culture marked at once by the provincial practices of *Landeskunde* and the transnational concerns of a multinational empire. This does not imply a direct influence from scientific practice to epistemology or vice versa. Rather, both the macroseismic surveys and Mach's epistemology reflected a peculiarity of Habsburg science: the function of translation.

Implicit in the work of the earthquake commission was a philosophical resistance to reduction. No single perspective on the event was privileged. Instead, the full array of impressions – visual, aural, tactile, instrumental – were taken as irreduceable elements of knowledge of the whole. The epistemic model was one of translation. Data derived from human, instru-mental, geological, historical and even animal sources were taken to be irreduceable yet comparable. This model both derived from and bolstered the research practices of the imperial commission, which built a multilin-gual seismic archive on the basis of questionnaires and intensity scales that were assumed to be unproblematically translatable between the various languages of the Monarchy. The goal was emphatically not to reduce the observational reports to a single 'language of nature', as the curves of self-registering instruments were often described in the nineteenth century.[84] To a North American seismologist of this period it seemed self-evident that 'reliable seismograms furnish us the data for all our computations about an earthquake…in short, the seismograms give the story of the earthquake, just as the spectrogram gives the story of the distant stars.'[85] For Mach, Suess and their colleagues, however, no seismogram could ever tell the story of an earthquake – most basically, because the earth was not a distant star. The goal was not to explain the earthquake in terms of a single physical cause, nor to reduce its description to a single variable. Instead, all possible perspectives on an earthquake, like the multiple languages of the Empire, were in principle granted equal status. 'Complete knowledge' of an earthquake corresponded not to a mathematical law, nor to an instrumental trace, but rather to a multilingual archive.[86] Yet Mach's epistemological fantasy hid the immense labour of constructing such an archive. He made translating between these various registers appear to be an instantaneous, effortless process. The history of the imperial Earthquake Commission demonstrates, on the contrary, that translation was labour – and it was labour that the Earthquake Commission was at times unwilling to fund. Commission members were not even aware of the extent of the work involved, since linguistic equivalences were constructed primarily through discussions between the reporters and local observers. For instance, the German, Croatian and Czech words 'Getöse', 'tutnjava' and 'rachot' became the standard descriptions for the rumbling noises accompanying earthquakes, rather than many other possibilities in each language.[87] Even less visible was the subsequent work of translating the official results of the Commission from German into languages in which they were more likely to be read by potential observers. Arguably, the invisibility of such labour is endemic in the history of science.[88] By eliding the labour and politics of translation, Mach contributed to the discipline's enduring tendency to view translation merely as an epistemological metaphor.[89]

Notes

1. Ernst Mach (1897) *Contributions to the Analysis of the Sensations*, trans. C.M. Williams (La Salle, Ill.: Open Court), 155. Previously, in his studies of the history of physics, Mach had cited earthquakes as events that disrupt the normal experience of gravity, producing a 'sensation of constant ascent', and rendering useless a terrestrial system of coordinates. In the *Analysis of Sensations* Mach was concerned with earthquakes as a problem not of mechanical explanation but of observation.
2. Jan Kozák and Axel Plešinger (2003) 'Beginnings of Regular Seismic Service and Research in the Austro-Hungarian Monarchy', *Studia Geophysica et Geodaetica*, 47, 99–119, 757–91, here 105; Martina Lehner (1995) *'Und das Unglück ist von Gott gemacht...': Geschichte der Naturkatastrophen in Österreich* (Vienna: Praesens), 143. Kozák and Plešinger briefly describe the immediate origins of the Austrian Earthquake Commission.
3. *Lotos: Zeitschrift für Naturwissenschaften* 11 (Nov. 1869), 165.
4. *'Da wird kein Stoß verspürt, keine Flüßigkeit wird verschüttet, ohne daß die Richtung, in der es geschah, am andern Morgen in der Zeitung bekanntgegeben würde ....Tagaus, tagein, heute, morgen ewig. Bis das Weltgebäude wirklich zusammenkracht...'* Karl Kraus (1908) 'Das Erdbeben', *Die Fackel*, 9, 16–24.
5. Mott Greene (1982) *Geology in the Nineteenth Century: Changing Views of a Changing World* (Ithaca: Cornell University Press); Jürgen Strehlau (2006) '"...Earthquakes occur on specific points and lines which...mostly coincide with traceable fracture lines...': Eduard Sueß' study of earthquakes in Lower Austria and southern Italy...', *Berichte der Geologischen Bundesanstalt*, 69, 67–68.

6. Melchior Neumayr (1887) *Erdgeschichte*, vol. 1 (Leipzig: Bibliographisches Institut), 305–06.

7. Other cases in which non-scientists played important roles in observation networks in the physical earth sciences are described in: Michael Reidy (2008) *Tides of History: Ocean Science and Her Majesty's Navy* (Chicago: Univ. of Chicago Press); Graham Burnett (2005) 'Matthew Fontaine Maury's "Sea of Fire": Hydrography, biogeography and providence in the tropics', in Felix Driver and Luciana Martins (eds.) *Tropical Visions in an Age of Empire* (Chicago: Chicago Univ. Press), 113–36 ; Fabien Locher (2008) *Le savant et la tempête* (Rennes: Presses Universitaires).

8. Gregory Clancey (2006) *Earthquake Nation: The Cultural Politics of Japanese Seismicity, 1868–1930* (Berkeley: University of California Press); Paul White (forthcoming) 'Darwin, Concepción, and the Geological Sublime', *Science in Context*.

9. Eduard Suess (1880) 'Ueber die Erdbeben in der österreichisch-ungarischen Monarchie', *Monatsblätter des Wissenschaftlichen Clubs in Wien*, 2/3, special supplement.

10. From 1867, the Monarchy had a complex 'dual' government, in which Austria and Hungary were bound by a joint army, common foreign policy and economic agreements, as well as by allegiance to the Emperor. Within each half of the Monarchy, burgeoning national movements competed for privileged positions, the Czechs and Poles in Austria and the Croatians in Hungary winning limited concessions to independence.

11. *Für Laibach: zum besten der durch die Erdbeben-Katastrophe im Frühjahre 1895 schwer betroffenen Einwohner von Laibach und Umgebung*, Die Genossenschaft der bildenden Künstler Wiens, 1895.

12. 'Naturlehre', *Der Floh*, 21 April 1895.

13. Reprinted in Ferdinand Hochstetter (1880) 'Ueber Erdbeben, mit Beziehung auf das Erdbeben vom 9. November 1880', *Monatsblätter des Wissenschaftlichen Clubs zu Wien*, 2/3, 1–14, here 11.

14. Neumayr, *Erdgeschichte*, 306.

15. Miksa (Max) Hantken (1882) *Das Erdbeben von Agram im Jahre 1880* (Budapest: Legrady).

16. See Dragutin Skoko and Josip Mokrović (1982) *Andrija Mohorovičić* (Zagreb: Školska knj.), 94–95.

17. Neumayr, *Erdgeschichte*, 306.

18. Franz Wähner (1883) *Das Erdbeben von Agram am 9. November 1880* (Vienna: Akademie der Wissenschaften), 5.

19. Wähner, *Das Erdbeben von Agram*, 288.

20. Eduard Suess (1873) *Die Erdbeben Nieder-Österreichs* (Vienna: k. k. Hof- und Staatsdruckerei).

21. *Getöse* is typically translated into English as 'roar', *tutnjava* as 'boom'. Wähner, *Das Erdbeben von Agram*, 153.

22. Wähner, *Das Erdbeben von Agram*, 289.

23. Rudolf Hoernes (1893) *Erdbebenkunde: Die Erscheinungen und Ursachen der Erdbeben, die Methoden ihrer Beobachtung* (Leipzig: Veit & Co.).

24. Lynn Nyhart (2009) *Modern Nature: The Rise of the Biological Perspective in Germany* (Chicago: University of Chicago Press), esp. 253–56.

25. J.F.J. Schmidt (1858) 'Untersuchung über das Erdbeben am 15. Jänner 1858', *Mittheilungen der k. k. Geographischen Gesellschaft*, 2. The Sillein quake should be recognized as a formative event for the monographic method in Austria.

See Jiři Vaněk and Jan Kozák (2007) 'First macroseismic map with geological background (composed by L.H. Jeitteles)', *Acta Geophysica*, 55, 594–606.

26. Rudolf Hoernes (1902) 'Erdbeben und Stoßlinien Steiermarks', *Mittheilungen der k. k. Erdbebenkommission* 7/1–2.

27. See, for example, Alan Confino (1997) *The Nation as a Local Metaphor: Württemberg, Imperial Germany, and National Memory, 1871–1918* (Chapel Hill: UNC Press); Celia Applegate (1990) *A Nation of Provincials: The German Idea of Heimat* (Berkeley: University of California Press); David Blackbourn and James Retallack (eds.) (2007) *Localism, Landscape, and the Ambiguities of Place: German-Speaking Central Europe, 1860–1930* (Toronto: University of Toronto Press).

28. These included Sime Ljubić and Ivan Kostrenčić. See Paola Albini (2004) 'A survey of past earthquakes in the Eastern Adriatic', *Annals of Geophysics*, 47, 675–703.

29. Quoted in Skoko and Mokrović, *Andrija Mohorovičić*, 96.

30. Wähner, *Erdbeben von Agram*, 4.

31. Richard Canaval (1882) 'Das Erdbeben von Gmünd am 5. November 1881', *Sitzungsberichte der kaiserlichen Akademie der Wissenschaften IIa*, 86, 353–409; Rudolf Hoernes (1880) 'Erdbeben in Steiermark während des Jahres 1880', *Mittheilungen des naturwissenschaftlichen Vereines für Steiermark*, 65–114. On the Swiss commission: Deborah Coen (forthcoming) 'The Tongues of Seismology in Nineteenth-century Switzerland', *Science in Context*.

32. Albert Heim (1879) *Die Erdbeben und deren Beobachtung* (Zurich: Zürcher & Furrer), 24.

33. Hoernes, 'Erdbeben in Steiermark', 114.

34. E.g. Suess (1874); Wähner, *Das Erdbeben von Agram*; A. Faidiga (1903) 'Das Erdbeben von Sinj am 2. Juli, 1898', *Mittheilungen der Erdbeben-Commission der Akademie der Wissenschaften Wien*, 17.

35. Canaval, 'Das Erdbeben von Gmünd', 354 (my emphasis).

36. Hoernes, 'Erdbeben in Steiermark', 73.

37. F.E. Suess (1897) *Das Erdbeben von Laibach am 14. April 1895* (Vienna: k. k. Geologische Reichsanstalt), 7.

38. Ibid., 203.

39. Andrew Herscher (2003) 'Städtebau as Imperial Culture: Camillo Sitte's Urban Plan for Ljubljana', *The Journal of the Society of Architectural Historians*, 62, 212–27.

40. Renato Vidrih and Jože Mihelič (2010) *Albin Belar: Pozabljen Slovenski Naravoslovec* (Ljubljana: Didakta).

41. (1903) *Biographisches Jahrbuch und deutscher Nekrolog*, vol. 7, ed. Anton Bettelheim (Berlin: Georg Reimer), 114.

42. See Belar's flattering obituary of Joseph Suppan in *Die Erdbebenwarte*, 2, (1902/3), 99–100.

43. Vidrih and Mihelič, *Albin Belar*, 127.

44. On the emergence of German and Slovenian identities through 1848 see Joachim Hösler (2006) *Von Krain zu Slowenien* (Munich: Oldenbourg).

45. Vidrih and Mihelič, *Albin Belar*, 135.

46. See, for example, his chart comparing three years of human and instrumental records in (1899) *Laibacher Erdbebenstudien* (Ljubljana: Verlag der Erdbebenwarte), Tafel II.

47. Albin Belar (1906/07) 'Was erzählen uns die Erdbebenmesser von den Erdbeben', *Die Erdbebenwarte*, 6, 101–10, here 101.

48. Ibid.

49. Vidrih and Mihelič, *Albin Belar*.
50. Kozák and Plešinger, 'Beginnings of Regular Seismic Service' (cit. Note 2).
51. Hans Commenda (1907) 'Aufruf zur Einsendung von Nachrichten über Erdbeben und andere seltene Naturereignisse', *Jahresber. Verein Naturkunde Linz*, 36, 3–15.
52. Wencel Laska (1902) 'Die Erdbeben Polens', *Mittheilungen der Erdbeben-Commission der kaiserlichen Akademie der Wissenschaften in Wien*, N.F. 8, 1–2.
53. Edmund v. Mojsisovics (ed.) (1901) 'Allgemeiner Bericht und Chronik der im Jahre 1900 im Beobachtungsgebiete eingetretenen Erdbeben', *Mittheilungen der Erdbeben-Kommission der Kaiserlichen Akademie der Wissenschaften*, Neue Folge no. 2, 6.
54. Mojsisovics, 'Allgemeiner Bericht und Chronik', 71.
55. Archive of the Austrian Academy of Sciences (AAAS), Erdbeben – Kommission, Kart. 1/Mp. II, Sitzungsprotokolle III Protokoll der Sitzung der Erdbeben – Commission (no. 41/1901), 10 January 1901.
56. AAAS, Erdbeben – Kommission, Kart. 1/Mp. II, Sitzungsprotokolle III Protokoll der Sitzung der Erdbeben – Commission (no. 1112/1900), 9 November 1900.
57. AAAS, Erdbeben – Kommission, Kart. 1/Mp. 2, Korrespondenz, Berichte III, Bericht Belars an die Akademie der Wissenschaften (no. 1286/1901), 29 December 1901.
58. Pieter Judson (2006) *Guardians of the Nation: Activists on the Language Frontiers of Imperial Austria* (Cambridge, Mass.: Harvard Univ. Press), chap. 3; Tara Zahra (2008) *Kidnapped Souls: National Indifference and the Battle for Children in the Bohemian Lands, 1900–1948* (Ithaca, NY: Cornell University Press).
59. Mojsisovics, 'Allgemeiner Bericht und Chronik', 168.
60. Jeremy King (2002) *Budweisers into Czechs and Germans: A Local History of Bohemian Politics, 1848–1948* (Princeton: Princeton University Press).
61. Mojsisovics, 'Allgemeiner Bericht und Chronik', 175.
62. '...nach anderen als nach sprachlichen Gesichtspunkten...', AAAS, Erdbeben – Kommission, Kart. 1/Mp. II, Sitzungsprotokolle III Protokoll der Sitzung der Erdbeben – Commission (no. 780/1901), 20 June 1901.
63. See Deborah Coen (2010) 'Climate and Circulation in Imperial Austria', *Journal of Modern History*, 82/4 (December), 839–75.
64. Alexander Supan (1900) 'Die Erdbebenforschung in Österreich', *Petermanns Geographische Mitteilungen*, 46, 143–45, here 144.
65. Alexander Supan (1889) *Österreich-Ungarn* (Vienna: Tempsky), 3.
66. Supan, 'Die Erdbebenforschung in Österreich', 144.
67. AAAS, Erdbeben – Kommission, Kart. 1/Mp. 2, Korrespondenz, Berichte III, Bericht Belars an die Akademie der Wissenschaften (no. 1286/1901), 29 December 1901.
68. Supan, 'Die Erdbebenforschung in Österreich', 144.
69. Ibid., 145.
70. Ferdinand Seidl (1899) 'Übersicht der Laibacher Osterbebenperiode für die Zeit vom 16. April 1895 bis Ende December 1898', *Mittheilungen der Erdbebencommission 12, Sitzungsberichte der kaiserlichen Akademie der Wissenschaften IIa*, 108, 395–430.
71. Mojsisovics, 'Allgemeiner Bericht und Chronik', 25.
72. Archive of the Austrian Academy of Sciences (AAAS), Erdbeben – Kommission, Kart. 1/Mp. II, Sitzungsprotokolle III Protokoll der Sitzung der Erdbeben – Commission (no. 1112/1900), 9 November 1900.

73. Massimo Mazzotti (2010) 'The Jesuit on the Roof', in David Aubin et al. (eds.) *The Heavens on Earth: Observatories and Astronomy in Nineteenth-Century Science and Culture* (Durham: Duke University Press), 58–85.

74. Giovanni Agamennone 'Gli studi sismici nell'Austria-Ungheria', *Bollettino della Societa sismologica italia* 7 (1901), 194–204, here 198.

75. Albin Belar (1901/2) 'Die Erdbebenforschung in Österreich-Ungarn', *Die Erdbebenwarte*, 1, 139–43.

76. Supan, 'Die Erdbebenforschung in Österreich', 145.

77. Hans Commenda (1900) 'Erdbeben und Erdbebennachrichten aus Oberösterreich', *Petermanns Geographische Mitteilungen*, 46, 143–45, here 143; reprinted in *Die Erdbebenwarte*, 6 (1907).

78. Kozák and Plešinger, 'Beginnings of Regular Seismic Service' (cit. Note 2).

79. Vidrih and Mihelič, *Albin Belar*, 87.

80. Vidrih and Mihelič, *Albin Belar*, 149.

81. Mach, *Contributions to the Analysis of the Sensations* (cit. Note 1), 19.

82. On the multiple possible reasons for Mach's resignation see John T. Blackmore (1972) *Ernst Mach: His Work, Life, and Influence* (Berkeley: University of California Press), 80. Blackmore points out that such resignations were not common.

83. On Suess's global tectonics see Andrea Westermann (2010) 'Overcoming the division of labor in global tectonics', paper presented at York University, 1 October.

84. Lorraine Daston and Peter Galison (2007) *Objectivity* (Cambridge, Mass.: Zone Books).

85. Otto Klotz (1920) 'Present status of seismological work in the Pacific', Bul*letin of the Seismological Society of America*, 10, 300–310, here 300.

86. Steve Fuller notes: 'If Planck's principle of unity was *reduction*, Mach's was *translation*.' Steve Fuller (2000) *Thomas Kuhn: A Philosophical History for Our Times* (Chicago: University of Chicago Press), 118.

87. E.g. Josef N. Woldřich (1897) 'Předběžná zpráva o zemětřesení v Pošumaví', *Rozpravy České akademie císaře Františka Josefa pro vědy, slovesnost a umění v Praze*, II/6, 1–6.

88. See Scott Montgomery (2000) *Science in Translation: Movements of Knowledge Through Cultures and Time* (Chicago: University of Chicago Press).

89. See Deborah Coen, 'Rise, *Grubenhund*: On Provincializing Kuhn,' *Modern Intellectual History* 9 (2012): 109–126.

9
Nationalizing Eugenics: The Hungarian Public Debate of 1910–1911

Marius Turda

Introduction

The period before the First World War abounded in national and international debates over the meaning and importance of eugenics. The Hygiene Exhibition, held in Dresden in 1911, the First International Eugenics Congress, held in London in 1912, and the First National Conference on Race Betterment, held at Battle Creek, Michigan in 1914, not only were seminal moments in the history of German, British and American eugenics, but also served to cement the fragile links existing between various eugenic movements throughout Europe and the United States, and functioned as important academic forums for debates on the role of eugenics and heredity within the natural sciences as well as in wider society.[1] These scholarly gatherings became a point of intersection between local and external factors shaping the emergence of eugenics concurrently as a scientific discourse about human improvement and as an ideology concerned with the biological transformation of society. To understand the interplay between these factors, one must demonstrate how eugenic knowledge was expressed in different geographic locations and the multiple networks of cultural and scientific power existing in these locations.

Recent scholarship on the transfer of eugenic knowledge has commendably tried to engage with different geographical locations, previously neglected or marginalized.[2] These scholarly efforts notwithstanding, the eugenic map of Europe in the twentieth century continues to be poorly drawn. The countries of Central, Eastern and South-eastern Europe have, for example, only just been sketched. We have but a preliminary understanding of those creative and complex eugenic cultures existing within this vast topography.[3] In expressing this growing interest in Central European eugenics, this chapter discusses the public debate on eugenics in Hungary, taking place over the course of two years, 1910 and 1911.[4]

The debate was important for several reasons. First, it created an auspicious environment for the nationalization of eugenic knowledge to occur in early twentieth-century Hungary. Although using the internationally recognizable language of evolutionary science, Hungarian eugenicists expressed specifically local imperatives; this was unquestionably a process that actualized the national context, serving as a major incentive for Hungarian academic knowledge about evolution and heredity to be expressed publicly. Eugenics, in this context, was seen as a mechanism capable of decoding particular social and national predicaments, an expression of the ideal of a healthy nation in the face of dramatic demographic and social changes.

Second, the debate illustrates the level of scientific sophistication achieved by Hungarian eugenicists at the time; in other words, their scholarly engagement with emerging European trends in heredity. During the first decades of the twentieth century, Hungarian physicians, anthropologists and sociologists struggled with dilemmas concerning the importance of social hygiene, heredity and eugenics in a similar way to their counterparts in Britain and Germany. Debates between biometricians and Mendelians, for instance, which caused many a methodological chasm amongst evolutionists in Britain at the beginning of the twentieth century, were widely echoed within the Hungarian scientific community, and dominated this public debate on eugenics.

The third reason why the Hungarian debate is important for the emergence of eugenics as an international movement in early twentieth-century Europe relates to its overall political and social message. Eugenic arguments presented at the debate were not explicitly political, but they were overtly social. Organized by the Society of Social Sciences (*Társadalomtudományi Társaság*) and publicized in the journal *Twentieth Century* (*Huszadik Század*), the debate was as much scientific as socio-political. The debate itself – it must be said – did not bring together all Hungarian eugenicists. But all participants were involved, directly or indirectly, with the social and cultural programme advocated by the Society of Social Sciences. This provided a shared intellectual identity, one which the Society of Social Sciences and its journal carved out of Hungarian public culture at the beginning of the twentieth century.

Finally, the Hungarian debate must be placed within its regional context. Considering that this is the first example of a public debate on eugenics in Austria-Hungary, its particulars may help us to understand other national eugenic movements in this region. Austrian, Czech and Polish eugenicists were likewise active at the beginning of the twentieth century, and we are gradually now discovering the innovativeness of many of these eugenicists and their personal and institutional connections as well as how eugenics, more broadly, interacted with its cultural, political and national contexts.[5] In this regional context, Hungarian eugenics gains heightened importance.

The Hungarian debate thus has a double significance for the nationalization of eugenics in the early twentieth century: on the one hand, it gave supporters of eugenics in Hungary the necessary opportunity to synthesize their views on heredity and articulate common programmes of hygiene and racial hygiene; on the other, it added a new dimension to general discussions on social and political transformation, which characterized the evolution of political thinking in Europe after 1900. This is not an overview of Hungarian eugenics at the beginning of the twentieth century, which I have endeavoured elsewhere;[6] rather, I have sought here to highlight this debate's major contribution to the nationalization of eugenic thought in Hungary, in particular.

Before I elaborate on the main eugenic themes discussed during the debate, it is important that the scientific framework be considered, so that the importance the Society of Social Sciences and the journal *Twentieth Century* had in preparing and organizing this event is clarified. I shall then look at some of the most important eugenic themes discussed during the debate, including biometry, Mendelism, degeneration and sterilization. These themes should lead to a more refined understanding of this debate, and should help us to position it not only within the history of European eugenics in general, but also within a more specific programme of social reform and biological improvement envisioned by a new generation of Hungarian intellectuals.

Science without frontiers

The beginning of the twentieth century corresponded in Hungary, and elsewhere in Europe,[7] with growing anxieties about social and national frontiers and the ways in which they ought to be managed politically, particularly in defining which groups were contained within them.[8] Undoubtedly, one cannot consider the emergence of eugenics in Hungary without considering cultural politics at the time. The eugenic debate accented the particular features of Hungarian society but, undoubtedly, the rationale behind it was shaped by wider developments within European science at the time.[9] Both the journal *Twentieth Century* (established in 1900) and the Society of Social Sciences (formed in 1901) were intimately connected to international developments in social and natural sciences.[10]

Moreover, there was an additional element that particularly characterized the relationship between the society and the journal: the rejection of Hungary's intellectual isolationism. As Ervin Szabó declared in his presidential address to the Society of Social Sciences in 1912:

[The Society of Social Sciences] can proudly claim that it was able to overcome its isolation in public life. Those who were formerly indifferent, and whose interest, friendship and help we have managed to win, nay, even former opponents often greet our initiates and accomplishments with pleasure and respect. Our membership, relatively large by Hungarian standards, the proliferation of our branches in the provinces, the large list of organizations which sympathize with our Society or derive directly from it, succeeded to place a social stratum of significant size and quality at the service of Hungarian society.[11]

Since its first issue, the journal expressed pride in presenting the Hungarian public with a different conception of scientific journalism, one that did not separate society from politics but engaged with both.[12] The journal attracted amongst its contributors some of the most promising Hungarian social and natural scientists at the time, including the lawyer and historian, Gusztáv Gratz; the lawyer and economist, Pál Szende; the sociologist and philosopher, Oszkár Jászi; the sociologist and legal scholar, Bódog Somló; the positivist philosopher, Gyula Pikler; the physician and naturalist, József Madzsar; and the Marxist philosopher, Ervin Szabó. Many of these intellectuals also pursued political careers and played important political roles in the following years, thus fully justifying the editorial credo behind the publication of the journal: namely, that politics and science should be part of the same cognitive effort to grasp social reality.[13]

In 1906 *Twentieth Century* moved towards a more radical cultural programme, one that would eventually inspire much of the political changes before and during the First World War. The public debate on eugenics was part of this general project of cultural awareness, envisioned by the Society of Social Sciences and advocated in the pages of *Twentieth Century*. As mentioned above, what is especially telling about this debate is the interdependency between science and politics. Through eugenics, supporters of social and national change sought to determine the relative degrees of reciprocity existing between those scientific theories of biological perfection and the language of societal reform utilized by them. The eugenic vocabulary that emerged during the debate thus embodied a distinctive scientific vision, one that was not merely biological but also social and historical. Correspondingly, the scientific jargon of eugenics needed to be popularized in order for it to address not only abstruse topics related to mechanisms of heredity but also general problems characterizing Hungarian society. Such an ambitious agenda was very much in accordance with the general strategy of intellectual engagement pursued by the Society of Social Sciences.

Eugenics aspired to offer a theory of racial development with a commitment to a new racial future. Indeed, eugenics was assumed to be flourishing in those countries – such as Britain, Germany, France and the US – that were considered the most advanced in terms of scientific achievement, and where public responsiveness to social and biological utopianism was well documented.[14] But, as the debate on eugenics in Hungary demonstrates, intellectual commitment to eugenics was widespread. By the late nineteenth century, the language of social and biological improvement was spoken as articulately in London and Berlin as in Vienna and Budapest. Together with the introduction of modern health policies and the professionalization of the public health service, eugenics was transformed from a specialized scientific discourse into a topic of public interest.[15]

In this context a key argument which organizers reaffirmed during the preparation of the Hungarian debate on eugenics, was a belief in the importance of contrasting opposing notions about the role of heredity, especially in view of the recent works by Francis Galton, August Weismann, Wilhelm Schallmayer, Karl Pearson and others. In the history of eugenics, the beginning of the twentieth century was a rather eclectic time, when the most diverse hereditarian trends – of which Lamarckism, Darwinism, Weismannism and Mendelism were the most prominent – intersected.[16] These debates not only coloured initial studies on eugenics, but also determined *Twentieth Century*'s editorial positioning with respect to its popularization in Hungary. Indeed, these initial Hungarian efforts to come to terms with eugenics were blessed by none other than Galton himself. His Herbert Spencer Lecture – entitled 'Probability, the Foundation of Eugenics' – which he delivered at the University of Oxford in 1907, was translated into Hungarian and published in *Twentieth Century* in the same year.[17] Moreover, biometry – which Galton enthusiastically advocated – shaped the eugenic orientation of most participants in the debate. Yet in this article Galton was preoccupied with more than just reinforcing his support for biometry. An essential corollary of this endorsement was Galton's commitment to the cultivation of individual abilities, talents and faculties; but he concentrated equally on the moral issues related to eugenics. He believed that the time for eugenics had not yet come, and that the favourable reception by the public was a matter of targeting society at large rather than individuals. In a visionary way, Galton concluded:

> Considering that public opinion is guided by the sense of what best serves the interests of society as a whole, it is reasonable to expect that it will be strongly exerted in favour of Eugenics when a sufficiency of evidence shall have been collected to make the truths on which it rests plain to all. That moment has not yet arrived. Enough is already known to those who have studied the question to leave no doubt in their minds about the general results, but not enough is quantitatively known to justify legislation or other action except in extreme cases. Continued studies will be required for some time to come, and the pace must not be hurried. When the desired fullness of information shall have been acquired then, and not till then, will be the fit moment to proclaim 'Jehad', or Holy War against customs and prejudices that impair the physical and moral qualities of our race.[18]

This was the most important part of Galton's eugenic philosophy: not only were public awareness and scientific investigation closely related, but in many ways the entire edifice of practical eugenics rested on as wide a dissemination of knowledge about racial improvement as possible. Equally importantly, these became the core principles guiding Hungarian supporters of eugenics before and during the debate. However, as so many eugenicists in Hungary recognized, there was no simple transition from seeing eugenics as an 'academic question' to implementing the 'Holy War' of practical eugenics.

Preparing the debate

Although the focus of this chapter is on one particular event in the history of Hungarian eugenics, some notion of the direct and indirect linkages and structures that bound this event to similar developments elsewhere is essential. After 1900, Budapest harboured a vibrant eugenic culture that placed it next to other European centres, London, Berlin and Paris.[19] Through their studies abroad and participation in international conferences, Hungarian eugenicists became familiar with various eugenic concepts and theories, which they then refined and adapted to their local realities. Of these, British eugenics was especially influential.

Defining the aims of eugenics, Francis Galton wrote in 1904: 'Firstly it [eugenics] must be made familiar as an academic question, until its exact importance has been understood and accepted as a fact; Secondly it must be recognised as a subject whose practical development deserves serious consideration; and Thirdly it must be introduced into the national conscience, like a new religion.'[20] Most Hungarian eugenicists of the early twentieth century followed this approach. In 1910 three seminal articles were published in *Twentieth Century*, whose objective was to prepare the ground for a more sustained discussion of eugenics. A physician and biologist, Lajos Dienes, authored the first. Entitled 'Biometrika' (Biometrics), this short commentary introduced Francis Galton's and Karl Pearson's contribution to the study of measurable biological characteristics. A convinced supporter of biometrics, Dienes repudiated previous assumptions that individual abilities were determined by the environment and relegated this formative role to heredity. Describing the importance of biometry to the study of heredity, Dienes agreed that Galton's and Pearson's 'methods are entirely new; they reveal the validity of things we could have thought of, but could never have demonstrated in the past.' On a more sceptical note, however, Dienes continued: 'The results of this method [biometry] will probably not lead to a complete theory of society, as neither did the exact methods introduced in physiology and psychology lead to a complete theory of life or of the psychological phenomena; yet they performed services which no other method could have performed.'[21] Dienes' scepticism was, however, rather conventional, as he felt that precaution was needed at that stage, particularly considering the novelty of the topic for the Hungarian public. A more general debate was to follow, he announced, on the topics of biometry and eugenics.

József Madzsar, vice-president of the Society of Social Sciences and a member of the editorial board of *Twentieth Century*, wrote the second article, entitled 'Gyakorlati eugenika' (Practical eugenics).[22] The text echoed Karl Pearson's *The Problem of Practical Eugenics* (1909), although, as in Dienes' case, Madzsar's account was meant to be a preface rather than a defining statement. Nevertheless, Madzsar was less ambiguous than Dienes with respect to the importance of biometry for eugenics. Hereditary factors, Madzsar continued, were paramount in the creation of a healthy individual, and biometry could help explain how certain hereditary traits, transmitted from one generation to another, differed in their range of variability. Madzsar was even more critical of attempts by state and social institutions to improve the health of the population which he saw as working against 'the goals of natural selection'. Madzsar's conception of heredity fused Darwinism and Weismannism and was diametrically opposed to other visions of public assistance and medical reforms, based on humanitarian principles. Thus: 'The present form of social charity is even more dangerous because in most cases it obstructs the perishing of elements which are most burdensome and dangerous for society and it encourages their proliferation.'[23]

Madzsar's radicalism did not stop at simply criticizing policies of 'social charity'; he went further and suggested radical policies of negative eugenics, including sterilization. Invoking Plato's argument that the 'disabled should be banished from the state', Madzsar maintained the necessity of introducing this practice in contemporary society. Eugenicists, he continued, should resist 'pseudo-humanism', and 'pursue the goal of preventing the proliferation of the unfit and promoting the proliferation of the fit'. In such a context, the state was invested with eugenic prerogatives. Indeed, according to Madzsar, 'If the state has the right to deprive citizens of their freedom, of their life even, it undoubtedly has the right to sterilize as well, especially when this can be executed without any other unpleasant consequences for the individual.'[24] This was an issue that each eugenicist needed to resolve. By demanding that the modern state should aim at the purification of its racial body whilst eugenically controlling the population, Madzsar outlined a radical programme of social and biological engineering. The eugenic state thus envisioned was seen not only as the symbol of modernity but, most importantly, as the repository of the biological qualities of the nation and their vigilant custodian. There were thus a 'multitude of queuing tasks' facing eugenic reformers, but, as Madzsar hoped, 'through the scientific organization of the research of heredity and through the establishment of eugenic laboratories we will have to create a scientific basis for all those tasks.'[25]

The ideological possibilities for eugenics were fully integrated by Madzsar into his understanding of 'practical eugenics'. In his conclusions, therefore, he positioned eugenics within the field of politics rather than within that of science. Following Galton, he predicted that 'a eugenic religion will take shape in the realm of ideologies. This religion will forbid all forms of senti-mental charity, which are damaging to the race; it will enhance kinship and increase love for the family and for the race. In brief: eugenics is a manly, promising religion which calls upon the noblest feelings of our nature.'[26] Madzsar's eugenic programme was a synthesis of science and politics. As elsewhere in Europe, in addition to worrying about the deterioration in health of future generations, Hungarian eugenicists were preoccupied with the major task of improving the racial quality of the population, whilst simultaneously maintaining its reproductive capacity.

In 'Eugenika' (Eugenics) Zsigmond Fülöp, the editor of Darwin's works in Hungary, expanded some of the arguments outlined by Madszar.[27] Fülöp's article is structured into three parts. The first includes general considerations about the history of eugenics and the importance of the revolutionary theory of natural selection. It is within the scientific revo-lution brought about by Darwinism that Fülöp placed the emergence of eugenics: 'Eugenics represents the application of Darwinism to society with the scope of improving the qualities of the race, especially social qualities.'[28] The second part discusses Galton's theories of heredity and the importance of English empiricism in shaping the eugenic principles initially developed by Galton and Pearson. Fülöp was also impressed with how Galton had reconciled his theories of heredity with Darwinism, and how methodical biometry was. 'To achieve the tasks of eugenics,' Fülöp argued, 'it is not enough just to establish which biological factors work in society. To understand the role of these factors we need to establish *their mathematical definition.*'[29] Galton usefully employed such exactness, Fülöp believed, in developing his theories of ancestral inheritance and filial regression, respectively. The popularization of eugenics was as important as scientific research, and Fülöp praised the pioneering work undertaken by *Biometrika*, the journal founded in 1901 by Galton, Pearson and the zoologist W.F.R. Weldon.

Fülöp also addressed the most problematic issue related to eugenics: its practical application to society. Based on his Darwinist convictions, and following Galton, Fülöp identified two ways of improving society: negative methods, centred on impeding the proliferation of degenerate and unfit individuals; and positive methods, based on encouraging healthy indi-viduals to reproduce. Moreover, the practical application of eugenics was hindered by a plethora of factors which, according to Fülöp, needed to be eliminated if the improvement of society as a whole was to succeed. Thus 'drafting' and 'war' should be abolished, for they maimed and killed the finest male individuals; 'mental asylums' should be closed down, for these institutions allowed the mentally ill to survive and eventually regain their place in society; urbanization was blamed for causing degeneration; 'reli-gious celibacy' was attacked for hindering the marriage and procreation of the educated elite; 'degenerate infants' should not be permitted to live;

and marriage amongst family relatives should be forbidden. Finally, Fülöp enumerated two further factors characterizing the Hungarian context: the 'one-child system' and emigration.[30]

Like Madzsar, Fülöp depicted the existing Hungarian social system as 'pseudo-humanism', deploring its 'charity', which favoured the proliferation of the unfit; and, in response, he proposed the following measures, including the 'ideological and emotional training' of society in order for it to accept negative eugenic measures. Fülöp's eugenic vision was interventionist, and the 'total state' was again invoked as the only efficient vehicle to supervise eugenic goals. Echoing Madzsar, Fülöp argued: 'If the state has the right – for the good of society – to behead the criminal, why should it not have the right – for the equally important biological development of society – to obstruct the reproduction of degenerates?'[31] The state, then, was to exercise complete control over the individual in the interest of improving society's racial health.

Fülöp's passionate eugenics resurfaced again towards the end of his article. Although he sympathized with English eugenicists like Galton and Pearson, especially their techniques for popularizing eugenics, such as the establishment of research teams and societies and lecturing, Fülöp was not persuaded that this form of propaganda would suffice to make eugenics 'the religion of the future'. The only way to achieve the eugenic transformation of society, Fülöp contended, was not to encourage technological progress or to support existing social institutions, but to radically change the social system itself. 'Only then', he claimed, 'would a healthy and unwavering socio-political eugenics be possible, namely the sacrifice of one's personal interests in favour of collective interest.'[32]

Claims about practical eugenics and racial scientism put forward by Madzsar and Fülöp were too radical to go unnoticed by other Hungarian scientists and intellectuals. Dienes, Madzsar and Fülöp were not marginal physicians and natural scientists. All this ensured that the appropriate conditions were met – scientific interest, public concern and enthusiastic supporters – for a more general debate on eugenics to occur.

The public debate

On 31 January 1911 the Society of Social Sciences sent an invitation to the Royal Medical Association of Budapest (*Budapesti Királyi Orvosegyesület*) announcing its intention to initiate a debate on the issues of eugenics (*eugenika*), racial improvement (*fajnemesítés*) and racial hygiene (*faji-higieniá*). Three speakers were announced: József Madzsar, Lajos Dienes and Zsigmond Fülöp. These lectures were subsequently published in *Twentieth Century*. In the first lecture, entitled 'Fajromlás és fajnemesítés' (Racial degeneration and racial improvement), József Madzsar presented a conceptual synthesis of various works on eugenics, including Francis Galton's *Natural Inheritance*,

Karl Pearson's *The Scope and Importance to the State of the Science of National Eugenics* and Leonard Doncaster's *Heredity in the Light of Recent Research*.[33] Agreeing with these authors, Madzsar explained his position as follows:

> First of all, following the example of the Galton Eugenics Laboratory, we need to work on the science of eugenics, examine the rules of heredity, collect good and bad pedigrees, thus creating a proper theoretical foundation for its later, practical implementation. In order to discover and implement biological views everywhere in our practical social policy, we need to reformulate our entire thinking on the bases of the accumulated data. It is neither enough to deal with economic aspects and external circumstances, nor to enhance education and individual hygiene: we need to consider the future generation as well.[34]

Next, Madzsar elaborated on some of the ideas presented in his previous paper, 'Gyakorlati eugenika' (Practical eugenics), of which the most important concerned degeneration and marriage. Fears of social and racial degeneration dominated the European cultural discourse at the beginning of the twentieth century; they also characterized much of Madzsar's earlier discussion of eugenics. Transposing these anxieties to the perceived decline of the racial qualities of the population confirmed Madzsar's belief: the proportion of the constitutionally weak and mentally disabled was increasing in Hungary. 'The institution of marriage' would ultimately be reformed. Reproducing familiar tropes of other European discourses on eugenics, Madzsar suggested that marriage between the mentally disabled caused profound social instability; eugenics should in turn consider regulating such marriages. The creation of 'a new biological aristocracy of fit families' – one based not on social class but upon hereditary qualities – was what Madzsar hoped to achieve in Hungary.[35]

The wider intellectual objectives of this lecture largely mirrored Madzsar's own ideological goals: an attempt to overcome divided attitudes about the implementation of social policies in Hungary, most notably in the spheres of medical care and social assistance; to educate a responsive audience in the virtues of eugenics, especially its heuristic and ideological potential as a source of individual liberation; and finally, to develop and clarify a policy of eugenic reform. Aware of the distance separating the profound transformations inherent in his eugenic doctrine from existing social and political contexts, Madzsar explicitly acknowledged that 'eugenics is still utopia today, but it might not be such a distant issue; after all, all theories have been utopias at a certain moment.'[36]

Contributions to the consolidation of eugenics as a scientific discipline by Francis Galton and Karl Pearson were further analysed by Lajos Dienes. In 'A fajnemesítés biometrikai alapjai' (The biometrical foundation of racial improvement), he explored the relationship between eugenics and biometrics.[37] On the one hand, Dienes accepted Madzsar's ideas about the importance of heredity, as well as his practical approach to eugenics; on the other, he maintained the importance of the biometrical approach for understanding heredity through statistical techniques. According to Dienes, differences in physical traits, health and intelligence could be explained by the statistical study of natural selection within the population. Dienes thus combined Pearson's population approach to Darwinist variation with Galton's hereditary conception of society in order to clarify his own concept of eugenics.

Biometry, Dienes further claimed, was not only mathematical and scientific scholarship *par excellence* – and thus less exposed to gratuitous speculation – it also convincingly explained how hereditary variability and correlation influenced natural selection. 'In any case,' Dienes noted, 'the most important fact is that the effect of biological factors upon society can be identified and compared quantitatively.' Based on this assumption, a much larger programme of eugenic rejuvenation of the nation could be envisaged: 'We cannot fail to realize the importance of these statements for the assessment of social factors and for the elaboration of our future social policy, and it is impossible not to discover the urgent need for continuing these investigations.'[38] Dienes nurtured the hope that degeneration could be countered if proper policies of public health and preventive medicine were introduced across society.

The last lecturer was Zsigmond Fülöp. In his 'Az eugenetika követelései és korunk társadalmi viszonyai' (The claims of eugenics and the social conditions of our age), Fülöp did not reciprocate Dienes' enthusiastic support of his quantitative eugenic policies and offered a more circumspect assessment of eugenics.[39] In parts, he reproduced some of the arguments presented in his earlier article 'Eugenika'; yet Fülöp clearly pursued a different strategy. First, he declared his hostility towards the quantitative aspect of eugenics: 'The best definition of eugenics I could think of, for the purposes of my lecture, is that eugenics is a demographic policy which takes not so much quantitative but qualitative aspects into consideration.'[40] Second, his attention turned towards some of the arguments raised by Madzsar's lecture, most notably the issue of degeneration. Such a topic necessitated additional expertise, and Fülöp enlarged his conceptual approach by including the work of the German eugenicist, Wilhelm Schallmayer. Under a veneer of concern for public opinion, Fülöp was nevertheless uncompromising in his insistence that degeneration constituted one of the most important issues facing eugenics, both in Hungary and elsewhere. Thus, 'as rapid degeneration is an undeniable fact', the question preoccupying eugenicists was 'whether we need to initiate a eugenic social policy, or in other words: is there a need for the creation of a human stock bodily and mentally stronger and more valuable than the present one? The answer to this question is nothing but yes.'[41]

Fülöp identified the following questions to be central to his eugenic agenda: '1. Does eugenic research offer enough positive results which could in turn lead to the practical realization of eugenics? 2. What are the most appropriate tools for this realization? 3. Can the identified goals of eugenics be realized within the present social framework or we should modify our social conditions first?'[42] In answering these questions, Fülöp engaged in a critical evaluation of biometry and eugenics. He first subtly refined his appreciation of the employment of statistical methods in studies of heredity: 'Not even the most perfect *statistical* research can elucidate *the causes of variability and heredity.*' Next, he explained his reticence on this issue as follows:

> I consider, from the point of eugenics, that it is much more important to decide whether acquired features are hereditary or not than to perform statistical research, because we will have to give eugenic social policy an entirely new direction if it turns out that we need not deal with such acquired characteristics or with the impact of the environment, but only with hereditary features instead.[43]

With respect to other concerns voiced at the beginning of Fülöp's lecture, especially the relationship between heredity and practical eugenics, in addition to its bearing upon the revolutionary change of the existing social order, he suggested a no less unorthodox perspective. On the one hand, Fülöp questioned the connection between research on 'heredity and eugenic social policy' whilst, on the other, he stressed the importance of Darwinism in shaping the knowledge of social selection. 'In the beginning,' Fülöp claimed, 'it will suffice for us to know the most important rules of general biology, which play an important role in the evolution of species, rules which have been for thousands of years applied to the breeding of animals and plants.'[44]

Fülöp's dissatisfaction with the state of eugenics research was in fact connected to one of his greater interests: the incompatibility between eugenics and existing social conditions, especially in the form of traditional Victorian values, which Fülöp identified to be permeating the work of British eugenicists. In other words, the form of eugenics advocated by Galton and his supporters was exaggeratedly non-interventionist and individualist. Mindful of important differences between the structures of society in Britain and Hungary, Fülöp realized that in the latter, without the resources of the state, eugenics would be undermined and eventually rendered ineffective in the wake of constant social changes. In addition to its potential for channelling social transformation, eugenics should prompt the creation of a new 'national ethics'; namely, an evolved form of social solidarity that would ensure the racial continuity of future generations. Fülöp prophesized that, with the emergence of a new morality and ethics based on biology, the old order, sanctified by religion and outdated political philosophies, would be replaced. With the predominance of science in society, eugenics would triumph.

These conceptualizations of eugenics provided the critical conditions for the emergence of scientistic language that both medicalized and politicized the demand for social change in Hungary. But how did contemporaries react to the public lectures delivered by Dienes, Madzsar and Fülöp? Was there any interest in eugenics outside the community of physicians and evolutionary biologists? As the Society of Social Sciences anticipated, the reaction to the lectures was immediate, and it prompted the first public discussion of eugenics within the Hungarian scientific community, the so-called 'Eugenika vita' (Debate on eugenics).

The debate had a clear aim: to clarify the scientific and social challenges posed by 'eugenika', 'eugenetika', 'fajromlás' (racial degeneration) and 'fajnemesítés' (racial improvement). *Twentieth Century* published the debate under the provocative title, 'A fajnemesítés (eugenika) problémái' (The question of racial improvement – eugenics).[45] The list of participants included representatives of diverse professions, such as lawyers, psychologists, veterinary surgeons, physicians, sociologists, social hygienists and natural scientists. Prominent scientists, like the zoologist István Apáthy and the immunologist Leó Liebermann, were also in attendance. Considering this heterogeneous intellectual background, it is perhaps unsurprising that the debate not only was characterized by scientific arguments concerning the application of eugenics to society, but also considered how theories of evolution could substantiate claims of biological control of the population. Several themes were addressed during these discussions, including biometry, Lamarckism, Mendelism, neo-Malthusianism, feminism, degeneration, sterilization and the protection of children. Some participants simply recapitulated arguments put forward by the three speakers; the majority, however, aspired to be innovative and combative. In particular, two issues deserve attention: the relationship between social environment and eugenics; and the creation of a new totalizing social order based on eugenic principles.

Social environment and eugenics

All participants in the debate concurred that heredity was paramount in shaping the life of the individual, but they disagreed on two claims: first, that alleged social 'afflictions' like alcoholism and sexually transmitted diseases were hereditary; second, that the transmission of hereditary traits could be affected by external influences, such as the environment and education. The neurologist René Berkovits, for instance, explicitly connected the issue of degeneration with discussions on heredity. Thus 'to employ a purposeful and practical form of eugenics, we need to determine the relationship which exists between two major issues: the heredity of acquired characteristics and degeneration. The matter of degeneration appears under a very different light depending on whether we acknowledge or deny the heredity of acquired characteristics.'[46]

Berkovits believed that this imbalance had to be remedied by appropriate social and family policies. The solution, as seen by supporters of his view, was one of artificial selection. The issue was not so much to improve the performance of the race but to encourage the reproduction of those individuals with 'good heredity' and to discourage the existence of those designated as 'unfit'. Berkovits condemned a social system that induced degeneration. 'It is of utter necessity', he continued 'to prevent the proliferation of the degenerated; this is the only way to extinguish haemophilia, inherited defective constitution, endogenous mental diseases, hereditary antisocial propensity, and so on.' Moreover, this interventionist policy, aimed at improving living conditions and social inequalities, was accepted by all participants; some, however, like Berkovits, argued that it would ultimately be ineffective. What was needed, Berkovits concluded, was 'a commission of eugenics to research the specific matters and come up with an evident recommendation for the legislation; the negative part of this legislation, namely the list of those who should be banished from marriage, has already grown enough to be formulated.'[47]

Berkovits was not the only participant to connect degeneration with modernity or to advocate negative measures for eugenic segregation. Imre Káldor, for instance, suggested that the transformations of society brought about by industrialization and urbanization had contributed to a weakening of human resistance to diverse degenerative factors, ranging from spiritual pollution to physical corruption and decreasing birthrates. The degeneration of the race was conterminous with modernity, Káldor maintained, as more and more artificial instruments were invented to correct – and thus normalize – hitherto conspicuous physical deficiencies. Káldor was not, however, against modern progress in general and he argued that, for example, a proper eugenic policy could balance the effects of industrialization and degeneration.

Which, then, was the eugenic policy to be recommended? Positive eugenics was, according to Káldor 'unscientific, because we cannot determine those features that are worth cultivating from the point of view of society'; negative eugenics was 'practically more manageable', especially in the case of 'natural born criminals and the mentally affected'. For those individuals who could not be reintegrated into society, sterilization was the only solution. Accordingly, 'the real eugenic movement is the one which strives to eliminate the non-human causes inflicting degeneration.' Káldor focused on the environmental causes of degeneration, but also believed that negative eugenic measures were necessary to protect society from 'undesirable elements'.[48]

István Apáthy, too, endeavoured to offer a more detailed analysis of both eugenics and its relationship to degeneration. Insisting that a terminological clarification was necessary in this context, Apáthy constantly employed the term *'faj egészségtana'* (race hygiene) for eugenics instead of *'eugenika'* (eugenics), although other participants in the debate extensively used the latter term. Apáthy viewed degeneration primarily as an innate process rather than the effect of external causes: 'The process known as racial degeneration is actually a malady of the species, a malady which can be cured by specific methods.' Well versed in hereditarian theories and a supporter of schemes for preventive hygiene, Apáthy adopted both as sources of inspiration for the 'specific methods' of combating the 'malady of modern degeneration'. Countering degeneration broadened the scope of eugenics. To this end, Apáthy proposed a more comprehensive programme of biological improvement: 'Public hygiene is concerned with the improvement of public life conditions and public health; eugenics fights against certain maladies which endanger not only the survival of isolated individuals but [also] the survival of the entire species.' The science of hygiene therefore needed to adapt to the challenges posed by eugenics, especially to the notion that benevolent hygienic schemes resulted in the multiplication of the 'unfit'. The importance of eugenics for rejuvenating society was thus perceived to be substantial. As Apáthy explained:

> The endeavours of these two sciences are in many aspects similar; furthermore, the improvement of public hygiene itself is one of the methods employed by eugenics. Yet the latter has also adopted methods which originated in sociology, on the one hand, and in ethics, on the other; and, finally, it adopted an entirely specific biological method as well. This method is the deliberate selection of those elements that protect the race and the impediment of the reproduction of certain individuals who might have a damaging effect on the future generation.[49]

Like other speakers, Apáthy placed eugenics at the confluence between the interests of the individual and the powers invested in the state. The eugenic policy advocated by Apáthy outlined new priorities for social hygiene, and he included these within his vision of state welfare. Biological degeneration and images of deteriorating social conditions only strengthened his conviction that liberal individualism was detrimental to biological improvement but – when the protection of the race was required – should be authorized to endorse the monitoring of the biological capital of the nation by the state. 'It is a great mistake', Apáthy noted, 'to believe that the interest of the individual is in conflict with state interest, or that private interest fights public interest in the problems of racial hygiene.'[50]

Apáthy advanced certain observational and experimental evidence in support of his theoretical perspective, such as animal breeding and the impact of Malthus on Darwin's thinking. His favourite examples were, however, drawn from the social sphere. As long as 'the task of producing the next generation' was left to 'to the best individuals in each generation', society was protected against degeneration. Whilst the acquired character was adaptive, evidence of its inheritance would support the claim that a 'healthy environment' was decisive for the improvement of social conditions. Classical social illnesses – such as prostitution, homosexuality and alcoholism – could be cured as long as individuals lived in a eugenically cleansed environment. In this context, 'the goal of eugenics is to call attention to the fact that each social order, each habit, fashion or morality which acts against the selection of the best individuals in fact undermines the future of the entire society and it trades evolution, the salvation of future generations, for the pleasures of the moment, for the individual comfort of the present generation.'[51]

Eugenics thus appealed to those concerned with the racial reconstruction and preservation of society. In conclusion, Apáthy reiterated this point:

> However, eugenics will be satisfied even with a possibility to prevent the proliferation of less desired individuals and to increase the proliferation of desirable individuals. Neither the morality of racial improvement nor state interference will make the accomplishment of these requirements easy. The two might initiate, in more favourable social conditions, a series of changes which might lead to the creation of a happier, better and nobler human species.[52]

Apáthy used an eclectic methodology to support his arguments, combining various elements, such as ideas about culture and biology, projects for social improvement, moral precepts and schemes of hygiene. Whilst it cannot be denied that degeneration played an important role in shaping the debate, eugenics represented something far more complex. Indeed, at a more general level, the overall goal was to provide the basis for a creative political biology, capable of protecting the Hungarian nation from degeneration. From this conviction arose those modernist projects of national and racial regeneration that I have recently described in *Modernism and Eugenics*. A new social order was heralded in Europe during the first decade of the twentieth century, and rarely was eugenics more frequently invoked as a vehicle for change than in Hungary.

The dawn of a new social order

Plans for a new social order based on scientific knowledge and in accordance with eugenic principles were constantly enunciated during the Hungarian debate on eugenics. All discussants took a keen interest in contemporary social problems, and their analyses of eugenics emphasized the biological rejuvenation of society that was inherent in philosophies of social progress.

Eugenic methods were thus meant to reinforce policies for social improvement, starting with alcoholism and prostitution. For example, the lawyer Zoltán Rónai described the situation in the following terms: 'The dominating society stubbornly clings to alcohol-consumption and prostitution, and in the present-day society the unavoidable conflict is not between the protection of the weak and the future of the race, but between the assault of the weak and the future of the race instead.'[53] Rónai's concern with the socially and racially degenerated was so ingrained in his pessimistic Lamarckism that he even doubted that negative eugenics could be efficient:

> ...The interdiction to marry and artificial sterilization, through analgesic surgical intervention, of those with venereal diseases, alcoholics and other degenerates threatening the future of the race, has little consequence. The syphilitic infection or the germ-intoxicating alcoholism can also intervene postnuptially, and it would hardly be possible to sterilize the huge numbers of syphilitics and alcoholics, or those degenerated under the influence of syphilis and alcohol.[54]

For Rónai, society was losing its battle against degeneration: 'The ultimate goals of racial hygiene could only be realized in a socialist society.' Accordingly, 'eugenicists must strive to encourage not new nobility, but a new culture, and in this respect we can say that the real ennoblement of the human race can be achieved only together with the real ennoblement of the society.'[55]

Rónai's socialist inclination was not unique amongst supporters of eugenics at the time. In fact, the majority of those participating in the debate were socialist sympathizers, if not activists. Neither was the relationship between eugenics and the Left uncommon at the time. As Diane Paul and Michael Schwartz had eloquently argued, it was rather common for Marxists and Fabians (in Britain) and social democrats (in Germany and Sweden) to entertain eugenic ideas and to suggest a fusion between socialism and eugenics.[56] Projects of biological engineering squared neatly with the socialist doctrine of transforming humankind. In many ways, socialist thinkers were quicker to grasp the importance of science for shaping projects of national rejuvenation than other political ideologues. The emerging scientism of the early twentieth century drew socialism and eugenics closer, and interestingly their fusion was not necessarily oriented towards the support of 'positive methods' of racial improvement but rather towards 'negative' ones, such as sterilization and social segregation.

Rónai concurred with other supporters of eugenics that schemes of racial improvement derived from a more general interest in natural sciences, biology and medicine than from social sciences. This is not to say that the biological improvement advocated by eugenics and the social transformation envisioned by social sciences were completely separated.

On the contrary, the two themes often became entwined. As the jurist, Dezső Buday, explained:

> Scientific research relates to the questions of eugenics in two ways. One is the biological method which analyses the rules of procreation and the heredity of the individual determined strictly on a biological basis; the other is the sociological approach which determines the effects of natural selection, of sexual selection and of heredity on the basis of wide research and using biometrical statistics. Both methods are equally necessary.[57]

Admittedly, there was a degree of conceptual overlapping between the two, for both eugenic and socio-political language stressed the importance of biological laws in shaping the character of the individual, as well as the race. Even those less inclined to adopt a social and political language – like Apáthy, for example – encoded their speculations about heredity into descriptions of the application of eugenics to society.

There were also participants who were openly critical of eugenics' ambiguous relationship with racial thought. Although the history of eugenics cannot be divorced from the history of racism, neither should they be considered identical.[58] Participants in the debate used 'race' both in its biological, Darwinian meaning and as synonym for nation, society and community.[59] This multifunctional usage of the term race illustrates the need to move away from a narrow interpretation of eugenics as simple racial biology to a more nuanced consideration of its many functions and its social (not only biological!) role within the national community.

Corrective eugenics

Hungarian eugenicists voluntarily narrowed the definition of what was recognized as the legitimate improvement of society and who could benefit from this improvement. According to Berkovits, for instance, the eugenic project in Hungary was too modest. For eugenicists supporting this view, nature rather than nurture provided the key to understanding how genetic material was transmitted from generation to generation. Disagreement persisted, however, and the willingness of some of the most prominent participants in the debate to accept the new racial scientism advocated by eugenics should not obfuscate the fact that other contributors adhered to competing forms of hygiene and social hygiene.

The histologist Tibor Péterfi criticized both the term race as employed by eugenicists and their claim that biological degeneration was equally identifiable in human societies and in nature: 'We must thus acknowledge that we cannot apply the regularities of racial transmittal and degeneration directly in our strivings aiming at the improvement of the composition of society. Not so much the laws of biology and phylogeny are applicable here as the means and results of hygiene and medical sciences.'[60] Race, as Péterfi conceived it, was understood in its Darwinian interpretation, namely as variety within the human species. As a corollary to this view, the impact of the environment on the race was seen as substantial, as some participants believed that improving nutrition, housing, sanitation and education was more effective than identifying the 'racially valuable' individuals and encouraging them to breed with each other.

Leó Liebermann advocated this view, whilst simultaneously disagreeing with negative eugenic policies. Remarking that some of the greatest minds of European culture, including Immanuel Kant, had fragile constitutions and suffered from hereditary diseases, Liebermann retorted that 'it would be far-fetched to try to produce the artificial or the drastic elimination of the weak or of those who have a poorer constitution for the furthering of the cause of eugenics.'[61] To those demanding 'the artificial termination of the reproductive ability for those deemed "unworthy"', Liebermann offered the example of colonization in North America and Australia, especially, by individuals who were not always of 'worthy stock'. 'It is not necessary to demonstrate the worth of their descendants', Liebermann continued, 'in order to prove that the civilization of the world would have indeed suffered great harm if these anti-social elements had been destroyed only because they did not find their place in the old Europe.'[62]

The feminist Vilma Glücklich added a new dimension to this critique of eugenics by emphasizing how greatly women were neglected by those designing schemes for biological improvement. She insisted that proper sexual education and the wide dissemination of basic knowledge of hygiene were required in order for more sophisticated eugenic ideas to emerge. Only a society in which women were not exploited exclusively for reproductive purposes could hope to achieve the level of free expression and political support that was necessary for the success of eugenics.

The psychologist Dezső Hahn was another sceptic. The discrepancy between the cultural level of Hungarian society and the public acceptance of eugenics in other European countries and the US was, for him, the principal obstacle preventing the practical application of eugenic principles. Indeed, he overtly rejected negative eugenic methods: 'In some American states certain recidivist criminals or alcoholics – only men – are violently and artificially, by way of operation, prevented from procreation; well, this is nothing but a distasteful caricature of eugenic activity.'[63] What is necessary, Hahn contended, was a more critical reception of eugenic ideas: 'The foundations of eugenics do not appear in such a light, and the great and most important task of serious eugenic research is to clarify its basic assumptions with the tools of objective and scientific criticism and not to waste its power on the utopian treatment of various ways of practical accomplishment.'[64] Artificial selection, whether in the form of negative or positive eugenics, was 'practically impossible to realize', Hahn further noted. Environmental factors, too, were not to be neglected. Even Richard Dugdale's 1877 classic study of the hereditary nature of anti-social behaviour, *The Jukes: A Study in Crime, Pauperism, Disease, and Heredity* – which he quoted approvingly – did not persuade Hahn to modify his view that a proper social environment was as important in shaping the individual character as hereditary heritage. Ultimately, however, he confessed his support for eugenics, and considered that there were two practical tasks for eugenics: combatting degeneration, and the creation of 'racial ethics which would introduce the feeling of responsibility for the race besides healthy self-ishness for the conscience and acts of man.'[65]

The general tendency was to accept the point of view suggesting that eugenics was of fundamental importance not only in assisting schemes of public health and social hygiene, but also in shaping new forms of polit-ical biology. Within this emerging context, eugenicists were elevated to a new status; there were now national shamans dealing with nation's lumi-nous racial future. It is therefore important to situate these eugenic repre-sentations of the nation within the epistemic possibilities that produced them.

'Not a Utopia anymore'

The sociologist and veterinary doctor, Jenő Vámos, published his 'Az alkalm-azott eugenika' (Applied eugenics) as a complementary text to the debate.[66] Based on his veterinary experience, Vámos was not surprised that eugenics had first emerged in Britain, a country renowned for its exquisite devotion to animal breeding. Such background aside he could not neglect the indi-vidual contribution of Francis Galton to a field that – although anticipated by many thinkers and traceable to the classic civilizations of Antiquity – was in fact a new discipline: 'Galton's biometrical studies provide a guar-antee that eugenics should not be considered hygiene, only treated in a new way, and mixed with social-politics – as here in Hungary many are already disposed to. Instead, it is to be considered a new science, which incorpo-rates all positive knowledge aiming at the purposive ennoblement of human races.'[67]

Eugenics, therefore, was intimately connected to racial development in Hungary. As such it was meant to serve practical purposes. In many ways, Vámos's paper followed in the footsteps of other participants in the 1911 debate on eugenics, especially in terms of harnessing eugenics with a social mission. In other, significant ways, however, Vámos strengthened a hith-erto criticized argument: that the practical application of eugenics should be combined with the racial development of society. Thus, 'for eugenics, the improvement of the individual is a means only, which in effect should lead to the improvement of the race.'[68]

Vámos identified a two-pronged eugenic agenda, which he based on '*a priori* eugenic goals'. The first measure was to facilitate 'the natural selection of healthy individuals', whilst the second was to encourage the growth of those 'healthy individuals who were fitter from a racial and sexual point of view'. Healthy individuals were defined on the basis of a rigorous racial and sexual profile. No exceptions were permitted, and Vámos insisted that: 'If we want to improve the Caucasian race, those individuals are *a priori* suitable who bear the typical characteristics of the race, provided they serve a social purpose. On the other hand, for instance, such individuals who show characteristics contrary to their sexual nature – like the homosexuals – are *a priori* unsuitable, even if, in other aspects, they do befit a social purpose.'[69]

One would have expected Vámos to endorse such a challenging description of eugenics with an equally sophisticated theoretical argument. However, he readily amalgamated biometry with ideas about artificial selection borrowed from his own area of expertise. In fact, Vámos set out a methodology based largely on a unified interpretation of biology and natural selection, one which asserted the universality of reproductive practices that was to be valid in the natural world. Such a view was widely disseminated after the publication of Charles Darwin's *Variations of Animals and Plants under Domestication* (1868), and it certainly influenced Galton's characterization of 'stirpiculture', a term that fascinated him before the coinage of eugenics in 1883 and one that was popular amongst eugenicists in the United States at the end of the nineteenth century.[70]

Vámos identified six principles adapted from animal breeding that could provide a useful underpinning for eugenics. The first was interbreeding: 'The task of eugenics would be to facilitate, with the help of the proper means (i.e. the restriction of reproducing possibilities of those individuals characterized by detrimental features), the consolidation of the good qualities and the elimination of the bad ones.' The second principle was based on the belief that racial qualities were mainly inherited from the mother. On this assumption, Vámos established a direct link between feminism and eugenics: 'Only when the woman is totally emancipated in spiritual and economic terms will the goals of eugenics be realized. The goal of feminism is the improvement of women, and the ennoblement of the race has to be grounded in the improvement of mothers.' According to the third principle, marriage should be allowed only between compatible partners, whose racial qualities were complementary. In this context, 'the main goal of biometric measurements is to determine which qualities – and in what quantity – are transmitted by the fathers'. The precision of biometry was also invoked in his fourth principle, which assumed that children inherited more from the parent of the opposite gender (boys from mothers; girls from fathers). Vámos then considered the relationship between environment and heredity, and argued that both played an equally important role in the history of the race. Nevertheless, 'creating proper life conditions is a major task for eugenics, and for social-politics, but by itself it is not eugenics.' Finally, the sixth principle identified by Vámos was that of 'saturation', meaning that in 'the case of multiple conceptions by the same father', it was the last child that 'inherited most of the father's qualities'. His naturalist analogies notwithstanding, and the fact that much of his eugenic language was more selectionist than hereditary and in tone, Vámos hoped to improve the quality of the race both in a biological and a social sense:

> I do not wish to elaborate here a detailed judgement about how this application would be imaginable. However, for those who, under the pretext of respecting the rights of the individual, want to impede the conscious application of eugenics, I have an argument. If the Church had, and in many places still has, the right – without respect to the interests of the individual – to prevent the procreation of desirable offspring for the race through the interdiction of mixed marriages, why would the future physicians of eugenics not have at least as many rights, in favour of the improvement of the human races, and in favour of the individual too?[71]

With the consolidation of eugenics as a scientific discipline, progress would be made towards understanding human nature, so that, according to Vámos, 'In today's society applied eugenics is not a utopia any more, and it will be even less so in the society of the future.'[72]

Conclusion

As elsewhere in Europe and the US, eugenics in early twentieth-century Hungary referred to a number of scientific and practical topics: heredity, biometry, Mendelism; but also degeneration, sterilization and marriage counselling. The interplay between scientific ideas and concepts of social change was propagated by individuals who, even when they did not share the same eugenic principles, were all preoccupied with the improvement of the racial health of the Hungarian nation. Some of assumptions formulated during the debate on eugenics were based on local, Hungarian experiences; others were drawn from similar debates in Germany and Britain. It was through this process of cultural transfer that participants in this public debate on eugenics found the concepts they needed to identity the social and medical problems characterizing Hungarian society and which they hoped they could solve through the development of a specifically Hungarian form of eugenics.

Moreover, supporters of eugenics in early twentieth-century Hungary believed that the ideal of regenerative biology – so central to the new theories of heredity – would inspire national values during social and political crises. At a time when theories of organic destiny enthralled political elites and intellectuals alike, Hungarian eugenicists offered the possibility of national regeneration, combining scientific dogmas with racial categories and glossing over what Galton described in 1904 as a 'new secular religion'.

Once eugenics had succeeded in attracting sufficient supporters, two dominant viewpoints emerged with respect to its application to Hungarian society. One category of eugenicists argued that economic reform and the improvement of social conditions should be directed by the precepts of racial hygiene. The advocates of this view of social transformation underlined the necessity for selective breeding policies designed to prevent those individuals with 'negative' characteristics from social interaction and, ultimately, reproduction. Yet such a radical position did not go unchallenged. There were supporters of eugenics who instead perceived their goals primarily in terms of social and medical reforms. For this category, eugenics was part of a broader hygienic *Weltanschauung*, which included diverse policies, such as improving the living conditions of the urban underprivileged classes in the suburbs of Budapest, the prevention of sexually transmitted diseases, and the social reintegration of prostitutes.

This public debate on eugenics highlighted and sharpened controversies about science, culture, race and social change that were to take a much more radical form in ensuing decades. Different models of practical eugenics were envisioned, and all were based on three principles: the importance of heredity; the link between biology, medicine and the health of the nation; and the connection between science and political power. These three principles would subsequently be tested during the systemic upheavals of the First World War, prefiguring the emergence of new forms of the biologization of national belonging during the interwar period and beyond. Integrating this Hungarian event into that of the larger European history of eugenics of which it was a part can only lead to historiographic enrichment. The next step is to acquire similar knowledge about other Central European eugenic movements, which are otherwise localized and fragmented, so that a more regionally encompassing perspective on the history of European eugenics can finally emerge.

Notes

Parts of this chapter have appeared in two previous publications: (2007) 'The first debates on eugenics in Hungary, 1900–1918,' in Marius Turda and Paul Weindling (eds.) *Blood and Homeland: Eugenics and Racial Nationalism in Central and Southeast Europe, 1900–1940* (Budapest: CEU Press), 185–221 and (2006) '"A new religion": eugenics and racial scientism in pre-World War 1 Hungary,' *Totalitarian Movements and Political Religions* 7, 3, 303–325. I would like to thank the publishers for permission to reprint these fragments in current form.

1. See Marius Turda (2010) *Modernism and Eugenics* (Basingstoke: Palgrave Macmillan). For an earlier treatment see Stefan Kühl (1997) *Die Internationalen der Rassisten: Aufstieg und Nidergang der internationalen Bewegung für Eugenik und Rassenhygiene im 20. Jahrhundert* (Frankfurt: Campus Verlag).
2. See Alison Bashford and Philippa Levine (eds.) (2010) *The Oxford Handbook of the History of Eugenics* (New York: Oxford University Press).
3. See Marius Turda and Paul Weindling (eds.) (2007) *Blood and Homeland: Eugenics and Racial Nationalism in Central and Southeast Europe, 1900-1944* (Budapest: CEU Press);

Christian Promitzer, Sevasti Trubeta, Marius Turda (eds.) (2011) *Healthy, Hygiene and Eugenics to 1945* (Budapest: CEU Press).

4. The debate has not received the attention it deserves, neither in the scholarship dealing with the history of medicine and the social sciences in Hungary, nor in the international scholarship on eugenics. Notable exceptions are Mária M. Kovács (1994) *Liberal Professions & Illiberal Politics: Hungary from the Habsburgs to the Holocaust* (Washington: Woodrow Wilson Center Press) and László Perecz (2005) '"Fajegészségtan", balról jobra. Az eugenika század eleji recepciójához: Madzsar és Pekár' ["Race Hygiene" from Left to Right: The Reception of Eugenics in Early Twentieth Century: Madzsar és Pekár] in *A totalitarizmus és a magyar filozófia* [Totalitarianism and Hungarian Philosophy] (Debrecen: Vulgo), 200–12.

5. Recent works on Austrian eugenics include Monika Löscher (2009) '... *der gesunden Vernuft nicht zuwider*...'? *Katholische Eugenik in Österreich vor 1938* (Innsbruck: Studienverlag); Paul J. Weindling (2009) 'A city regenerated: eugenics, race, and welfare in interwar Vienna', in Deborah Holmes and Lisa Silverman (eds.) *Interwar Vienna: Culture between Tradition and Modernity* (Rochester, NY: Camden House), 81–113; Britta I. McEwen (2010) 'Welfare and Eugenics: Julius Tandler's *Rassenhygienische* Vision for Interwar Vienna', *Austrian History Yearbook* 41 (2010): 170–90. For Polish eugenics see Magdalena Gawin (2003) *Rasa i nowoczesność: historia polskiego ruchu eugenicznego, 1880–1952* [Race and Modernity: The History of Polish Eugenics Movement, 1880-1952] (Warsaw: Neriton); for eugenics in the Czech Lands see Michal Šimůnek (2007) 'Eugenics, Social genetics and racial hygiene: plans for the regulation of human heredity in the Czech Lands, 1900–1935', in Marius Turda and Paul J. Weindling (eds.) *Blood and Homeland: Eugenics and Racial Nationalism in Central and Southeast Europe, 1900–1940* (Budapest: CEU Press), 145–66.

6. See Marius Turda (forthcoming) *The Ideal of a Healthy Nation: Eugenic Ideas of Social and Biological Improvement in Early Twentieth-century Hungary.*

7. For a comprehensive overview see Daniel Pick (1989) *Faces of Degeneration: A European Disorder, c. 1848–1918* (Cambridge: Cambridge University Press).

8. The literature on social problems and nationalism in Hungary is extensive. For a recent Hungarian perspective see János Gyurgyák (2007) *Ezzé lett magyar hazátok. A magyar nemzeteszme és nacionalizmus története* [This is what your Hungarian homeland has become. A History of Hungarian nationalism and national idea] (Budapest: Osiris).

9. See, for example, Woodruff D. Smith (1991) *Politics and the Science of Culture in Germany, 1840–1920* (Oxford: Oxford University Press); John W. Burrow (2000) *The Crisis of Reason: European Thought, 1848–1914* (New Haven: Yale University Press).

10. As eloquently demonstrated by György Litván and László Szűcs (eds.) (1973) *A szociológia első magyar műhelye: A Huszadik Század köre*, [Sociology's First Hungarian Workshop: The *Twentieth Century*'s Circle] 2 vols (Budapest: Gondolat); Attila Pók (1990) *A magyarországi radikális demokrata ideológia kialakulása. A 'Huszadik Század' társadalomszemlélete (1900–1907)* [The Development of Hungarian Radical Democratic Ideology: The *Twentieth Century*'s View of Society, 1900-1907] (Budapest: Akadémiai Kiadó).

11. See Ervin Szabó (1982) 'The tasks of the sociological society (Presidential Address)', in György Litván and János M. Bak (eds.) *Socialism and Social Science: Selected Writings of Ervin Szabó* (London: Routledge and Kegan Paul), 197.

12. See Attila Pók (1989) 'The social function of sociology in fin-de-siècle Budapest', in György Ránki (ed.) *Hungary and European Civilization* (Budapest: Akadémiai kiadó), 265–83.

13. As eloquently discussed by György Litván (2006) *A Twentieth-Century Prophet: Oscar Jászi* (Budapest: CEU Press).
14. See Pauline M.H. Mazumdar (1992) *Eugenics, Human Genetics and Human Failings: The Eugenics Society, Its Sources and Its Critics in Britain* (London: Routledge); Paul Weindling (1989) *Health, Race and German Politics between National Unification and Nazism, 1870–1945* (Cambridge: Cambridge University Press); William H. Schneider (1990) *Quality and Quantity: The Quest for Biological Regeneration in Twentieth-Century France* (Cambridge: Cambridge University Press); Paul A. Lombardo (ed.) (2011) *A Century of Eugenics in America: From the Indiana Experiment to the Human Genome Era* (Bloomington, IN: Indiana University Press).
15. See, in particular, Dorothy Porter (1999) *Health, Civilization and the State: A History of Public Health from Ancient to Modern Times* (London: Routledge).
16. Sándor Sóos (2008) 'The scientific reception of Darwin's work in nineteenth-century Hungary' and Katalin Mund, 'The reception of Darwin in nineteenth-century Hungarian society', in Eve-Marie Engels and Thomas F. Glick (eds.) *The Reception of Charles Darwin in Europe*, vol. 2 (London: Continuum), 430–40 and 441–62, respectively.
17. Francis Galton (1907) 'A valószínűség, mint az eugenetika alapja' [Probability, as the Foundation of Eugenics], *Huszadik Század*, 1013–29.
18. Francis Galton (1909) 'Probability, The Foundation of Eugenics', in Francis Galton, *Essays in Eugenics* (London: Eugenics Education Society), 98–99; Galton, 'A valószínűség, mint az eugenetika alapja', 1029.
19. See Marius Turda (2009) 'The biology of war: eugenics in Hungary, 1914–1918', *Austrian History Yearbook*, 40, 238–64.
20. Francis Galton (1909) 'Eugenics: its definition, scope and aims', in Francis Galton, *Essays in Eugenics* (London: The Eugenics Education Society), 42.
21. Lajos Dienes (1910) 'Biometrika' [Biometrics], *Huszadik Század*, 51.
22. József Madzsar (1910) 'Gyakorlati eugenika' [Practical eugenics], *Huszadik Század*, 115–17.
23. Ibid., 116.
24. Ibid.
25. Madzsar, 'Gyakorlati eugenika', 116.
26. Ibid., 117.
27. Zsigmond Fülöp (1910) 'Eugenika', *Huszadik Század*, 161–75.
28. Ibid., 162.
29. Ibid., 163. [emphasis in the original]
30. Ibid., 170–71.
31. Ibid., 173.
32. Ibid., 175.
33. József Madzsar (1911) 'Fajromlás és fajnemesítés' [Racial degeneration and racial improvement], *Huszadik Század*, 145–60.
34. Ibid., 158–59.
35. Ibid., 159.
36. Ibid., 160.
37. Lajos Dienes (1911) 'A fajnemesítés biometrikai alapjai' [The biometrical foundation of racial improvement], *Huszadik Század*, 291–307.
38. Ibid., 307.
39. Zsigmond Fülöp (1911) 'Az eugenetika követelései és korunk társadalmi viszonyai' [The claims of eugenics and the social conditions of our age], *Huszadik Század*, 308–19.
40. Ibid., 308.

41. Ibid., 309.
42. Ibid.
43. Ibid. [emphasis in the original]
44. Ibid., 310.
45. (1911) 'A fajnemesítés (eugenika) problémái' [The question of racial improvement – eugenics], *Huszadik Század*, 29–44, 157–170, 322–336 and 694–709.
46. Ibid., 39–40.
47. Ibid., 44.
48. Ibid., 158–59.
49. 'A fajnemesítés (eugenika) problémái', 700.
50. Ibid., 701.
51. Ibid., 708–09.
52. Ibid., 709.
53. Ibid., 168.
54. Ibid.
55. Ibid., 170.
56. See Diane Paul (1984) 'Eugenics and the Left', *Journal of the History of Ideas*, 45/4, 567–90; Michael Schwartz (1995) *Sozialistische Eugenik: Eugenische Sozialtechnologien in Debatten und Politik der deutschen Sozialdemokratie, 1890–1933* (Bonn: J.H.W. Dietz). See also Richard Weikart (1998) *Socialist Darwinism: Evolution in German Socialist Thought from Marx to Bernstein* (San Francisco: International Scholars Publications) and Gunnar Broberg and Nils Roll-Hansen (eds.), *Eugenics and the Welfare State: Sterilization Policy in Denmark, Sweden, Norway, and Finland* (East Lansing, MI: Michigan State University Press, 2005).
57. 'A fajnemesítés (eugenika) problémái', 29.
58. See Marius Turda (2010) 'Race, science, and eugenics in the twentieth century,' in Bashford and Levine (eds.) *The Oxford Handbook of the History of Eugenics*, 62–79.
59. This is very much in line with most Hungarian social and political literature at the time. See Marius Turda (2005) *The Idea of National Superiority in Central Europe, 1880–1918* (New York: Edwin Mellen).
60. Ibid., 157.
61. Ibid., 323.
62. Ibid., 324.
63. Ibid., 327.
64. Ibid.
65. Ibid., 332.
66. Jenö Vámos (1911) 'Az alkalmazott eugenika' [Applied eugenics], *Huszadik Század*, 571–77.
67. Ibid., 571.
68. Ibid.
69. Ibid., 572.
70. See Francis Galton (1876) 'A theory of heredity', *The Journal of the Anthropological Institute of Great Britain and Ireland* 5, 329–48.
71. Vámos, 'Az alkalmazott eugenika', 573–76.
72. Ibid., 577.

10
The Politics of *Fin-de-siècle* Anatomy

Tatjana Buklijas

Introduction

In 1922, children of the recently deceased former Chair of the Second Anatomical Department in Vienna, Carl Toldt, published their father's autobiography. In the preface, they described him as a 'German-feeling and freedom-loving man' (*deutschfühlender und freiheitsliebender Mann*) as well as 'a true Tyrolean' (*ein echter Tiroler*).[1] Indeed, hardly any obituary of Toldt failed to mention his affection for the German nation and his Tyrolean homeland – often linking them to his attachment to German science.[2] In contrast, around 1910 obituaries for Emil Zuckerkandl, Toldt's contemporary and Chair of the First Anatomical Department, said little about Zuckerkandl's place of birth and national politics. The obituarists portrayed him as a successor to an older tradition, founded by Joseph Hyrtl, which saw anatomy as the basis of and contributor to clinical knowledge.[3] In this article, I take these views seriously.[4] I demonstrate the differences between two anatomical disciplinary orientations practised in fin-de-siècle Vienna and their close links with the political views and social networks with which the two anatomists allied themselves. I show how anatomical divergences can be understood only if we place them back into the context of contemporary Austrian politics and society, and of the growing middle-class rift along ethnic and religious lines.

The rift began to open after 1848 and widened manifestly after the Empire's reorganization into a Dual Monarchy, consisting of Austrian Cisleithania and Hungarian Transleithania, in 1867–68.[5] The burst bubble of the stock market in 1873 revealed the Empire's economic instability. The imperial court turned to Slavs and the Church for support and to the European south-east as its main sphere of interest, but the population that identified with the German nation felt threatened. In 1879, the Liberal government fell from power when trying to resist further engagement in the Balkans after the Austrian occupation of Bosnia in 1878. An era of non-party governments, supported by the court and an assortment of political groups (the

Church with clericalists, conservatives and Slavs), ensued. The first government survived for fifteen years, but the succeeding administrations proved less durable and each fell on an issue related to the cultural autonomy of the non-German ethnic groups.

Nowhere was the division between Germans and other ethnic groups as manifest as in multinational, diverse Vienna and the Empire's microcosm, the university. In the 1860s student societies admitted all German-speaking students regardless of their ethnic or religious identity.[6] But by the 1870s students, disillusioned with the liberalism of their fathers, moved closer to radical German nationalism. To be modern was to be German: Carl Schorske cited the historian of liberalism Richard Charmatz saying, 'he who placed his Austrianism above his Germanism was "old".'[7] The increasingly assimilated Jewish students threw themselves wholeheartedly into the national revival but soon found themselves disenfranchised. By the 1880s, membership of student societies was closed to all but Christian ethnic Germans.[8] Barred from the largest student societies, non-German students formed their own national associations.

The professoriate was not immune to these centrifugal processes. From the early 1860s the introduction of Hungarian and Polish languages at the universities of Budapest and Cracow, respectively, and the subsequent departure of German-speaking professors and students gave rise to some criticism.[9] But the split of Prague University into a Czech and a German part in 1882 was perceived by many as the 'Czechization' of 'the oldest German university' and thus provoked the loudest outcry. The majority of Viennese professors and the medical public rose to defend its German character, arguing that German was not just the language of one of the imperial ethnic groups but also a precious link between Austro-Hungarian academia and the German scientific powerhouse.[10] Viennese medical journalists and professors warned their Czech colleagues of the examples of Cracow and Budapest, which had declined into second-rate universities after they had relinquished the 'universal languages', German and Latin, in favour of Polish and Hungarian, respectively.

In the medical Vienna of the 1870s, the gradual departure, by death or retirement, of the 'Second Vienna School' generation opened space for new directions and alliances, but also for conflicts.[11] In 1876, the Prussian-born surgeon and star of the medical school Theodor Billroth published a 'cultural history' of medical education.[12] The book, which advocated elitism favouring the German middle classes over Jews from the rural East, received acclaim from nationalist students but also criticisms from the liberal press.[13] Billroth's colleague, the aristocrat-surgeon Johann von Dumreicher, responded with a booklet that extolled the democratic inclusiveness of Austrian universities and the empire's civilizing mission to the European south-east.[14] This conflict deepened towards the end of the century, informed the appointments to surgical chairs and directions in surgical research and teaching, and contributed to divisions within the faculty.[15]

In the medical universe of 1880s Vienna, German nationalism became linked to appreciation for German science and the German model of higher education, while support for the official policies of ethnic and cultural autonomy often went hand in hand with the idea of a distinctively Austrian style of medical studies. Billroth feared the downfall of the university if it lost its German character, while Dumreicher believed that the German model was not suitable for a largely rural and ethnically diverse empire. Negotiating academic appointments and curriculum alterations with these constraints in mind became a complex matter where disciplinary orientation was perceived as intrinsically political.

Anatomy presents a uniquely suitable case through which to study the political and scientific divisions of the Viennese professoriate around the turn of the twentieth century. First, it was a well established discipline. Appointments to anatomical chairs merited considerable attention from the faculty. Clinicians and non-clinicians alike wanted to make sure that students received the best possible instruction on the structure of the normal body. Second, it had two chairs, both of which came up for appointment during the turbulent 1880s. Decisions made in 1884 and 1888 cemented two political and scientific directions within Viennese anatomy, which would grow increasingly apart in the following decades. Third, anatomy's research tools and questions attracted public interest and controversy worldwide. Throughout the nineteenth century, the passing of anatomy acts across the Anglo-American world were in many cases preceded by body snatching scandals and surrounded by widespread public unease.[16] In mid-nineteenth-century London, the debate over the evolution and development of animal and human bodies, much of which took place within anatomy's disciplinary remit, was tightly linked to political fights between social reformers and conservatives.[17] By the 1870s anatomists came to play a prominent role in the nascent discipline of anthropology, which aimed to elucidate human origins and the relationships between ethnic groups.[18] These issues were of substantial national political interest: I have discussed elsewhere how in Vienna the rise of mass political parties around 1900 made the ethnic and religious identities of dissected cadavers a topic of parliamentary debate.[19] Here I shall show how interpretations of anatomical variation, inter-individual and intergroup, corresponded with political standpoints.

The article is divided into three sections. The first examines anatomical appointments in 1884 and 1888. How and why were anatomical orientations inscribed with political meanings? Lynn Nyhart used files on professorial appointments to assess the standing of diverse research orientations within anatomy and zoology at nineteenth-century German universities.[20] I rely on the same type of sources to understand what kinds of anatomy, and politics, the Viennese medical faculty endorsed.

The subsequent two sections are portraits of two appointed professors, Carl Toldt and Emil Zuckerkandl. I look for patterns in their research, other professional activities, politics and personal life, using an approach similar to Jonathan Harwood's take on 'styles of scientific thought'.[21] Harwood started from Fritz Ringer's division of the Wilhelmine German professoriate into traditionalist 'mandarins', who had sought to preserve their privileged but increasingly precarious social position by defending the value of *Bildung*, and minority progressives, who stood in favour of industrialization and modernization.[22] Harwood studied the German genetics community in the early twentieth century to show how different social experiences and backgrounds gave rise to two distinct 'styles of thought', which pervaded scientists' lives, from their national politics to their views of science and approaches to research problems.

The key questions that divided the Austrian professoriate were the viability of the Empire, ethnic differences and the cultural positions of ethnic groups. Their split may be seen as an extension of the broader crisis in the Austrian educated middle class. Historians headed by Carl Schorske attributed that crisis to the failure of the 'overly rationalist' Viennese liberal middle class to confront rising populist movements, nationalism and anti-Semitism.[23] The new generation of Viennese bourgeois, he argued, turned away from public life into cafés and salons, to create an ahistorical, modernist culture of subjectivity and uncertainty across fields such as the fine arts, literature, architecture and psychoanalysis. Schorske's critics pointed out that nationalism had been part of liberal politics all along.[24] They noted, further, that Schorske and his followers had failed to address the high proportion of Jews in the new culture.[25] More recently, Schorske's sharp division between liberal public and modernist private worlds, as well as between liberal objectivity and determinism on the one hand and modernist subjectivity and uncertainty on the other, has been criticized as well.[26] I take this scholarship on board to show how divisions in the professoriate were not between 'old' liberals on one side and young 'modernists' and 'nationalists' on the other, but rather between two liberal groups, of *staats-* and of *kulturnational* provenance.[27] Neither withdrew into private worlds but each spoke to different audiences; and while neither broke away from history, they saw themselves as successors to different traditions.

The politics of anatomical appointments

The post-1849 education reforms changed the procedure for professorial appointments in the Habsburg Monarchy after the German model. While previously, civil servants and courtiers had selected candidates, professors now proposed candidates for appointment.[28] A faculty would elect an appointments committee among professors whose disciplines were close to that of the post in question. The committee would then usually make a list of three candidates (*Terna*) in the order of preference *primo, secundo* and *tertio loco,* and in some cases with one name only, *unico loco.* In addition to teaching experience and research output, a key criterion was the candidate's area of expertise, as he was expected either to fill an existing gap or to develop a new field. Once the faculty had decided, the list would be sent to the minister of religion and education for consideration. If rejected, the list would be returned to the faculty, sometimes accompanied by suggestions on the kind of candidate the ministry preferred. If the list was accepted, the minister would start negotiations with the proposed candidate, and if the ministry could meet the candidate's demands on salary, working conditions, institute funding and accommodation, the chair would be filled.

The process was fraught with difficulties at every step: even if the appointments committee could agree on three names that fitted the demands of the faculty, the ministry could still refuse them if the candidates were deemed politically uncongenial or simply too expensive, and nominate a candidate from outside the *Terna.* The faculty and the ministers were all too painfully aware that the Austro-Hungarian Empire could not compete with the lavishly furnished and endowed institutes of Germany. Furthermore, even if the University of Vienna could offer an attractive remuneration package, the cost of living in the Habsburg capital often exceeded that of the cities of the German Empire.[29] Nonetheless, Vienna and Prague aimed high and frequently proposed eminent German or Swiss names, befitting the reputation of the schools. The ministry occasionally agreed, even at high cost. For example, in 1870, after failing to win over Ernst Haeckel for the Vienna zoology chair, the ministry appointed Göttingen professor Carl Claus.[30] But frequently the state authorities insisted on an Austrian candidate. When in 1874 Wilhelm Henke, a mechanical anatomist, left Prague for Tübingen, the faculty composed a list consisting of three non-Austrian candidates: Christian Aeby (Bern), Carl Gegenbaur (Jena) and Carl Hasse (Breslau). They reasoned that, because in Prague (descriptive) histology and embryology were already represented by an *Extraordinarius,* Walther Flemming, the school needed an anatomist of 'morphological orientation'. While the ministry considered the proposal, Flemming left for Kiel, so the school suddenly found itself in need of a different profile for the post. Instead of asking the faculty to update the list of names, the ministry stated that no established histologist and embryologist would come to Prague without the promise of an institute, so the best they could do was to appoint a young local talent with experience in anatomy, histology and embryology. In short, the best person for the job was the extraordinary professor of anatomy in Vienna, Carl Toldt.

Reasons for favouring either an Austro-Hungarian or a German candidate went beyond disciplinary profile and cost. The experimental physiologist Ernst Brücke, hired in 1849, was the first and for almost twenty years the only Protestant in the Vienna medical faculty.[31] It is no coincidence that the second German Protestant in the faculty, Theodor Billroth, was hired in 1867, in the aftermath of another important conflict with German lands. Hiring a Prussian after the Austrian military defeat at Sadowa (Sadová) was controversial but also regarded as a splendid revenge.[32] The satisfaction was further increased in the following decades when Billroth's fame soared and he refused all of the many calls he received to German universities.

The appointment to the Second Chair in 1884 was the first time the 'German' procedure was used to elect a professor of anatomy at the University of Vienna.[33] Although the university hired two anatomists between 1849 and 1884, in neither case was the regular procedure followed. The appointment committee consisted of Brücke, Billroth, the anatomist Carl Langer, the surgeon Eduard Albert (recently appointed successor to Billroth's opponent Johann von Dumreicher) and the pathologist Hans Kundrat.[34] Billroth and Brücke almost immediately withdrew and were replaced by the internist Heinrich von Bamberger and obstetrician Joseph Späth.[35] The committee briefly considered several German candidates of diverse research orientations, such as Wilhelm Henke, still in Tübingen, and Gustav Schwalbe in Strasbourg, only to compose a list of two names: Toldt and Zuckerkandl.

At first glance, the two were very similar. Both men had begun their anatomical careers in Vienna in the 1870s, and both now held professorships elsewhere in the empire: Toldt in Prague, and Zuckerkandl in Graz. But the similarities stopped here. The Tyrolean Roman Catholic Toldt, who was almost a decade older, had started his career as a military doctor in the early 1860s and specialized in histology. Zuckerkandl, a Jew raised in today's Hungary, was trained in pathological histology by none other than the old master Carl Rokitansky. He then turned to gross anatomy, but remained closely connected to the Vienna General Hospital, working with doctors to propose anatomical solutions to clinical problems.

The committee members, and then the whole faculty, clashed over the *primo loco* candidate.[36] The majority voted for Toldt, but two members of the committee, Albert and Kundrat, supported by 6 out of 21 faculty members present at the meeting (Albert, Kundrat, the psychiatrist and neuroanatomist Theodor Meynert, the chemist Ernst Ludwig, the forensic medicine professor Eduard Hofmann and the experimental pathologist Salomon Stricker), objected. Albert accused the dean of the Medical Faculty, Gustav Braun, of having engineered the election process by replacing Billroth and Brücke with two members whose expertise in anatomy was inferior to Meynert's and Stricker's.

In his *Separatvotum*, Albert protested against all the arguments in Toldt's favour: that he had more and better publications; represented two fields, anatomy and histology; was more senior, and had experience in directing a large anatomical institute. Albert stressed that Zuckerkandl was regarded as a great talent and had been promoted to the position of an *Extraordinarius* unusually rapidly, while Toldt had not earned such a reputation. And Zuckerkandl's research by no means fell behind Toldt's, but rather followed a different path, focusing on hitherto unresolved issues in gross anatomy, such as variations in humans.[37] Rokitansky had taught him how to use a microscope. In any case, Albert stressed, expertise in histology alone did not recommend someone for this job: small schools should and must choose teachers who represented two orientations, but Vienna required excellence in a single orientation; a professorship in anatomy did not require expertise in histology.[38] The size of the institutes the candidates had directed did not matter either, because, in Albert's view, for anatomy, as opposed to physics and chemistry, the size of the institution was irrelevant: dissection of the human body depended more on the skill of the man than on the building.[39] Zuckerkandl was praised for the manual skill he had acquired by dissecting many corpses in his longer career as a teacher and researcher of anatomical variations. Finally, he was renowned as an exceptionally gifted lecturer, a talent that would be of great use in Vienna. He could teach topographical anatomy, a subject that the reform of 1872 had wrongly left out of the curriculum in Vienna but was still taught widely at smaller schools such as Innsbruck and Graz.[40]

Albert's *Separatvotum* revealed an emerging difference in opinion on what road anatomy should take and how it should relate to other disciplines. Much of it had to do with institutional traditions. The large Vienna medical school could support more independent disciplines than smaller universities. Physiology and anatomy had been separate from 1805[41] and histology represented by an *Extraordinarius*, Carl Wedl, from as early as 1853, rising to an *Ordinariat* in 1871.[42] Embryology received a chair and institute under the directorship of *Extraordinarius* Samuel Schenk in 1873.[43] Pathological anatomy became an *Ordinariat* in 1844 but the department itself went back to the early nineteenth century and preceded the oldest German chair, in Berlin, by several decades.[44] Finally, experimental pathology, which at German universities was accommodated either together with pathological anatomy or at clinical chairs, had been an independent chair under Salomon Stricker from 1872.[45] This division of labour encouraged anatomists to focus on gross normal anatomy.

The disciplinary organization of the Vienna School of Medicine strongly influenced smaller Austrian medical schools to establish independent chairs of pathology, histology and embryology: indeed, separate professorships in histology and embryology seem to have been an exclusively Austrian phenomenon.[46] In contrast, at many German universities new chairs were founded later; not only was histology taught at anatomical (or, in some cases, physiological or pathological) departments in the 1870s, but combined anatomy/physiology chairs were common as late as the 1850s.[47] Issues of institutional organization thus further encouraged German anatomists to pursue research orientations beyond gross – or indeed human – anatomy.[48]

But the size of the school and the related tradition of specialization explain only part of Albert's preference for a clinically inclined anatomy. For anatomical appointments, German and Austrian medical faculties alike often preferred 'practically' oriented candidates. Yet German professors practising or supporting less immediately applicable orientations often stressed that the student 'desired to become not simply a medical craftsman, but a *Wissenschaftler*'.[49] They argued that the value of teaching anatomy as a *Wissenschaft*, rather than a mere practical art, lay not only in honing students' scientific skills and way of thinking, but also in helping them memorize facts. Even in German universities they often failed to persuade, but in Austria clinicians were additionally armed with the argument of an old 'practical' orientation going back to Joseph II. The emperor's vision of an efficient factory for the quick production of health personnel came to life in the Josephinum, but also influenced the university. The political support that propelled Vienna to a leading place among centres of medical education in the 1830s and 1840s was directed to the General Hospital, where clinical knowledge was grounded in dissections at Carl Rokitansky's institute of pathological anatomy.[50] Rokitansky's contemporary, the professor of normal human anatomy Joseph Hyrtl, was a comparative anatomist of Cuvier's bent. As a teacher, Hyrtl, a political conservative with connections in the clergy, preached anatomy in service of clinical medicine.[51] But as the nineteenth century drew to a close and Rokitansky and his contemporaries gradually withdrew, many believed that the school sank into a crisis caused by its departure from a successful tradition.

The strong clinical tradition and the apparent crisis it faced in the 1880s may explain why Albert and his group supported Zuckerkandl. This, however, leaves unclear why many Austrian-born professors who had spent their entire careers at Austrian universities favoured Toldt. One argument is that they voted for Toldt for pragmatic reasons, because in the same year the professor of histology, Carl Wedl, retired and Toldt was qualified to teach histology in the interim period. But it was Brücke who replaced Wedl and who protested against the reappointment when the ministry decided to fill the vacated histological post.[52] I suggest that it was a rift in ethnic and cultural (including scientific) alliances that caused the faculty to divide. Albert's faction wanted a continuation of tradition, embodied in a person who was, happily, trained by both Hyrtl and Rokitansky. Others voted for a departure from established ways, an anatomy closer to dominant trends in German science.

In terms of institutional politics, the debate around the anatomical appointment could be seen as a follow-up to the controversy around the election of Dumreicher's successor in 1881. The faculty proposed Billroth's former student, Vinzenz von Czerny, a Heidelberg professor famous for new methods of laryngeal extirpation, oesophageal resection and uterine extirpation via the vaginal route. The minister, however, decided that Czerny was too similar to Billroth and that Eduard Albert would represent better the 'Austrian school of surgery'.[53] Eduard Albert (1841–1900), a lower-middle-class Bohemian and Rokitansky's protégé, had been Dumreicher's assistant before his appointment to a professorship in Innsbruck.[54] In contrast to Billroth's focus on the abdomen, Albert, like Dumreicher, had been a keen orthopaedic surgeon even before orthopaedics became an independent discipline.[55] While Billroth had an eye for research talent and built a highly successful and tightly knit school around him, he was less interested in lecturing to general student audiences.[56] Adolf Lorenz – the famous surgeon and father of the even better known Konrad – recollected that although Albert's technical skill was inferior to Billroth's, he filled his auditorium while Billroth often spoke to empty benches.[57] Yet it remains unclear what the minister meant by the 'Austrian school of surgery'.[58] It seems that it first and foremost signified an orientation in national and university politics. Albert was seen as a successor to Dumreicher, who represented the 'Austrian' style of medical education. Czerny, although admittedly an excellent surgeon, was perceived as too German.

Zuckerkandl thus could have been seen as a reinforcement to the 'Austrian' faction within the faculty. Albert was neither the first nor the last to have been installed at the request of the minister. For example, the career of the young Graz pathologist Hans Kundrat received a strong and surprising boost when he was appointed to Rokitansky's former chair after the ministry's failed attempt to attract a famous name: both Julius Cohnheim (Leipzig) and Friedrich von Recklinghausen (Strasbourg) had declined the offer.[59] For all of Zuckerkandl's talents, his greatest recommendation was probably his claim to Hyrtl's tradition. For 'pro-Austrians', the old master was not only a skilled human anatomist but also a symbol of the old golden era of Viennese medicine, of a time without ethnic divisions. On the occasion of the installation of Hyrtl's bust in the courtyard of the new university building in 1889, Albert said of Hyrtl: 'He greets the children of all tribes of the Austrian state who come to the old imperial city.'[60]

Albert's efforts to populate Viennese medical chairs with Austrians of the same political persuasion failed on this occasion.[61] The ministry, perhaps trying to appease the opposing faction after the Czerny case, appointed Toldt. But only four years later, in 1888, the faculty again had a chance to appoint an anatomist, this time to the First Anatomical Chair, vacant after Langer's death. The list included Gustav Schwalbe of Strasbourg in *primo loco*, Zuckerkandl, still in Graz, in *secundo loco*, and Carl Rabl, the junior but talented recently appointed Prague professor, in *tertio loco*.[62] Toldt's preferred candidate was Schwalbe, while Rabl was supported by those who favoured an Austrian candidate but did not want a Jew.[63] Zuckerkandl enjoyed strong support: practically the entire medical public expected him to obtain a Vienna professorship.[64] The widely read medical weekly *Wiener Medizinische Wochenschrift* attacked his competition, arguing that

Schwalbe was 'more histologist than anatomist' at a small university where he had to teach two disciplines, and Rabl a person 'who had worked in the more remote field of zoology', 'can show just one anatomical paper' and was considered only because he was an Austrian of the 'right' religion.[65] The minister again concluded the discussion: 'Professor Dr. Schwalbe is without doubt the more important, he towers above the candidate offered in the second place, Professor Dr. Zuckerkandl, especially in that he [Schwalbe] represents in the most excellent way the newer orientation in anatomy, so-called morphological anatomy, which seeks to make comparative zoology and histology, as well as embryology, fruitful for anatomy', but these fields were already represented in Vienna by Toldt.[66] Rabl was not considered. Zuckerkandl was described as a first-class anatomist and, even more importantly, an outstanding teacher 'with excellent talent for observation, perseverance in work and love for his discipline as well as with complete mastery of the methods of so-called gross anatomy – the ability to explain skilfully difficult anatomical relations.'[67] The objections concerning Zuckerkandl's Jewish background and family connections with journalistic circles via his wife Bertha were deemed insufficiently important to justify giving precedence to a foreigner.

Discussions around anatomical appointments to the Vienna medical faculty in the 1880s thus, to a large extent, followed those in Germany, revolving around the question of the kind of anatomy that was best suited to the education of students and to the reputation of the school. But it also had a dimension that was characteristic of Austria. Research orientations were seen to correspond to positions in university and national politics. The next two sections will show how politics and anatomy came together in the careers of Carl Toldt and Emil Zuckerkandl.

A German anatomist

In 1897, the government of Cisleithania under Count Kasimir Badeni added Czech to German as the official language of the Bohemian public administration. Just two years earlier, Badeni's predecessor had fallen when his government allowed the use of Slovenian in secondary-school instruction in the Styrian town of Cilli (Celje). Now Badeni's decision brought on the biggest parliamentary crisis in the history of the country. In this period, languages came to be configured as tools of nationalist political activism.[68] As long as education and public administration were all conducted in German, upward social mobility resulted in assimilation into the German ethnic corpus; linguistic emancipation, by contrast, fostered the appearance of a middle class with a different national identity. Violent protests against Badeni's decision erupted especially among the German nationalist studentship, threatening to shake the Habsburg throne. The compromise that brought the students back into the lecture halls was reached at the cost of Badeni's fall; unusually, the chief negotiator for students and their informal leader was the Rector of the University of Vienna, the anatomist Carl Toldt. Who was this professor, the high imperial official, who protected and helped German nationalist students?

Carl Toldt was born in 1840 into a middle-class family in the small town of Bruneck (Brunico) in South Tyrol.[69] His civil servant father expected him to follow in his footsteps and study law, but his premature death, a fall in the family fortunes and an interest in natural sciences prompted Toldt to choose medicine. The choice fell upon the Josephinum, where he could study medicine at the expense of the state. Toldt left for Vienna in 1858, but his attachment to Tyrol remained strong. After graduation, the compulsory ten-year military medical service took him to the Verona Military Hospital, where his thorough training in new diagnostic and treatment methods of eye, ear and throat diseases earned him the position of Head of the Department of Eye Diseases. He later fondly remembered the (German) chief physician of the hospital, but wrote that 'the rest [of the physicians], predominantly Italians, clung to medically obsolete views, and assistant doctors were completely incompetent.'[70] After a transfer to the Mantua Infantry Regiment, which fought in the 1866 battle of Custozza, Toldt decided that his true interests lay in science. He obtained an assistantship at the Josephinum Chair of Physiology under Ewald Hering, who put him in charge of microscopy practicals for medical students.[71] By 1869 Toldt had habilitated as *Dozent* in microscopical anatomy, and was lecturing on histology. In this period, his research focused on microscopical studies of the structure of the lymphatic vessels, fat tissue and intestines.

After the dissolution of the Josephinum, Toldt, still an active military officer with no experience in anatomical research or teaching, was placed in charge of the Courtyard of Corpses (*der Leichenhof*) in the Military Hospital (*Garnisonspital*). He supervised all handling of military corpses, pathological autopsies and anatomical dissections at the new Second Anatomical Department. In 1875 the army commissioned him to determine whether torso girth was a valid measure of military fitness.[72] Toldt compared the lung volume of the cadavers he received from the Military Hospital with their torso girth, and concluded that these two measures were not related; soon afterwards the Austrian army abandoned this method of fitness assessment.[73] This study was his first venture into the field of gross anatomy and anthropology.

In 1876 Toldt was appointed Professor of Anatomy at the University of Prague. Fortunately for him, the faculty had lost a histologist, the ministry had a preference for Austrian candidates, Langer supported him, and his administrative experience in running the Courtyard of Corpses counted in his favour.[74] As a Prague professor (1876–84), Toldt oversaw the building of a new anatomical institute and served as the dean of the Medical Faculty in 1881/82 and a member of the University Senate in 1883/84. His early experiences in university administration thus coincided with the fission of the Charles University into Czech and German universities in 1882, and with rising tensions between Germans and Czechs.[75] The strained Prague atmosphere may have been an additional reason for Toldt to welcome his 1884 invitation to the Vienna Second Anatomical Chair.

His inaugural lecture, 'On education in morphological sciences at medical faculties' [*Über den Unterricht der morphologischen Wissenschaften an den medizinischen Facultäten*] encouraged students to begin their careers as natural scientists by independent observation in the dissection hall.[76] Toldt's support for teaching science to medical students was especially strongly expressed in his 'Special report on the position of natural sciences in the medical curriculum and examinations' [*Sondergutachten über die Stellung der naturgeschichtlichen Fächer in der medicinischen Studien- und Prüfungsordnung*]. There he argued persuasively – and not unlike Billroth before him – for the crucial importance of science in developing independent observational skills, against relegating science education to *Gymnasia* and for retaining mineralogy, botany and zoology as examinable subjects in medical curricula.[77]

From the second half of the 1870s onwards Toldt's own research took a turn away from the studies of the microscopic structure of the gastrointes-tinal system, via the structure and development of bone tissue, to forensic methods of dating human remains on the basis of the stage of ossification.[78] He finally arrived at anthropology, and in particular the anthropometry of Tyrol and Vorarlberg. In the 1870s and 1880s this field was the battleground between ethnographers, anatomists and linguists supporting one of the two opposed theories.[79] The Tyrolean scholar Franz Tappeiner claimed that the Alpine populations on the border with Switzerland had skulls markedly wider than Germans and so, despite of their blond hair and blue eyes, consti-tuted a distinct race, which he named Rhaetian (*Rhätisch*), thus stressing their Latin/Roman origins.[80] Tappeiner's craniometric observations fitted perfectly with the historical ethnography of Carl von Czörnig, director of the Büro für administrative Statistik, who claimed that Tyroleans were by origin Etruscans – the people that founded Rome – pushed northwards by the Gauls and then overrun by the Germans, who left them their language but in exchange adopted their culture.[81]

The Tappeiner-Czörnig hypothesis exploited the special status of Tyrol, a region of turbulent history, to reaffirm the connection between the Roman Empire and the Habsburgs, who had ruled the Holy Roman Empire until 1804, and to establish Austrian racial and cultural independence from, and superiority over, Germans. The theory was, however, hotly disputed by the German middle classes, whose sense of cultural dominance over other ethnic groups of the Empire was based precisely on their connection to the German nation. For them, Tyroleans were not proto-Austrians but the purest Germans.

Figure 10.1 Toldt family summer residence, known as '*das Genspichler Gut*' or '*Gut am Rasitten*', in Vahrn near Brixen; woodcut by M. Strele, published in Karl Toldt (1939) *Geschichte der Familie Toldt, verfasst von weil. Univ.-Prof. Dr. Carl Toldt, 1893, bearbeitet und ergaenzt von seinem Sohn Dr. Karl Toldt* (Innsbruck) 123

Toldt is an excellent representative of this group. Although he left Tyrol at eighteen, he returned regularly. In 1872, he married Kreszenz Pfaundler (b. 1847), a daughter of a Brixen (Bressanone) merchant. The family, which soon increased to five with the births of Anton (1873), Karl (1875) and Marie (1876), never failed to spend their summer holidays in Vahrn (Varna) with their Brixen family.[82] In 1883, they purchased an old farm that would become the family summer residence for generations (Figure 10.1).

For the Austrian liberal bourgeoisie, summer residences in the countryside were much more than holiday retreats: they were places where, within the family's supportive circle, new ideas were explored and concepts created, free from the constraints of urban schedules and institutions.[83] In Tyrol, Toldt spent time with his family, planted trees and conducted anthropological studies of the local population, first using the military recruit lists that included anthropometric data and then in the field. He found that the Grossglockner region was populated by very tall people, while the inhabitants of the Ortler region were short, and explained this finding not by invoking the influence of ecological factors, such as nutrition or the altitude, but by pointing to their different ethnic origins, the former being predominantly Germans and the latter Italians.[84] This view won out in the end even in court circles: Toldt, and not advocates of the 'Rhaetian' race, contributed a chapter on the physical properties of the Tyrol and Vorarlberg populations to 'The Austro-Hungarian Monarchy in Words and Pictures' [Die österreichisch-ungarische Monarchie in Wort und Bild], a 24-volume book project started in 1886 under the auspices of Crown Prince Rudolph and completed in 1902 to celebrate the diversity of the peoples, cultures and nature of the Empire.[85]

Toldt's professional network consisted mainly of other anatomists, his students or senior scholars, who were almost exclusively German speakers.[86] Most of them came together in the 'Anatomical Society' [Anatomische Gesellschaft], a society founded in 1886 from a section of the 'Society of German Naturalists and Physicians', [Gesellschaft Deutscher Naturforscher und Ärzte]. Toldt was one of its first chairmen and a co-president with Wilhelm His and Albert von Kölliker from 1891 to 1894.[87] The society organized meetings and published the Anatomischer Anzeiger, a journal that would soon gain much respect and influence, but its largest and most ambitious project was the reform of anatomical nomenclature. The goal of this project, started at the 1889 Anatomical Society meeting when Toldt was elected to the nomenclature committee, was to standardize the names for the body parts and gradually eliminate often controversial eponyms.

Toldt invested much effort and hope in the nomenclature project, published in Basel in 1895 as Basel Nomina Anatomica (BNA). Together with his long-term assistant, Tyrolean Alois dalla Rosa, in 1900 Toldt completed Anatomical Atlas for Students and Physicians [Anatomischer Atlas für Studierende und Ärzte], a three-volume masterpiece of detailed preparation presented in the form of sober yet highly precise wood engravings. The foreword to the first edition stated that its goal was to disseminate the new nomenclature and in this it succeeded as in the forthcoming decades it would become one of the most popular anatomical atlases.[88]

Although initially intended for the German medical community only, in subsequent decades the nomenclature gained acceptance worldwide. In 1908, Toldt was succeeded by Ferdinand Hochstetter (1861–1952) in the Second Anatomical Chair. Hochstetter continued to maintain close contacts with German anatomists, which included participation in the Anatomical Society as a co-president (1920–24) and member and one-time chair of the nomenclature reform committee (1923–35).[89] He furthermore took over the editorship of the Anatomischer Atlas from the twelfth edition in 1921; like his predecessor, he used it to publicize a new anatomical nomenclature when the BNA was revised around 1930. In contrast to Toldt's, however, Hochstetter's research was focused on descriptive embryology and comparative anatomy. Besides his papers on the development of the central nervous system, heart and circulation, he became known for his innovations in the art of preparation.[90]

The political views and affiliations of the successors to Toldt's chair constitute one of the best known chapters of the Viennese anatomical history. Under Ferdinand Hochstetter, German nationalist students of his department clashed with socialists and Jews of the First Anatomical Chair, while Eduard Pernkopf, the Second Anatomical Chair 1932–45, was the most prominent Nazi of the Medical Faculty and the author of an atlas that notoriously used the cadavers of executed political prisoners as models.[91] Toldt's political opinions are, by contrast, less well known. As a military officer of the imperial army he probably kept his opinions private. But as a professor in Vienna, he clearly positioned himself with the German nationalists. For example, when in 1890 police broke up a gathering of a German nationalist student society, the student and anatomical demonstrator Franz Hoffmann punched a policeman on duty.[92] He was fined and reprimanded by the university vice-chancellor. But three months later, the ministry of internal affairs was surprised to discover that, in spite of the incident, Hoffmann had been reappointed to the demonstrator post at the Second Anatomical Department. Toldt submitted an application in Hoffmann's support, stating that his help was essential for the regular course of instruction because the extraordinary professor, Alois dalla Rosa, rarely ventured into the dissection hall and skilled help was hard to find. But the ministry was not convinced and Toldt had to find a new demonstrator.[93] His sympathies became even more evident in 1897/98, when he served as the Rector of Vienna University.[94] In his inaugural speech 'On the history of the university medical departments' [*Zur Geschichte der Medicinischen Universitätsinstitute*], he emphasized the autonomous rights of the university and the authority of the vice-chancellor. Such assertions of independence would inevitably resonate strongly in the historical moment when the authority of the court and dynasty came under threat from various sides. In April 1897, Emperor Franz Joseph finally yielded and sanctioned the election of the controversial populist leader of Christian Socials, Karl Lueger, after almost two years of refusing to do so.[95] In the same year, the Badeni language ordinances weakened the linguistic, and consequently cultural, hegemony of Germans in Cisleithania and provoked an explosion of student protests throughout the country.

Even after the ordinances had been revoked and the government had fallen in November, the riots continued, but now the immediate reason was a prohibition on the public display of student association insignia in Prague. The rectors of all Austrian universities held an emergency meeting on 2 February 1898, but the next day German students started a strike. The government, however, returned the blow and officially ended the winter term with effect from 10 February, ordering the beginning of the summer term on 21 February. For enrolment in the next term, students were ordered to submit a written application. Many students and part of the professoriate saw this as intentional humiliation, bound to reawaken the protests at the beginning of the summer term. But Toldt managed to work out a compromise that satisfied the studentship and saved the face of the government: the written application was exchanged for a form that students just needed to sign. Although the early twentieth century was hardly a peaceful period for the university, student protests were not repeated with the same force as in 1897/98.

Toldt's obituarists described him as 'an upright German man' (*ein aufrechter Deutscher Mann*)[96] and 'a stalwart German' (*ein strammer Deutscher*).[97] Typically for the German-speaking Austrian middle classes, his allegiance to the larger German nation provided him with a sense of identity. His national feeling was shaped by his upbringing in Tyrol, a border region of mixed ethnic make-up, and then education at the Josephinum at the time when Austria hoped to head the unification of the German lands yet had to face the reality of military failure.

Toldt's politics were by no means extreme. He was a typical middle-class liberal, whose belief in progress in the polarized atmosphere of late nineteenth-century Austria metamorphosed into a belief in progress under the leadership of the 'most civilized' ethnic group. The fact that he had started his career as a histologist and outside the university certainly influenced his professional choices. Yet the most important element in his biography was his German cultural identity, which, as shown by examples of his professional networks, research projects and university politics, deeply informed his personal and professional life.

Fin-de-siècle anatomy? Between the university tradition and anti-semitism

Emil Zuckerkandl belonged to the lucky generation that enjoyed a brief moment of (relative) freedom and unity at Austrian universities. Born into a Jewish family in the Hungarian town of Raab (Győr) on 1 September 1849, he was educated in Pest, where the family had relocated.[98] The family followed a typical route to Jewish assimilation, propelled by the liberal legislation of 1867. New generations moved from the trades to the professions, especially medicine, to which all three Zuckerkandl brothers were connected: the youngest, Otto, was a highly successful urological surgeon, while Viktor, the only businessman of the three, purchased a psychiatric hospital, the Purkersdorf Sanatorium, in 1902.[99] Emil matriculated at the University of Vienna in 1868.[100] Studying hard and working as an anatomical demonstrator did not stop him from taking part in the boisterous activities of student fraternities, including drinking parties and duels.[101] Yet just a few years after his graduation, young Jews were painfully awakened to the fact that, in spite of their cultural identification with the German nation, they were unwanted outsiders. For the medical student Sigmund Freud the wake-up call came in 1876 with the controversy over Billroth's book, discussed above.[102] Zuckerkandl's professional choices were determined on

Figure 10.2   Sculpture of Emil Zuckerkandl by 'the sculptor of Red Vienna', Anton Hanak (1923). Copper alloy cast, dimensions: H 135 × W 63 × D 50 cm; with the stand: H 220 × W 63 × D 50 cm. The sculpture, funded by money raised by an American, 'Dr. Ash', was unveiled in the *Arkadenhof* of the University of Vienna's main building on 5 April 1924. Reproduced by permission of the University Archives of the University of Vienna.

the one hand by his education under Joseph Hyrtl and on the other by his precarious position as a Jewish academic in an anti-Semitic setting.[103]

Information on Zuckerkandl's early life is scarce and conflicting. According to his successor and obituarist Julius Tandler, he initially wanted to study engineering.[104] But his widow, Bertha Szeps, wrote that he had passionately studied nothing but violin until the age of 16, when he suddenly abandoned it and decided to devote himself to 'learning'.[105] In the first term of his medical studies, his exceptional dexterity at dissection attracted the attention of the highly critical Hyrtl, who rarely built close relationships with his students and never established a school. When demonstrating preparations in lectures, he described them as coming from the 'expert hands of his favourite student Zuckerkandl': perhaps the young man's skilled hand and artistic inclination reminded Hyrtl of his own youth.[106]

On Hyrtl's recommendation, Zuckerkandl was appointed to a temporary prosectorship at Amsterdam's Athanaeum while still a medical student until the 'sad circumstances of that anatomy, the inability to work in the anatomical subject at that place' brought him back to Vienna.[107] He then briefly worked as Rokitansky's assistant (1873/74) and, after graduation in 1874, moved to the Anatomical Institute, where he stayed until his appointment to the Anatomical Chair in Graz in 1880. Stories about his duels and nights on the town abounded during his prosectorship years, and probably contributed to his popularity with students. Zuckerkandl's romantic aura was further heightened by his black robe, reminiscent of his teacher Hyrtl and seventeenth-century Dutch anatomy (Figure 10.2).

His *Burschenschaft* past came in handy when on his first day in Graz he was booed by anti-Semitic students.[108] Zuckerkandl challenged the protesters to a duel.[109] This clever act secured his favourable reception in the conservative milieu of the Styrian capital. Although his Jewish background was repeatedly brought up as he climbed the professional ladder, his career progressed fairly smoothly.[110] He became Extraordinary Professor in 1879, at the age of 31, without having previously habilitated as *Dozent*.[111] In 1888, he returned to Vienna, where he chaired the First Anatomical Department until his death in 1910.

Zuckerkandl's research interests spanned several fields – descriptive and physiological anatomy, histology, embryology and anthropology – and the entire human body: the ear (temporal bone, Eustachian tube), tongue, feet, cranial nerves and peritoneum.[112] His first anthropological papers date from the early 1870s, when he analysed skulls collected during the Austrian frigate *Novarra*'s exploratory circumnavigation, 1856–59.[113] But in contrast to Toldt's work on torso girth and Tyrolean populations, Zuckerkandl was not interested in ascribing characteristic physical properties to a race or ethnic group. Instead, he focused on inter-individual differences, and preferred a qualitative comparison to statistics.[114] His approach was anatomical rather than anthropological, as anthropologists not only focused on determining typical characteristics of a group and inter-group differences but also, from the 1860s, relied on increasingly complex statistical methods.[115]

Zuckerkandl's anatomy drew on the topographical (applied) tradition established by Hyrtl. Topographical anatomy divided the body into regions demarcated by natural boundaries, such as muscles and large blood vessels. This approach, also called 'surgical', differed from systematic or descriptive anatomy, which divided the body into functionally and histologically unified systems. Hyrtl published a widely used textbook of topographical

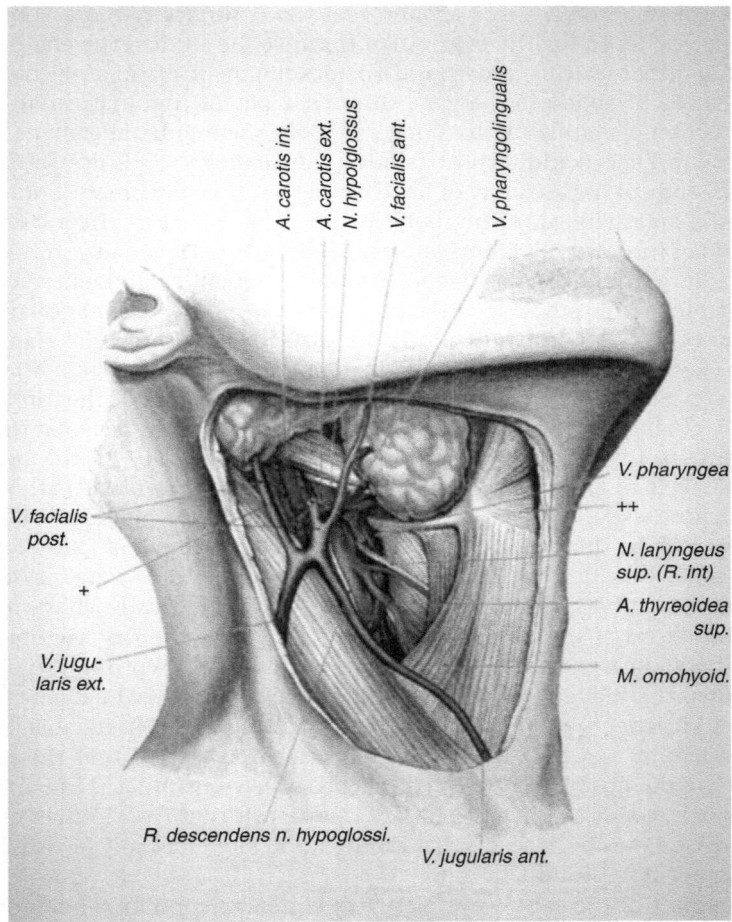

Figure 10.3 Plate showing *Fossa carotica* and *Fossa submaxillaris*, drawn by Bruno Keilitz, in Emil Zuckerkandl (1899) *Atlas der topographischen Anatomie* (Vienna: Wilhelm Braumüller), vol. 1, 186, fig. 193. 25.2 × 17.5 cm.

anatomy; some 50 years later, Zuckerkandl produced a topographical atlas (Figure 10.3).[116]

Yet Zuckerkandl was by no means a carbon copy of his predecessor. Hyrtl's research, largely within the comparative anatomy of Cuvier's school, was separate from his clinically oriented teaching. For Zuckerkandl, by contrast, there was no clear boundary between the two. The purpose of anatomical knowledge was to literally open up new bodily regions for the clinic and offer solutions to existing problems. He experimented with dissection to propose new, visually more advantageous and surgically more conservative approaches to bodily regions. For instance, he studied the anatomy of the abdominal wall to understand the development of inguinal hernia – the descent of abdominal organs through a patent inguinal canal – and construct an operating method that cut above the inguinal ligament.[117]

Probably Zuckerkandl's most famous contribution to clinical medicine was his study of the anatomy of nasal cavities and the pneumatic bones that surround and delineate them, published in a book that in effect launched the field of rhinology.[118] From the early 1800s, Vienna had had a strong clinical tradition in the disease diagnostics and treatment of all the head organs engaged in communication with the external world – eye, ear and larynx – yet the nose remained unexplored. In 1870, the dermatologist Hans von Hebra described the unusual growth on a patient's nose as *rhinoscleroma*; four years later laryngologists found similar formations inside the throat and mouth.[119] The following decade saw a flurry of research targeted at finding the cause of this chronic disease that produced deformity; in the 1880s it was ascribed to a Gram-negative bacillus. Research into other pathologies of the nose expanded apace, and Zuckerkandl's two-volume publication may be seen in this context. Dissections of 300 normal and 'pathological' adult noses were used to set the anatomical boundaries of nasal cavities as well as to describe the morphology of the bones, specifically, characteristic inflammations of the mucosal lining and polyps in the first volume. The second volume was dedicated to anatomy of the nasal septum. Erna Lesky attributed the sudden prominence of the nose in Viennese medicine to the onrush of patients from Galicia, Poland, Wallachia and Bessarabia, where rhinoscleroma was endemic. But it has also been argued that the plastic surgery of the nose (rhinoplasty) developed as Jews assimilated in European and North American societies in late nineteenth century.[120] Whatever the main reason may be, it seems that immigration and especially Jewish immigration played a prominent role.

Being part of the early wave of Jewish academic appointments probably influenced Zuckerkandl's career in other ways as well. In the early twentieth century, about 50 per cent of Viennese doctors were Jews.[121] While they could qualify as physicians, open private practices or, with some difficulty, get hospital jobs, they often faced obstacles in their academic careers. In both the Austro-Hungarian Empire and Germany, professors were in effect high-ranking civil servants, and Jews were denied civil service posts. This led to a disproportionately high number of unsalaried *Privatdozenten* in comparison with professors of Jewish background: at the University of Berlin 44 out of 113 *Privatdozenten* were Jewish, but there was not a single Jew among the 19 full professors (*Ordinarii*).[122] In Vienna, the situation was somewhat better; between 1848 and 1900 around 10 per cent of professorial appointments in Vienna were to Jews.[123] Yet, reaching the highest positions was difficult and the proportion of Jews among *Ordinarii* was smaller than among *Extraordinarii* and *Dozenten*. Zuckerkandl's own brother, Otto, never advanced further than an extraordinary professorship.[124] The career woes of *Privatdozent* Sigmund Freud are well known. While *Privatdozenten* in clinical disciplines supported themselves with private practice as they climbed the academic ladder, for those in preclinical subjects, such as anatomy or physiology, a salaried position was a necessity. For that reason few Jews dared to pursue an academic career in a preclinical discipline. Zuckerkandl was an exception to the general rule in the period when few German-speaking anatomical professors were Jews.

Yet in spite of his apparently successful assimilation, being a Jew may have informed Zuckerkandl's career, for instance by shaping his professional network. In contrast to Toldt's exclusively anatomical alliances and collaborations, the audience for Zuckerkandl's papers largely consisted of clinicians. In the absence of letters and memoirs, this is best understood by looking at the journals in which he published. While his anthropological articles were published in German and Austrian anthropological journals, the rest of his work largely appeared in clinical periodicals. For example, he published in *Monatschrift der Ohrenheilkunde* (20 articles), *Monatsschrift für H-N-O Krankenheiten* (2), *Archiv der Ohrenheilkunde* (2), *Langenbecks Archiv für Klinische Chirurgie* (1), *Wiener Klinische Wochenschrift*, *Deutsche Zeitschrift für Chirurgie*, *Arbeiten des Neurologischen Instituts in Wien*.[125] His collaborators were almost exclusively clinicians. For the book on nasal cavities he collaborated with the pathologist Kundrat and a laryngologist; in his later years he co-authored several articles on muscular movement with Wilhelm Erb, the German neurologist who became famous for his work on muscular dystrophies. And in further contrast to Toldt, Zuckerkandl was not an active member of the Anatomische Gesellschaft. The reason for his absence from his own professional community may have been his lack of interest in the topics researched by other anatomists or, conversely, anatomists' lack of interest in his work.[126] But it may also be that this strong national professional association looked unfavourably at a Hungarian-born Jew.

Perhaps precisely because of his exposed position as a Jewish professor at the medical faculty, Zuckerkandl was not nearly as active in university politics as Toldt, and abstained from any public demonstration of his opinions. The only issue on which he took a clear position in a public debate was his support for women in medicine. Zuckerkandl was the second professor of the medical faculty (and the University of Vienna) to employ a female demonstrator, Gertrude Bien, who then became the first woman to advance to the position of university assistant, in spite of the ministry's reluctance.[127] When at her very first lecture she took anatomical preparations of sex organs around the lecture hall and male students greeted her with shouts and protests, Zuckerkandl sent for the porters to throw the rebels out.[128] The First Anatomical Department continued to pioneer women under Zuckerkandl's successor, Julius Tandler, whose female demonstrator, Marianne Stein, went on to become his assistant in the early 1920s.[129] The Second Department, by contrast, appointed female demonstrators only when forced to do so by the shortage of male students due to the First World War; likewise, its first female assistants were employed in the 1940s after the male ones had been drafted into the army.[130]

Zuckerkandl's belief in universal access to education did not stop at middle-class women but extended to the proletariat. He was a founding member of the independent left-wing movement for popular education (*Volkshochschulbewegung*).[131] From its beginnings in a modest rented house in the proletarian heartland of Ottakring, this initiative grew into a network of 'people's universities' (*Volkshochschulen*) strategically distributed through working-class neighbourhoods, which provided workers with edifying content for their leisure time. This public educational activity made a political statement but also provided an important economic incentive to the poorly paid junior university staff.[132] After 1919 these institutions would be taken up by the new Social Democratic municipal establishment as the central places of the proletarian intellectual and psychological transformation.

Outside the university, Zuckerkandl moved in highly educated, well-off, liberal and largely Jewish circles. Around 1880, a diplomatic correspondent of the liberal *Neues Wiener Tagblatt* introduced him to his future wife, Bertha, the younger daughter of newspaper editor Moritz Szeps.[133] Szeps was a well known Viennese opinion-maker, Francophile – his elder daughter Sophie would marry Paul, the brother of French Prime Minister Georges Clemenceau – and confidant of Crown Prince Rudolph, while Bertha herself was a leading art journalist. For art historians, the Zuckerkandl family is significant for its crucial support of the Vienna Secession. Bertha's journalistic pieces promoted the new movement and especially the career of Gustav Klimt, while the wealthiest brother, Viktor, was an important patron whose collection included Klimt's *Pallas Athene* and many of his best landscapes.[134] It was through Bertha's connections that Viktor engaged Josef Hoffmann, the cutting-edge architect and founder of the applied arts group Wiener Werkstätte, to design the Purkersdorf sanatorium in a modern, strictly functional and minimalist fashion.[135] Viktor Zuckerkandl probably met the artists in Emil's and Bertha's house in the leafy suburb of Unterdöbling. In the salon, Johann Strauss and Ernst Mach mingled with the radical young artists and intellectuals such as Max Reinhardt, Hugo von Hofmannsthal, Arthur Schnitzler and Hermann Bahr, as well as medical professors such as Wagner-Jauregg and Billroth.[136] The best known Viennese *fin-de-siècle* love story, between the composer Gustav Mahler and the budding musician and socialite Alma Schindler, began, reputedly, in the Zuckerkandl household.[137]

Cafés and salons as the meeting places of academics, artists and their patrons have been identified as the feature that fostered the cohesiveness

Figure 10.4   Gustav Klimt, *Danaë* (1907), oil on canvas, 77 × 83 cm. Detail.

of the Viennese cultural elite, in contrast to the independent, mutually poorly communicating professional circles in London, Paris or Berlin.[138] People from diverse fields came together around the central problems of life, consciousness, neuropsychological processes, social relations, and the history and future of humankind. Historians have noted the striking similarities between Ernst Mach's psychology and Hugo von Hofmannsthal's literary impressionism, as well as between Freudian psychological doctrine and characterizations in Arthur Schnitzler's novels and plays.[139] Bertha Zuckerkandl later recollected Emil's discussions of the nature of death with Hermann Bahr and Ernst Mach.[140] This interdisciplinary environment probably stimulated Zuckerkandl to engage with questions outside the remit of his own discipline.

The exchange went both ways. Josef Hoffmann drew on contemporary ideas that located the cause of psychiatric – nervous – illness in the modern environment when designing the previously mentioned Purkersdorf sanatorium around 1903.[141] The European iconoclastic art movements of 1900 took inspiration from nature and transformed natural motifs into abstract decorative forms: most famously, Ernst Haeckel's marine organisms decorated *Jugendstil* furniture.[142] The right corner of Gustav Klimt's famous painting *Danaë*, depicting the mythical moment when Zeus in the shape of golden rain surreptitiously impregnated the princess of Argos and Eurydice, is scattered with the shapes of a blastocyst (Figure 10.4).[143]

This early embryonic structure, first described in humans in 1895, was still a novelty; Klimt must have seen it in embryological literature or, more likely, attending Zuckerkandl's anatomical lectures in the summer of 1903.[144]

In her memoirs, Bertha recalled the exhilarating atmosphere of these evening science lectures for artists. Dressed in his flowing black gown, Zuckerkandl dazzled the gathered painters, writers and musicians with slide projections of '*wohlausgeklügelte Färbungen von Gefäßen, von einem Stückchen Epidermis, einer Arterie, einem Blutstropfen, ein wenig Gehirnsubstanz* [ingenious colourings of blood vessels, of a small piece of the epidermis, of an artery, a drop of blood, a bit of brain tissue]'.[145] Bertha argued that both Klimt's palette and the ornamental repertoire of the Wiener Werkstätte borrowed from nature's treasure trove. Indeed, a close look at Klimt's work in the years immediately after 1903 reveals an abundance of shapes that look like cells: see, for instance, on the left side of the *Wasserschlangen* (1904–07) shapes reminiscent of epithelium, with black 'nuclei' in whiteish 'cytoplasm', and, hugging the hips of the woman on the right, a fabric made of dividing cells. Cell division (mitosis), too, was a novel concept, first described by the former Prague professor Walther Flemming in the early 1880s.

Zuckerkandl's anatomy, then, should not be understood as a straightforward continuation of Hyrtl's approach, but as an act of negotiation between tradition and modernity. Not only was he a student of Hyrtl; he was so perceived by the faculty that wished to maintain continuity. Yet this was not possible. Zuckerkandl's position as a rare Jew within the anatomical community may have directed him towards less homogenous and more open clinical professional networks and thus further pushed his research in the direction of the clinic. The social circles in which his family moved exposed him to knowledge exchanges that took place within the city rather than within the university or the anatomical professional community. These interdisciplinary yet highly localized and city-centred exchanges would become even more manifest for Zuckerkandl's successor Julius Tandler – a Jew and Social Democrat who took the practical approach to an extreme by becoming a politician and a social reformer intent on transforming proletarian bodies.[146]

Conclusion

Through the nineteenth century, anatomical appointments were sites of tension between clinicians, who supported 'medically relevant' approaches, and anatomists, who favoured orientations perceived as cutting-edge science within their professional community. In Vienna, these discussions carried strong political overtones because the approaches favoured by the clinic – in this case, topographical/clinical anatomy – were seen to follow the Austrian educational tradition and were favoured by professors supporting the court politics of maintaining a hoped-for, albeit fragile ethnic stability. In contrast, those in favour of the current trends in German anatomical science were also inclined to forge stronger links with Germany, as an assurance of the continuing cultural dominance of Germans in the empire.

The crisis of the academic profession around 1900 was not unique to the Austro-Hungarian Empire, but the division lines here ran differently from those in Germany. The split in anatomy reflected the division in the professoriate, which in turn was related to a corresponding process in the middle class. The major political issue was the future shape of the empire; the major political force was nationalism. Positioning Austrian professors on the conventional left–right spectrum would lead only to confusion: first, because of the specificities of the political scene, where the left really appeared only in the twentieth century, and second, because each of the two factions was diverse in any case. Thus, although the 'Austrian' faction with its rhetoric of tradition and loyalty to the dynasty may be regarded as something of a counterpart to the German 'mandarins', such parallels would only distort the picture.

This division differs from the generational one proposed by Schorske. Nationalism did not defeat and succeed liberalism, but rather the two co-produced one another. The example of anatomy and, as indicated in the section on anatomical appointments, medicine more generally, indicates that the split that emerged in the liberal era extended seamlessly into the 1900s. Each camp drew on a different set of resources and talked to a different audience. To reconstruct them, I sketched synoptic portraits of two anatomists, bringing together their research publications, teaching careers, university and national politics, and professional networks as well as their upbringing and private lives. Neither of these anatomists withdrew from the public, but rather their publics were different. And if the private world where Zuckerkandl's anatomy developed was a glamorous salon in the foothills of the Vienna woods, Toldt's was a *Sommerfrische* in South Tyrol. Perhaps against our expectations, it was the urbanite Zuckerkandl who chose to build on the tradition, rather than breaking away from it.

Acknowledgements

An earlier version of this essay was delivered at the ESHS conference in Vienna in September 2008. Parts of it were presented in talks in Cambridge, Ljubljana, Manchester, Melbourne and Vienna between 2007 and 2009. I am grateful to all of these audiences for their questions and comments. I furthermore thank Nick Hopwood, Hans-Georg Hofer, Emese Lafferton, Johannes Feichtinger and the editors of this volume for their comments on this essay.

Names of places are given as they appear in original documents, in German, with their commonly used names in other languages in brackets. All translations from German are mine unless otherwise stated.

Notes

1. Carl Toldt (1922) *Autobiographie* (Vienna: Urban & Schwarzenberg), 45.
2. See Ferdinand Hochstetter (1921) *Bericht über die Rektors-Inauguration der Wiener Universität* (Vienna: Universität Wien); Rudolf Pöch (1920) 'Karl Toldt', *Wiener Klinische Wochenschrift*, 33; idem (1921) 'Carl Toldt (gestorben am 13. November 1920)', *Mitteilungen der Anthropologischen Gesellschaft in Wien*, 51, 77–93; Siegmund von Schumacher (1921) 'Carl Toldt †', *Anatomischer Anzeiger*, 54, 82–91. He also received a long and affectionate obituary in a local Tyrolean journal: Raimund von Klebelsberg (1921) 'Carl Toldt. † am 13. November 1920', *Der Schlern: Südtiroler Halbmonatsschrift für Heimatkunde und Heimatpflege*, 2/2, 25–33.
3. Carl Toldt (1911) 'Emil Zuckerkandl', *Almanach der Österreichischen Akademie der Wissenschaften*, 61, 364–71. Julius Tandler (1910) 'Emil Zuckerkandl', *Wiener Klinische Wochenschrift*, 23, 798–800; idem (1910) 'Emil Zuckerkandl', *Wiener Medizinische Wochenschrift*, 60, 1360–62; Julius Tandler (1910) 'Emil Zuckerkandl', *Anatomischer Anzeiger*, 37, 86–96.
4. Obituaries and eulogies have long been used in historical research, but in recent decades they have been reappraised by historians interested in the relationship between history and memory, in particular in the construction of collective memory. For collective memory and science, see Pnina Abir-Am (1999) 'Introduction', in idem (ed.) *Commemorative Practices in Science: Historical Perspectives on the Politics of Collective Memory*, Osiris, 2nd Series, 14, 1–33.
5. The names referred to the river Leitha/Lajta/Litava, which until 1921 marked the border between Austria and Hungary. For overviews of Austrian politics in the nineteenth and early twentieth centuries see Erich Zöllner (1979 [1961]) *Geschichte Österreichs: von Anfängen bis zur Gegenwart* (Munich: Oldenbourg); Alan J.P. Taylor (1981 [1941]) *The Habsburg Monarchy 1809–1918: A History of the Austrian Empire and Austria-Hungary* (Harmondsworth: Penguin); Helmut Rumpler (1997) *Eine Chance für Mitteleuropa: bürgerliche Emanzipation und Staatsverfall in der Habsburgermonarchie* (Vienna: Ueberreuter).
6. On student associations see Michael Gehler (1994) 'Männer im Lebensbund. Studentenvereine im 19. und 20. Jahrhundert unter besonderer Berücksichtigung der österreichischen Entwicklung', *Zeitgeschichte*, 21, 45–66.
7. Carl Schorske (1998) *Thinking with History: Explorations in the Passage to Modernism* (Princeton, NJ: Princeton University Press), 142.
8. Gary Cohen (1996) *Education and Middle-class Society in Imperial Austria 1848–1918* (West Lafayette, IN: Purdue University Press), 237–40.
9. Cohen, *Education and Middle-class Society in Imperial Austria*, 47. See also, on medical professors, (1861) 'Miscellen', *Vierteljahrschrift für die practische Heilkunde*, 18/2, 1.
10. This topic filled the pages of the *Wiener Medizinische Wochenschrift*, *Medizinische Presse* and daily papers around 1880. See, for example (1881), 'Die Czechisirung der Prager Universität', *Wiener Medizinische Wochenschrift*, 31/15, 425–26.
11. See Erna Lesky (1965) *Die Wiener Medizinische Schule im 19. Jahrhundert* (Vienna: Böhlau).

12. Theodor Billroth (1876) *Über das Lehren und Lernen der medizinischen Wissenschaften an den Universitäten der deutschen Nation nebst allgemeinen Bemerkungen über Universitäten: eine culturhistorische Studie* (Vienna: Carl Gerolds Sohn).

13. On Billroth's politics, its origins and how these played out in the Viennese context, see Felicitas Seebacher (2006) '"Der operierte Chirurg": Theodor Billroths Deutschnationalismus und akademischer Antisemitismus', *Zeitschrift für Geschichtswissenschaft*, 54, 317–38.

14. Johann Freiherr von Dumreicher (1878) *Über die Nothwendigkeit von Reformen des Unterrichtes an den medicinischen Facultäten Österreichs* (Vienna: Alfred Holder).

15. Tatjana Buklijas (2007) 'Surgery and national identity in late nineteenth-century Vienna', *Studies in History and Philosophy of Biological and Biomedical Sciences*, 38, 756–74.

16. Anatomy acts were legislative documents that regulated the supply of cadavers to anatomical institutes. While prior to ca. 1800 most cadavers came from executed prisoners, these new acts named hospitals and welfare institutions as the chief and often sole sources. Ruth Richardson (2001) *Death, Dissection and the Destitute*, 2nd ed. with a new afterword (London: Phoenix Press); Michael Sappol (2002) *A Traffic of Dead Bodies: Anatomy and Embodied Social Identity in Nineteenth-century America* (Princeton: Princeton University Press); Helen MacDonald (2006) *Human Remains: Dissection and its Histories* (New Haven: Yale University Press).

17. Adrian Desmond (1989) *The Politics of Evolution: Morphology, Medicine and Reform in Radical London* (Chicago: University of Chicago Press).

18. There is much literature on the history of anthropology but hardly anything substantial that explores the relationship between anatomy and anthropology. The new project on 'Anatomies of Empire: race, evolution, and scientific networks in the twentieth-'century British world', funded by the Australian Research Council, promises to remedy this omission, at least for the British Empire. On anthropology in the Austro-Hungarian Empire see Brigitte Fuchs (2003) *'Rasse', 'Volk', Geschlecht: Anthropologische Diskurse in Österreich* (Frankfurt: Campus) and Emese Lafferton (2007) 'The Magyar moustache: the faces of Hungarian state formation, 1867–1918', *Studies in History and Philosophy of Biological and Biomedical Sciences*, 38, 706–32.

19. Tatjana Buklijas (2008) 'Cultures of death and politics of corpse supply: anatomy in Vienna, 1848–1914', *Bulletin of the History of Medicine*, 82, 570–607.

20. Lynn Nyhart (1995) *Biology Takes Form: Animal Morphology and the German Universities, 1800–1900* (Chicago: University of Chicago Press).

21. Jonathan Harwood (1993) *Styles of Scientific Thought: the German Genetics Community, 1900–1933* (Chicago: University of Chicago Press).

22. Fritz Ringer (1969) *The Decline of the German Mandarins. The German Academic Community, 1890–1933* (Cambridge, Mass.: Harvard University Press).

23. Carl Schorske (1979) *Fin-de-siècle Vienna: Politics and Culture* (London: Weidenfeld and Nicolson).

24. Pieter M. Judson (1996) *Exclusive Revolutionaries: Liberal Politics, Social Experience, and National Identity in the Austrian Empire, 1848–1914* (Ann Arbor: University of Michigan Press); Steven Beller (ed.) (2001) *Rethinking Vienna 1900* (New York: Berghahn Books).

25. Steven Beller (1989) *Vienna and the Jews, 1867–1938: a Cultural History* (Cambridge: Cambridge University Press).

26. Deborah R. Coen (2007) *Vienna in the Age of Uncertainty: Science, Liberalism and Private Life* (Chicago: University of Chicago Press).

27. On this important distinction see also Johannes Feichtinger, '"Staatsnation", "Kulturnation", "Nationalstaat". The role of national politics in the advance of science and scholarship in Austria from 1848 to 1938', in this volume.

28. On professorial appointments in the Vormärz see Irene Montjoye (ed.) (1989) *Oscar Wildes Vater über Metternichs Österreich: William Wilde – ein irischer Augenarzt über Biedermeier und Vormärz in Wien, Studien zur Geschichte Südosteuropas* (Frankfurt am Main: Peter Lang). On the politics of the Viennese medical professoriate concerning university appointments (focusing on Polish candidates) see Marja Wakounig (1993) 'Wissenschaft und Karriere? Polnische Mediziner an der Wiener Uni zwischen 1870 und 1914', in Walter Leitsch and Stanisław Trawkowski (eds.) *Wiener Archiv für Geschichte des Slaventums und Osteuropas Bd. 16. Polen im alten Österreich. Kultur und Politik* (Vienna: Böhlau), 107–15.

29. The inability to offer comparable salaries played a crucial role in the rejection of German professors, as shown in Jan Surman (2008) 'Supranational? Die habsburgischen Universitäten im Spannungsfeld zwischen "république des lettres" und "république des nations"', in Alexandra Millner, Helga Mitterbauer and Katharina Scherke (eds.) *Moderne. Kulturwissenschaftliches Jahrbuch*, Themenschwerpunkt Migration (Innsbruck: Studienverlag), 213–24. Nonetheless, the University of Vienna could occasionally step up and put together an attractive offer. For instance, after Billroth's death, in 1894, Vincenz Czerny, a former student of Billroth and a Heidelberg professor, was offered a 5000-florin salary, the title of *Hofrat*, a new building for the Second Surgical Department and a minimum of twenty more years in service. In comparison, his Heidelberg salary was 4700 florins while Billroth, just before his death, had been paid 4220 florins per annum. See Buklijas, 'Surgery and national identity in late nineteenth-century Vienna', 771.

30. (1870) 'Notizen', *Wiener Medizinische Wochenschrift*, 20, 912.

31. Lesky, *Die Wiener Medizinische Schule*, 258–73.

32. Lesky, *Die Wiener Medizinische Schule*, 436; Seebacher, '"Der operierte Chirurg": Theodor Billroths Deutschnationalismus und akademischer Antisemitismus', 318–19.

33. Anatomy had a single chair until the closure of the military medical anatomy Josephinum in 1870, when its professor, Carl Langer, and its buildings in Sensengasse were transferred to the University as the Second Anatomical Chair. When the Chair of the First Department, Joseph Hyrtl, retired in 1873, Langer moved to the First Department and Christian August Voigt, who had been a professor without an institute since the 'Polonization' of the University of Cracow in 1860, was appointed the Chair of the Second Department. After his retirement in 1878, the chair remained vacant as the military authorities re-appropriated the buildings. It was advertised only after the funding for a new anatomical institute had been approved in the early 1880s. For more on anatomy in nineteenth-century Vienna see Tatjana Buklijas (2005) 'Dissection, discipline and urban transformation: anatomy at the University of Vienna, 1845–1914' (PhD thesis, University of Cambridge).

34. Minutes of the committee meeting, ministerial opinion and Albert's *Separatvotum* may be found in the Österreichisches Staatsarchiv (ÖStA), Allgemeines Verwaltungsarchiv (AVA), Unterricht: Allgemeine Reihe (1848–1914), Universität Wien, Sig. 4 Med, Professoren, Fasz. 606, Carl Toldt, Z. 4.588 (1884).

35. Billroth and Brücke probably withdrew because of their failure to get their preferred candidate appointed to the vacated surgical chair; see below.

36. ÖStA, AVA, Unterricht: Allgemeine Reihe (1848–1914), Universität Wien, Sig. 4 Med, Professoren, Fasz. 606, Carl Toldt, Z. 4.588 (1884).

37. Albert divided Zuckerkandl's research into two major groups: 1) anatomical discoveries: *corpus callosum* of the brain, the vascular network between pulmonary and systemic veins; suprathyroid gland; and 2) issues that he either corrected or explained. In human anatomy, these included the connections between bronchial and pulmonary blood vessels; blood vessels and the biliary tree of the liver; the vestibular aqueduct; the aetiology of skull asymmetry and the anatomy of the ethmoidal region; the fixation of kidneys during development; and the development of external inguinal hernia. In comparative anatomy, the *tensor tympani* muscle; the morphology of the facial skull; homologies in the growth of jaw bones between anthropoid monkeys and peoples at lower cultural levels; and his major project, the anatomy of the nasal cavity.

38. ÖStA, AVA, Unterricht: Allgemeine Reihe (1848–1914), Universität Wien, Sig. 4 Med, Professoren, Fasz. 606, Carl Toldt. Z. 4.588 (1884).

39. Ibid.

40. Ibid.

41. Hans-Heinz Eulner (1970) *Die Entwicklung der medizinischen Spezialfächer an den Universitäten des deutschen Sprachgebietes* (Stuttgart: Enke), 553, 662.

42. On histology see ÖStA, AVA, Unterricht: Allgemeine Reihe (1848–1914), Universität Wien, Sig 4 Med, Professoren, Fasz. 607, Carl Wedl, Z. 14.770 (1884); also Lesky, *Die Wiener Medizinische Schule*, 513.

43. Lesky, *Die Wiener Medizinische Schule*, 521.

44. Technically, the first professor of pathological anatomy in German-speaking countries was Johann Lobstein in Strasbourg (1819), but he also taught obstetrics and internal diseases. See Eulner, *Die Entwicklung der medizinischen Spezialfächer*, 103.

45. Lesky, *Die Wiener Medizinische Schule*, 549–50.

46. Eulner, *Die Entwicklung der medizinischen Spezialfächer*, 554–55.

47. Nyhart, *Biology Takes Form*, 368–69.

48. Nyhart, *Biology Takes Form*.

49. Ibid., 226.

50. The Second Vienna School in medicine was close to the strongly empirical reaction to *Naturphilosophie* that developed in medical schools in German lands in the 1830s. See Johanna Bleker (1988) 'Biedermeiermedizin – Medizin der Biedermeier? Tendenzen, Probleme, Widersprüche 1830–1850', *Medizinhistorisches Journal*, 23, 5–22. The support it received in Vienna may be related to the success of positivism in this period. See Johannes Feichtinger (2005) 'Positivismus und Machtpolitik. Ein wissenschaftliches Programm und dessen Transfer nach Österreich/Zentraleuropa. Zu einem Beispiel von Wissenstransfer', in Helga Mitterbauer and Katharina Scherke (eds.), *Entgrenzte Räume. Kulturelle Transfers um 1900 und in der Gegenwart* (Studien zur Moderne 22) (Vienna: Passagen Verlag), 297–319.

51. For more on Hyrtl's anatomy and politics, see Buklijas, 'Dissection, discipline and urban transformation: anatomy at the University of Vienna, 1845–1914', especially Chapter 3 'Conservative anatomy and the Second Vienna School, 1845–70'.

52. ÖStA, AVA, Unterricht: Allgemeine Reihe (1848–1914), Universität Wien, Sig. 4 Med, Professoren, Fasz. 607, Carl Wedl, Z. 14.770 (1884) and Fasz. 607, Emil Zuckerkandl, Z. 18.821 (1888).

238   *Tatjana Buklijas*

53. The decisive vote in Albert's favour was that of Eduard Hofmann, professor of forensic medicine, who submitted a minority vote for Albert. See (1880) 'Notizen', *Wiener Medizinische Wochenschrift*, 30, 1407; (1881) 'Notizen', *Wiener Medizinische Wochenschrift*, 31, 24.
54. Lesky, *Die Wiener Medizinische Schule*, 437, 449–50; Buklijas, 'Surgery and national identity in late nineteenth-century Vienna'.
55. Buklijas, 'Surgery and national identity in late nineteenth-century Vienna', 770.
56. On Billroth's recruitment of surgical and research talent among the Viennese studentship, his views on surgical education and the building of a school, see Buklijas, 'Surgery and national identity in late nineteenth-century Vienna', 762.
57. Adolf Lorenz (1936) *My Life and Work: the Search for a Missing Glove* (New York: Charles Scribner's Sons), 73–74. For more on the comparison between Albert and Billroth as lecturers see Buklijas, 'Surgery and national identity in late nineteenth-century Vienna', 768–69.
58. A question that divided surgeons worldwide in the 1870s was the causation of postoperative fever and the use of antisepsis. Yet it seems that the viewpoint of 'Austrians' and 'Germans' on this matter depended on their generation rather than on their national politics. Dumreicher was opposed to antisepsis, Billroth at first unconvinced but then converted, and Albert an enthusiastic supporter from the start. See Buklijas, 'Surgery and national identity in late nineteenth-century Vienna'.
59. (1881) 'Zur Besetzungsfrage der Lehrkanzel der pathologischen Anatomie', *Wiener Medizinische Wochenschrift*, 31, 1158–59.
60. Felicitas Seebacher (2000) '"Primum humanitas, alterum scientia". Die Wiener Medizinische Schule im Spannungsfeld von Wissenschaft und Politik' (PhD dissertation, Universität Klagenfurt), 363.
61. But this was not his only attempt. For instance, when the faculty named Hermann Nothnagel, professor of pharmacology and medical polyclinics at Freiburg University, *unico loco* candidate for the vacated Chair of Internal Medicine, Albert submitted a *Separatvotum*. A *Wiener Medizinische Wochenschrift* journalist wondered about his consistently negative attitude towards German professors. He wrote: 'Are we at war with Germany, or should not Germany be regarded as the truest ally of Austria? Should not German colleagues be trusted to teach at Austrian universities?' (1882) *Wiener Medizinische Wochenschrift*, 32, 800–01. Albert generally had a strong opinion on who should be allowed to join the Viennese medical faculty: in 1888, with his friend Stricker, he objected the *Habilitation* of the Polish-speaking otologist Abraham Eitelberg on the grounds of his supposedly poor German. See Wakounig, 'Wissenschaft und Kariere? Polnische Mediziner an der Wiener Uni zwischen 1870 und 1914'.
62. (1888) *Wiener Medinizische Wochenschrift*, 38, 375–77.
63. Ibid. The most vocal of Zuckerkandl's opponents was Heinrich von Bamberger, who had sat on the 1884 appointment committee and voted for Toldt.
64. (1888) *Wiener Medizinische Wochenschrift*, 38, 88–89, 375–77, 1263–64.
65. The WMW wrote that Schwalbe was not among the three anatomists considered by the University of Jena following Oskar Hertwig's departure for Berlin. But Schwalbe had taught in Jena between 1873 and 1881, and then in 1883 moved to the Imperial University of Strassburg (Strasbourg), a new and prestigious institution founded right after German unification, in 1872. It is not clear why the journal thought that Schwalbe would return to a smaller university. See (1888) *Wiener Medizinische Wochenschrift*, 38, 376, 1263.

66. ÖStA, AVA, Unterricht: Allgemeine Reihe (1848–1914), Universität Wien, Sig. 4 Med, Professoren, Fasz. 607, E. Zuckerkandl, Z. 18.821 (1888).
67. Ibid.
68. For a discussion of the emergence of 'language frontiers', such as the southern Bohemia and southern Styria, in the late nineteenth century see Pieter M. Judson (2006) *Guardians of the Nation: Activists on the Language Frontiers in Imperial Austria* (Cambridge, Mass.: Harvard University Press).
69. For biographic information on Toldt see Toldt, *Autobiographie*; Pöch, 'Carl Toldt (gestorben am 13. November 1920)'.
70. Toldt, *Autobiographie*, 12.
71. On Ewald Hering's physiology of vision see Steven R. Turner (1994) *In the Eye's Mind: Vision and the Helmholtz-Hering Controversy* (Princeton, N.J.: Princeton University Press).
72. Carl Toldt (1875) *Studien über die Anatomie der menschlichen Brustgegend mit Bezug auf die Messung derselben und die Verwertung des Brustumfanges zur Beurteilung der Kriegsdiensttauglichkeit* (Stuttgart: Ferdinand Enke). Anthropological studies on military recruits had been conducted in the Austrian army from at least the early 1860s, when the pathologist Augustin Weisbach compared the head circumferences of military recruits from different ethnic groups; see Fuchs, *'Rasse'*, *'Volk'*, *Geschlecht* (cit. Note 18), 142. To my knowledge, Toldt's was the first anthropological study in Austria to use soft-tissue material and to take place in an anatomical institute.
73. Pöch, 'Carl Toldt (gestorben am 13. November 1920)', 81.
74. The first two factors were decisive. ÖStA, AVA, Unterricht: Allgemeine Reihe (1848–1914), Universität Prag, Sig. 5, Professoren, Fasz. 1123, Carl Toldt, Z. 5026 (1876).
75. Karl Toldt (1939) *Geschichte der Familie Toldt, verfaßt von weil. Univ.-Prof. Dr. Carl Toldt 1893, bearbeitet und ergänzt von seinem Sohn Dr. Karl Toldt* (Innsbruck: Wagner).
76. Pöch 'Carl Toldt (gestorben am 13. November 1920)', 83. On inaugural lectures as occasions for academic elites to shape the public image of their disciplines, see Coen, *Vienna in the Age of Uncertainty*, 138, on Adolf Exner.
77. This undated manuscript, printed in 50 copies, was prepared by Toldt to distinguish his position from the conclusion of the committee on natural sciences that recommended that these subjects remain in the curriculum but without examination. The committee was appointed in relation to the ministerial decree of 16 February 1891, concerning the change in the medical *Studien-* and *Prüfungordnung*. Handschriftensammlung, Sammlungen der Medizinischen Universität Wien.
78. Michael Stober (1971) *Personalbibliographien der Professoren und Dozenten der Anatomie an der Medizinischen Fakultät der Universität Wien im ungefähren Zeitraum von 1845 bis 1969, mit biographischen Angaben und Überblick über die Hauptarbeitsgebiete* (Med. diss., Medizinische Fakultät, Universität Erlangen-Nürnberg).
79. Fuchs, *'Rasse'*, *'Volk'*, *Geschlecht*, 160–64.
80. Franz Tappeiner (1883) *Studien zur Anthropologie Tirols und der Sette Comuni* (Innsbruck: Wagner). See also Gustav Sauser (1938) *Die Ötztaler: Anthropologie und Anatomie einer Tiroler Talschaft* (Innsbruck: Naturwissenschaftlich-medizinischer Verein), especially 'Frühere anthropologische Untersuchungen in Tirol', 413.
81. In 1858, Carl Freiherr von Czörnig (1804–99) published the *Ethnographische Karte* of the Habsburg Monarchy, accompanied by a comprehensive text on the

Ethnographie Österreichs. The ethnographic map visually argued that the Empire had no ethnically homogenous territories, so the solution to the national questions could not be in the separation of nation-states. See Fuchs, *'Rasse', 'Volk', Geschlecht,* 152–58.

82. Toldt, *Geschichte der Familie Toldt.*

83. Coen, *Vienna in the Age of Uncertainty.*

84. Pöch, 'Carl Toldt (gestorben am 13. November 1920)', 88; Toldt, *Autobiographie.*

85. Carl Toldt (1893) 'Physische Beschaffenheit von Tirol und Vorarlberg', in *Die österreichisch-ungarische Monarchie: Tirol und Vorarlberg,* vol. 13 (Vienna: Kaiserlich-königliche Hof und Staatsdruckerei), 229–40.

86. Stober, *Personalbibliographien der Professoren und Dozenten der Anatomie an der Medizinischen Fakultät der Universität Wien.*

87. Robert Herrlinger (1965) 'Kurze Geschichte der Anatomischen Gesellschaft', *Anatomischer Anzeiger,* 117, 1–60.

88. Carl Toldt (1900) *Anatomischer Atlas für Studierende und Ärzte* (Vienna, Berlin: Urban & Schwarzenberg), 1; Eliane Rautenberg (2002) 'Die jüngere Firmengeschichte des Urban & Fischer Verlags anhand seiner anatomischen Publikationen' (Diploma diss., Ludwig-Maximilians-Universität München), 50–52.

89. ÖStA, AVA, Unterricht: Allgemeine Reihe (1848–1914), Universität Wien, Sig. 4 Med, Professoren, Fasz. 600, Ferdinand Hochstetter. See also Lesky, *Die Wiener Medizinische Schule,* 508.

90. See the Hochstetter Nachlass in the Handschriftensammlung, Sammlungen der Medizinischen Universität Wien.

91. See, for example, Daniela Claudia Angetter (1999) 'Die Wiener Anatomische Schule', *Wiener Klinische Wochenschrift,* 111, 764–74; Michael Hubenstorf (1989) 'Medizinische Fakultat 1938–1945', in Gernot Heiss et al. (eds.), *Willfahrige Wissenschaft: Die Universitat Wien 1938–45* (Vienna: Verlag fur Gesellschaftskritik), 233–82; Peter Malina (1998) 'Eduard Pernkopf's anatomischer Atlas oder: die Fiktion einer "reinen Wissenschaft"', *Wiener Klinische Wochenschrift,* 110, 193–201.

92. ÖStA, AVA, Unterricht: Allgemeine Reihe (1848–1914), Universität Wien, Sig. 4 Med, Fasz. 620, Hospitanten, Präparatoren, Mechaniker, Demonstratoren (1848–1928), Z. 24.254 (1891).

93. Ibid., Z. 2.262 (1892).

94. (1897) *Die feierliche Inauguration des Rectors der Wiener Universität für das Studienjahr 1897/98 am 28. October 1897* (Vienna: Selbstverlag der k. k. Universität).

95. Richard S. Geehr (1990) *Karl Lueger: mayor of fin-de-siècle Vienna* (Detroit: Wayne State University Press), 79–99.

96. Klebelsberg, 'Carl Toldt. † am 13. November 1920', 29.

97. Toldt, *Geschichte der Familie Toldt,* 123.

98. See the biographical references in note 3.

99. On Viktor Zuckerkandl and Purkersdorf, see Leslie Topp (1997) 'An architecture for modern nerves: Josef Hoffmann's Purkersdorf Sanatorium', *Journal of the Society of Architectural Historians,* 56, 414–37.

100. His *National* file (1868) gives his religion as '*mosaisch*' and his father as 'Leon, Kaufmann, Pest'. In 1871, the rubric 'Religion' is crossed out. UA, Medizinische Fakultät, Personalakten, Emil Zuckerkandl, f. 1.

101. For Arthur Schnitzler's memories of Zuckerkandl as a young prosector around 1880 see Therese Nickl and Heinrich Schnitzler (eds.) (1968) *Jugend in Wien: eine Autobiographie/Arthur Schnitzler* (Vienna: Molden), 127–28.

102. Seebacher, *'"Primum humanitas, alterum scientia"*. Die Wiener Medizinische Schule im Spannungsfeld von Wissenschaft und Politik', 327–32; John M. Efron (2001) *Medicine and the German Jews: a History* (New Haven: Yale University Press), 242.

103. On Jews in Vienna, their immigration, participation in the social and economic life, assimilation and obstacles to it see Marsha M. Rozenblit (1983) *The Jews of Vienna, 1867–1914: Assimilation and Identity* (Albany: State University of New York Press); Beller, *Vienna and the Jews, 1867–1938*.

104. Julius Tandler (1910) 'Emil Zuckerkandl', *Wiener Klinische Wochenschrift*, 23, 798–800.

105. Bertha Zuckerkandl (1938) *My Life and History* (London: Cassel and Company).

106. Julius Tandler (1910), 'Emil Zuckerkandl's Nachruf', in *Feierliche Inauguration des Rektors der Universität Wien für das Studienjahr 1910/1911* (Wien: Selbstverlag), 39–52.

107. UA, Medizinische Fakultät, Personalakten, Emil Zuckerkandl, Z. 566 (1873/74).

108. His mention of Darwin's theory and common ancestry of humans and apes in a lecture at the University of Graz apparently did not bother the Ministry but was severely criticized in the Styrian conservative press. See Bertha Zuckerkandl (1981) *Österreich intim. Erinnerungen 1892–1942, ergänzte und neu illustrierte Ausgabe mit 30 Abbildungen* (Vienna: Amalthea) (original edition: Frankfurt: Ullstein, 1970), 132–34.

109. Seebacher, *'"Primum humanitas, alterum scientia"*. Die Wiener Medizinische Schule im Spannungsfeld von Wissenschaft und Politik', 355.

110. I have no record of Zuckerkandl's conversion to Christianity, although it is likely that he did convert. His son Fritz was baptized: see Emil Zuckerkandl, Verlassenschafsabhandlungen, Stadtarchiv Wien.

111. This succeeded largely thanks to the efforts of Carl Langer, who commended his prosector's 'technical, literary and didactic skills'. See (1879) 'Notizen', *Wiener Medizinische Wochenschrift*, 29, 1297. Freud, for example, became *Extraordinarius* at the age of 45.

112. Stober, *Personalbibliographien der Professoren und Dozenten der Anatomie an der Medizinischen Fakultät der Universität Wien*, 41–54.

113. Fuchs, *'Rasse', 'Volk', Geschlecht* (cit. Note 18), 137–42.

114. Zuckerkandl (1875), quoted in Fuchs, *'Rasse', 'Volk', Geschlecht*, 140.

115. Andrew Zimmerman (2001) *Anthropology and Antihumanism in Imperial Germany* (Chicago: University of Chicago Press), 87–88, 277–88. On the ways in which the multiethnic environment informed the direction of anthropology in Austria-Hungary see Fuchs, *'Rasse', 'Volk', Geschlecht*; Lafferton, 'The Magyar moustache: the faces of Hungarian state formation, 1867–1918'.

116. Joseph Hyrtl (1847) *Handbuch der topographischen Anatomie und ihrer praktische medicinisch-chirurgischen Anwendungen* (Vienna: J.B. Wallishauser).

117. Otto Zuckerkandl (1897) *Atlas und Grundriss der chirurgischen Operationslehre* (Munich: J.F. Lehmann), 503.

118. Emil Zuckerkandl (1882) *Normale und pathologische Anatomie der Nasenhöhle und ihrer pneumatischen Anhänge* (Vienna: Wilhelm Braumüller) (the second edition was published in 1893). The book contained 22 black-and-white lithographic plates, drawn by Julius Heitzmann.

119. Hans v. Hebra (1870) 'Ueber ein eigenthümliches Neugebilde an der Nase; Rhinosclerom; nebst histologischem Befunde vom Dr. M. Kohn', *Wiener Medizinische Wochenschrift*, 20, 1–5.

120. Sander L. Gilman (1999) *Making the Body Beautiful: a Cultural History of Aesthetic Surgery* (Princeton: Princeton University Press), 124–37, 186–99; idem (1994) 'The Jewish nose: are Jews white? Or, the history of the nose job', in Laurence J. Silberstein and Robert L. Cohn (eds.) *The Other in Jewish Thought and History: Constructions of Jewish Culture and Identity* (New York: New York UP), 364–401. Zuckerkandl's only nod in this direction is his discussion of the anatomy of the septum in the second volume of his 1882 book (Note 118). It included anthropological considerations on the percentage of physiological septum deviation significantly lower in 'prognate peoples' than in Europeans.

121. Efron, *Medicine and the German Jews*, 235.

122. Ibid., 239.

123. Cohen, *Education and Middle-class Society in Imperial Austria*, 232–33.

124. Lesky, *Die Wiener Medizinische Schule*, 465.

125. Stober, *Personalbibliographien der Professoren und Dozenten der Anatomie an der Medizinischen Fakultät der Universität Wien*, 41–54.

126. Most German anatomists, even if employed to teach a 'clinically relevant' orientation in anatomy, pursued research of no immediate clinical value. For example, Wilhelm Waldeyer in Berlin worked in pathology and topographical anatomy but also (and more famously) in embryology and on cell structure.

127. The first female demonstrator in the medical school at the University of Vienna was Elsa Friedland, employed from 1 January 1903 at the Heinrich Obersteiner's Institute of Anatomy and Physiology of CNS; see ÖStA, AVA, Unterricht: Allgemeine Reihe (1848–1914), Universität Wien, Sig. 4G, Hospitanten, Mechaniker, Präparatoren, Demonstratoren (1848–1928), Fasz. 620, Z. 37957 (1902); 42292 (1903); 41892 (1904). On Gertrude Bien's demonstratorship from 1 May 1906 see ibid., Z.25311 (1906), 2. I am grateful to Dr. Ingrid Arias for locating the document that confirms that it was Gertrude Bien in 1907, and not Marianne Stein after the First World War, who was the first female assistant at the medical faculty. UA, Dekanatsakten der Medizinischen Fakultät, Z. 1112 (1907/08), Z. 1404 (1906/07), Z. 494 (1907/08).

128. Zuckerkandl, *My Life and History*, 133. Gertrude Bien (1881–?) published extensively in the early twentieth century but soon left anatomy. She then worked as an assistant physician at the *Carolinenspital*, a Viennese children's hospital (1913–18), and afterwards as *Primaria* in the Reception Centre for Children (*Kinderübernahmestelle*). It is probably no coincidence that Bien was employed in this innovative observation station for abused and neglected children, founded and funded by the Social Democrat government of Vienna, as the councillor for health and welfare was Bien's colleague and sometime superior, the anatomist Julius Tandler. Bien had other ties to the Zuckerkandl circle: she was the house physician of Klimt's famous model Adele Bloch-Bauer. In 1934, she retired and emigrated to the United States. See Ingrid Arias (2000) 'Die ersten Ärztinnen in Wien: Ärztliche Karrieren von Frauen zwischen 1900 und 1938', in Birgit Bolognese-Leuchtenmüller and Sonia Horn (eds.) *Töchter des Hippokrates: 100 Jahre akademische Ärztinnen in Österreich* (Vienna: Verlag der Österreichischen Ärztekammer), 55–78, here 70; Martina Gamper (2000) '"…so kann ich nicht umhin mich zu wundern, das nicht mehr Ärztinnen da sind". Die Stellung weiblicher Ärzte im "Roten Wien" (1922–1934)', in idem, 79–96, 83.

129. See Gamper '"…so kann ich nicht umhin mich zu wundern, das nicht mehr Ärztinnen da sind"', 92; ÖStA, AVA, Unterricht: Allgemeine Reihe

(1848–1914), Universität Wien, Sig. 4G, Hospitanten, Mechaniker, Präparatoren, Demonstratoren (1848–1928), Fasz. 620, Z. 37233 (1911).

130. The first female demonstrator at the Second Chair was Helena Maslowski in 1916; see ibid., Z.25431 (1916).

131. See Mitchell G. Ash and Christian Stifter (eds.) (2002) *Wissenschaft, Politik und Öffentlichkeit: von der Wiener Moderne bis heute* (Vienna: Wiener Universitätsverlag); Klaus Taschwer (1997) 'People's universities in a former metropolis: interfaces between the social and spatial organisation of popular adult education in Vienna, 1890–1930', in Barry J. Hake and Tom Steele (eds.) *Intellectuals, Activists and Reformers. Studies of Cultural, Social and Educational Reform Movements in Europe, 1890–1930* (Leeds: University of Leeds Press), 175–202, and idem (2000) 'Wissenschaft für viele: zur Wissenschaftsvermittlung im Rahmen der Wiener Volksbildungsbewegung um 1900' (PhD thesis, Fakultät für Human- und Sozialwissenschaften, Universität Wien).

132. Emil Zuckerkandl was already older and in poor health, but the active Tandler taught a range of courses in the early 1900s. I am grateful to Dr. Christian Stifter from the *Volkshochschularchiv* in Vienna for the information on anatomical courses.

133. Zuckerkandl, *My Life and History*, 64–65.

134. On the fate of Viktor's collection and that of the members of Zuckerkandl family in the Second World War see the excellent chapter on the family Zuckerkandl in Natter, T.G. (2003) *Die Welt von Klimt, Schiele und Kokoschka* (Cologne: Dumont).

135. For more on Purkersdorf Sanatorium, see Topp, 'An architecture for modern nerves: Josef Hoffmann's Purkersdorf Sanatorium'.

136. Zuckerkandl, *My Life and History*, 101; Beller, *Vienna and the Jews, 1867–1938*, 27.

137. Zuckerkandl, *My Life and History*.

138. William M. Johnston (1972) *The Austrian Mind: an Intellectual and Social History, 1848–1938* (Berkeley: University of California Press); William J. McGrath (1974) *Dionysian Art and Populist Politics in Austria* (New Haven: Yale University Press); Carl Schorske, *Fin-de-siècle Vienna* (cit. Note 23).

139. Johnston, *The Austrian mind*; Schorske, *Fin-de-siècle Vienna*.

140. Bertha Zuckerkandl (1981) 'Hermann Bahr, Ernst Mach und Emil Zuckerkandl im Gespräch, Wien 1908', in Gotthard Wunberg (ed.), *Die Wiener Moderne. Literatur, Kunst und Musik zwischen 1890 aund 1910* (Stuttgart: Reclam), 171–77.

141. Topp, 'An architecture for modern nerves: Josef Hoffmann's Purkersdorf Sanatorium'.

142. See Rainald Franz (1998) 'Stilvermeidung und Naturnachahmung: Ernst Haeckels "Kunstformen der Natur" und ihr Einfluß auf die Ornamentik des Jugendstils in Österreich', in Erna Aescht (ed.) *Welträtsel und Lebenswunder: Ernst Haeckel – Werk, Wirkung und Folgen* (Linz: Oberösterreichisches Landesmuseum), 475–80; Tatjana Buklijas and Nick Hopwood (2008) 'Evolution: Visual culture', *Making Visible Embryos* [http://www.hps.cam.ac.uk/visibleembryos/s4_3.html] (University of Cambridge); see also Marsha Morton (2009) 'From *Monera* to man: Ernst Haeckel, Darwinismus and nineteenth-century German Art', in Barbara Jean Larson (ed.) *The Art of Evolution: Darwin, Darwinisms and Visual Culture* (Lebanon, NH: Dartmouth College Press), 59–91.

143. See Emily Braun (2007) 'Ornament and evolution: Gustav Klimt and Berta Zuckerkandl', in Renée Price (ed.) *Gustav Klimt: The Ronald S. Lauder and Serge*

Sabarsky Collections (New York: Neue Galerie), 144–69; Scott F. Gilbert and Sabine Brauckmann (2011), 'Fertilization narratives in the art of Gustav Klimt, Diego Rivera and Frida Kahlo: repression, domination and Eros among cells', *Leonardo*, 44(3), 221–227.

144. According to Lajos (Ludwig) Hevesi, the Viennese journalist, author and art critic, 'even the nightmarish new piece 'From the realm of death', for which he [Klimt] tirelessly attended Professor Zuckerkandl's anatomical lectures this summer, has been sold' (*'selbst das schauerliche Novum "Aus dem Reich des Todes" für das er diesen Sommer rastlos die anatomischen Vorlesungen des Professors Zuckerkandl besucht hat, ist verkauft'*). See Hevesi in *Kunstchronik*, Leipzig, NF, 15 (1903–04), 136.

145. Zuckerkandl, *Österreich intim*, 132–34.

146. For Tandler's biography see Karl Sablik (1983) *Julius Tandler: Mediziner und Sozialreformer* (Vienna: A. Schendl).

Index

Abegg, Richard (1869–1910), 144
  Handbuch der anorganischen Chemie
    (1905–1913), 144
Académie française, 106
Academy of Sciences and Arts in
  Cracow, 10
activism, national, 62
Adelung, Johann (1732–1806), 33
Admiral Tegetthof (ship), 11
Ady, Endre (1877–1919), 116
Aeby, Christian (1835–1885), 213
Agamennone, Giovanni
    (1858–1949), 174–5
agreement, culture of, *see* consensus,
  culture of
Albert, Eduard (1841–1900), 44, 214–17,
    222, 237–8
  and appointment of Hermann
    Nothnagel, 238
  and appointment of Carl Toldt,
    214–17
Anatomische Gesellschaft, 222
Anatomischer Anzeiger (journal), 222
anatomy, 209–44
  and anthropology, 211
  appointments in Vienna, 212–18
  v. histology, 216
  topological, 226–7
  as *Wissenschaft*, 216
Anderson, Benedict (1936–), 2
Andrássy, Gyula (1823–1890), 119, 131
Andrian-Werburg, Victor Franz Freiherr
  von (1813–1858), 119
anti-Semitism, 71, 107, 217, 226
  and university appointments, 71,
    228–9
  see also Karl Lueger; Emil Zuckerkandl
*Anzeiger der Akademie der Wissenschaften
  in Krakau* (journal), *see Bulletin
  international de l'Académie des
  sciences de Cracovie* (journal)
Apáthy, István (1863–1922), 195–8
Arany, Dániel (1863–1945), 129–30
Arany Dániel országos matematika verseny
  (journal), 130

*Arbeiten des Neurologischen Instituts in
  Wien* (journal), 229
Archiv der Ohrenheilkunde
    (journal), 229
Archives slaves de biologie (journal), 45
art and medicine, 230–1
assimilation
  in Hungary, 105, 117
  see also Jews and assimilation
Athaenaeum (Amsterdam), 226
Ausgleich (1867), *see* Compromise
Aussee, 93
Austrian Earthquake Commission, 160,
    169–72
Austrian School of Surgery, 217
Austro-Hungarian Monarchy
  collapse of, 71
  and geology, 176
  language conflicts in, 218, 223
  university reforms in, *see* reforms
Austro-Prussian War, 120
autonomy
  v. autarchy, 75, 82
  and heteronomy 59–60
  relative, 59, 63
  of science, 60
  of universities, 65, 75

Bach, Alexander von (1813–1893),
    68, 140
Badeni, Kasimir (1846–1909),
    218–19, 223
  language ordinances, 218–19, 223
Bahr, Hermann (1863–1934), 230–1
Balázs, Béla (1884–1949), 126
Bamberger, Heinrich von (1822–1888),
    214, 238
Banská Štiavnica, 121
Baraniecki, Marian (1848–1895), 46
Basel Nomina Anatomica, 222
Batthyány, Kázmér (1807–1854), 115, 131
Batthyány, Lajos (1807–1849), 119, 131
Beethoven, Ludwig van (1770–1827),
    115, 124
Beketov, Nikolay (1827–1911), 142, 148